Tools for Structured and Object-Oriented Design
An Introduction to Programming Logic

Sixth Edition

Marilyn Bohl

Maria Rynn
Northern Virginia Community College

Prentice
Hall

Upper Saddle River, New Jersey
Columbus, Ohio

Library of Congress Cataloging-in-Publication Data

Bohl, Marilyn.
    Tools for structured and object-oriented design : an introduction to programming logic /
Marilyn Bohl, Maria Rynn. -- 6th ed.
        p. cm.
    Includes index.
    ISBN 0-13-049498-4
    1. Structured programming.   2.   Object-oriented programming (Computer science)   I.
Rynn, Maria, 1953-   II. Title.

QA76.64 .B63   2003
005.1'17--dc21

2002032653

Editor in Chief: Stephen Helba
Assistant Vice President and Publisher: Charles E. Stewart, Jr.
Assistant Editor: Mayda Bosco
Production Editor: Alexandrina Benedicto Wolf
Production Supervision: Custom Editorial Productions, Inc.
Production Manager: Matthew Ottenweller
Design Coordinator: Diane Ernsberger
Cover Designer: Ali Mohrman
Cover Image: SuperStock
Marketing Manager: Ben Leonard

This book was set in Times Roman by Custom Editorial Productions, Inc., and was printed and bound by The Banta Company. The cover was printed by Phoenix Color Corp.

Earlier edition © 1978 by Science Research Associates.

Acknowledgments
    Figure 2–14 reprinted by permission from GX20-8021-2, flowcharting worksheet by International Business Machines Corporation.
    Figure 2–17 reprinted by permission from GS20-8020-1-U/MO10, flowcharting template by International Business Machines Corporation.

Pearson Education Ltd.
Pearson Education Australia Pty. Limited
Pearson Education Singapore Pte. Ltd.
Pearson Education North Asia Ltd.
Pearson Education Canada, Ltd.
Pearson Educación de Mexico, S.A. de C.V.
Pearson Education—Japan
Pearson Education Malaysia Pte. Ltd.
Pearson Education, *Upper Saddle River, New Jersey*

10   9   8   7   6   5
ISBN: 0-13-049498-4

Preface

Tools for Structured and Object-Oriented Design: An Introduction to Programming Logic, Sixth Edition, teaches program design in a well-thought-out, language-independent manner. This text assumes no previous programming background. It can be used as a main text in a programming logic class or as a supplement in any beginning programming class.

Our approach is to start with simple concepts and build on these concepts as new topics are introduced. We use a sequential, step-by-step approach that introduces, by way of example, only one new concept at a time. Sample problems are included throughout the chapters to illustrate the use of program design tools in practical situations. Enrichment sections are included in many chapters to illustrate the program design concepts in Basic and Visual Basic. Exercises are given at the end of each chapter to help you apply what you are learning.

Our objective is to analyze a problem and express its solution in such a way that the computer can be directed to follow the problem-solving procedure. With simple language and frequent examples, this book explains how to understand and how to use important problem-solving tools. We begin with system and program flowcharts. Flowcharting guidelines approved and published by the American National Standards Institute (ANSI) and its international counterpart, the International Standards Organization (ISO), are explained and applied to solution planning. Emphasis is placed on maintaining an overall structure in program design. We show how to use pseudocode as an alternative or supplement to flowcharting in planning the logic of a well-structured program. We analyze techniques of top-down, modular program development by describing how to read and how to develop structure charts that show the hierarchical relationships of modules within a program. Our coverage of object-oriented design is expanded significantly in this edition, as indicated by the new title for this edition.

Enrichment sections are included in many chapters to illustrate some of the sample problems in the programming languages Basic and Visual Basic. Basic illustrates how the design of a program can be implemented using a procedural approach. Visual Basic illustrates how the design of a program can be implemented using an event-driven approach. It is important to note that our approach to teaching program design remains language independent. All program design concepts are covered prior to the enrichment sections in each chapter. The enrichment sections are optional and are included as a supplement to further illustrate some of these concepts.

The sixth edition of this book offers the same pedagogical features as the fifth edition. Each chapter includes objectives and a list of key terms. The sixth edition also includes an index and is supported by an *Instructor's Manual*. The *Instructor's Manual* contains the objectives and key terms listed in the textbook as well as suggested teaching strategies. Solutions to some end-of-chapter exercises are included in the text. Solutions to all end-of-chapter exercises are included in the *Instructor's Manual*. A CD containing PowerPoint slides is available to instructors. In addition, a CD containing the source code for all the Basic and Visual Basic examples in the enrichment sections of the text is packaged with the textbook.

The text is organized into three parts as follows:

> Part One–Structured Programming Concepts (Chapters 1 through 9) introduces the theory of structured programming and includes a chapter on each control structure as well as a chapter on array fundamentals. These chapters should be covered in sequence.

> Part Two–Object-Oriented Programming Concepts (Chapters 10 through 12) introduces many fundamental concepts of object-oriented design and programming. These chapters should be covered in sequence after Part One is completed.

> Part Three–Applications (Chapters 13 through 15) illustrates more complex applications, building on material previously introduced. These chapters should be covered in sequence after Part One is completed.

The specific content of each chapter and appendices follows.

Chapter 1 describes the system development life cycle and how program design fits with it. Computer-assisted software engineering (CASE) tools are introduced in this chapter and are referred to throughout the book, where appropriate, to increase your awareness of current tools and trends in the industry. The history of structured programming is also introduced in this chapter. Several nontechnical examples illustrate the basic control structures to give you a sense of what structured programming entails. The concepts of object-oriented design, event-driven programming, and graphical user interfaces are also introduced in this chapter.

Chapter 2 introduces the SIMPLE SEQUENCE control structure. Chapter 3 introduces the IFTHENELSE control structure and teaches simple, sequential, and nested IFs. Chapter 4 introduces the DOWHILE control structure, focusing on simple counter loops and header record logic. Chapter 5 focuses on trailer record logic and also includes a discussion of automatic end-of-file processing and multiple-heading logic. Chapter 6 introduces modularization. In this edition, the material on modularization is introduced after both header record logic and trailer record logic are discussed.

Chapter 7 introduces the CASE control structure. Chapter 8 introduces the DOUNTIL control structure. Chapter 9 introduces one- and two-dimensional arrays with many short, simple examples. Chapter 10 introduces concepts of object-oriented design. Unified Modeling Language (UML) class diagrams and pseudocode examples are used to illustrate object-oriented design concepts such as classes, objects, data members, methods, encapsulation, driver programs, and overloading. Chapter 11 introduces generalization/specialization relationships, inheritance, overriding, polymorphism, and abstract classes. Chapter 12 introduces additional types of relationships: association, aggregation, and composition. Inner classes are explained. The advantages of object-oriented design and programming are summarized.

Chapter 13 concentrates on more advanced array applications such as searching (both sequential and binary) and sorting. Chapter 14 illustrates the design of a sequential master file update program. Chapter 15 covers control-break processing.

Appendix A contains ANSI-approved symbols for program flowcharting, and Appendix B summarizes the basic control patterns of structured programming. Appendices A and B should be referred to whenever you are in doubt about which symbols to use in flowcharts. Responses to selected end-of-chapter exercises are provided in Appendix C to help you evaluate your understanding of the material.

Acknowledgments

This book would not exist today if it were not for the hard work of many people. We would first like to thank Charles E. Stewart, Publisher, and Mayda Bosco, Assistant Editor, for all their help in putting this book together. We would also like to thank Megan Smith-Creed for her support and assistance during the production of this manuscript. In addition, the following reviewers provided valuable feedback: Phillip Davis, Del Mar College; Lee Rosenthal, Fairleigh Dickinson University; and Costas Vassiliadis, Ohio University.

Maria Rynn would like to thank all of her colleagues at Northern Virginia Community College for their continuing support and advice throughout this revision. Finally, she wishes to thank her husband, Tedd, for his constant encouragement, patience, and loving support during this endeavor.

Marilyn Bohl would like to thank Vinh Nguyen and Jeff Tassin, whose timely Internet responses to specific questions were invaluable in the preparation of Part Two of this edition.

We hope that all who use this book will find that it provides a clear, systematic, and direct approach to problem solving. We welcome your comments and suggestions.

Marilyn Bohl
Maria Rynn

To My Family—
Tedd, Vanessa, Teddy, Donna, and Michael
From Maria Rynn

Contents

Part One—Structured Programming Concepts

Part Two—Object-Oriented Programming Concepts

Part Three—Applications

Part

Introduction to Structured Design

Objectives

Upon completion of this chapter you should be able to

- Name and identify the six steps in the system development life cycle.
- Define a computer-based information system.
- State four objectives of computer-assisted software engineering (CASE) and give examples of CASE tools.
- Name and identify the five steps in the program development cycle.
- Name and identify some of the tools and methodologies used in the design of well-structured programs.
- Define what is meant by a graphical user interface (GUI).
- Distinguish between procedure-oriented languages and object-oriented languages.
- Define what is meant by event-driven programming.
- Distinguish between syntax errors and logic errors in a program.
- Distinguish between unit testing and system testing.
- Name some forms of documentation needed in a computer-based information system.
- Name the three basic control structures of structured programming.

Introduction

We live today in a business world. Goods and services are bought and sold, distributed, produced, and created worldwide at incredible rates. Some businesses are international conglomerates; others are small mom-and-pop shops. The success of any business depends, for the most part, on how well the business is run. Every business uses one or more systems to produce its end products or services. A **system** is a combination of people, equipment, and procedures that work together to perform a specific function. A system can be manually operated or computer-assisted. A **computer-based information system** is a system in which some of the procedures are performed by a computer. Since desktop computers have become easily accessible, even small businesses are using computers or are looking into converting their manual systems to computer-based ones. Such a conversion is not an easy task, but it can be simplified by following a series of well-defined steps.

System Development Life Cycle

The **system development life cycle (SDLC)** (see Figure 1–1) is a series of well-defined steps that should be followed when a system is created or changed. The SDLC represents the big picture of what happens during system creation or modification. In this chapter we outline the steps in the SDLC and identify the ones on which this book focuses.

Analyze the Current System

Assume the owners of a local stationery store have decided to computerize the store's ordering and inventory control procedures. By doing so they hope to know the quantities of inventory in stock at any time, what products are selling the best, when to reorder to avoid being out of stock, and so on. Where might they begin? All aspects of the current system need to be studied carefully. This is usually done with the help of an information system professional known as a **system analyst**. The analyst studies every aspect of the existing system to get a clear understanding of what things are done and how. He or she also attempts to identify any problems associated with the system.

There are several ways the analyst can go about this task, but the most effective technique is to talk to users of the system. The **users** are people who are directly involved with the system in their day-to-day activities. (In our example, the store buyer, the clerks, and so on are users.) They are the ones who can most clearly define each system function and any problems associated with that function.

Define the New System Requirements

Once the existing system is understood, the requirements of the new or changed system are defined in the second step of the SDLC. The analyst should specify what needs to be done, not how to do it. These requirements should state which changes are necessary to eliminate the problems identified in the initial analysis. For example, concerning the required system outputs, the analyst might ask: What types of reports are required? What information should each report contain? How should the reports be formatted? What headings, spacing, and so on should be used? If users of the system will interact directly with it, the analyst may need to define screen formats for display on terminals, desktop computers, laptop computers, and even personal digital assistants (PDAs).

In addition to input and output requirements, all storage and processing requirements must be defined. For example, what files will be needed? What data should be contained in those files? After all the requirements are worked out by the analyst, he or she prepares a report outlining these

Figure 1–1
System Development Life Cycle (SDLC)

1. Analyze the current system.
2. Define the new system requirements.
3. Design the new system.
4. Develop the new system.
5. Implement the new system.
6. Evaluate the new system.

requirements for management. Management can then decide whether to proceed. If the decision is yes, the next step of the SDLC can begin.

Design the New System

In the third step of the SDLC, either the analyst or a coworker known as a **system designer** uses the requirements defined in the preceding step as a basis for designing the new or modified system. He or she determines how the system will be constructed. Any of several tools may be used to illustrate the system design.

One such tool is a **system flowchart**, which is a graphical representation detailing all of the programs within a system and how they interrelate. For each program, the system flowchart shows all the major inputs and outputs. Figure 1–2 illustrates a system flowchart for a payroll system. Each symbol in the system flowchart has a special meaning. These symbols will be discussed in more detail in Chapter 2.

In this example the data on time sheets is entered into a payroll program, together with data from the payroll master file. (Master files will be discussed in Chapter 7 and Chapter 14.) The payroll program produces two reports and the actual paychecks. This example shows what data the program needs to begin its processing and what output the program actually produces. No detail is given as to how the program actually works. This will be done when the new system is developed.

In recent years much emphasis has been placed on automating the tasks of system development. The goals underlying such automation are to gain more control over the processes of system development (to avoid large budget overruns; to avoid projects that are months or years late or never completed at all); to improve programmer productivity (especially in businesses with not enough trained people to do all the necessary system

Figure 1–2
System Flowchart

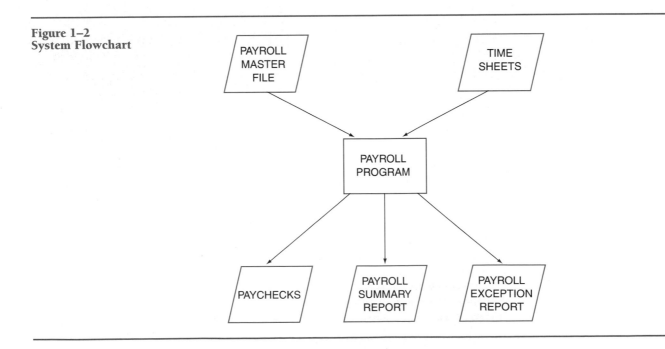

development); to make the programmer's job easier (manual drawing of system flowcharts, for example, can be very tedious); and to improve program quality (a program that is used to direct airplanes in flight or to maintain a hospital's life-support system must be right). The term **computer-assisted software engineering (CASE)** has been coined to refer to the automation of tools, methods, and procedures for system development. The widespread availability of desktop computers has accelerated the trend toward the use of CASE. By increasing your basic analysis and design skills through study of this book, you can enable yourself to make effective use of the ever-increasing array of CASE tools in the marketplace.

Develop the New System

In the fourth step of the SDLC, each of the programs called for in the system design is constructed. For example, the details of the payroll program shown in the system flowchart in Figure 1–2 will be specified. This development stage is composed of a series of well-defined steps called the **program development cycle (PDC)** (see Figure 1–3). The steps of the PDC are carried out by a **programmer**. A description of the types of activities in each step follows.

Review the Program Requirements   In the first step of the program development cycle, the programmer reviews the requirements defined in the second step of the SDLC. If anything is unclear at this point, the programmer asks for more information from the system analyst who wrote the original requirement, from the system designer, or even from a future user of the program (such as the store buyer). The programmer must not make unfounded assumptions. It is just as useless to solve the wrong problem correctly as it is to solve the right problem incorrectly.

Develop the Program Logic   In this step the actual processing steps within each program in the system are developed. We shall focus on this step throughout this book. Just as there are tools to use in the design of a system, there are tools to use in the design of a program. Two common tools discussed in detail in this book are **program flowcharts** and **pseudocode**. Like system flowcharts, program flowcharts contain standard symbols. These symbols graphically depict the problem-solving logic within a program. Pseudocode consists of English-language statements that describe the processing steps of a program in paragraph form. Figure 1–4 illustrates both a program flowchart and pseudocode to compute and print an employee's paycheck.

Figure 1–3
Program Development Cycle (PDC)

1. Review the input, processing, output, and storage requirements.
2. Develop the logic for the program.
3. Write the program using a programming language.
4. Test and debug the program.
5. Complete the program documentation.

Figure 1–4
Program Flowchart and
Pseudocode

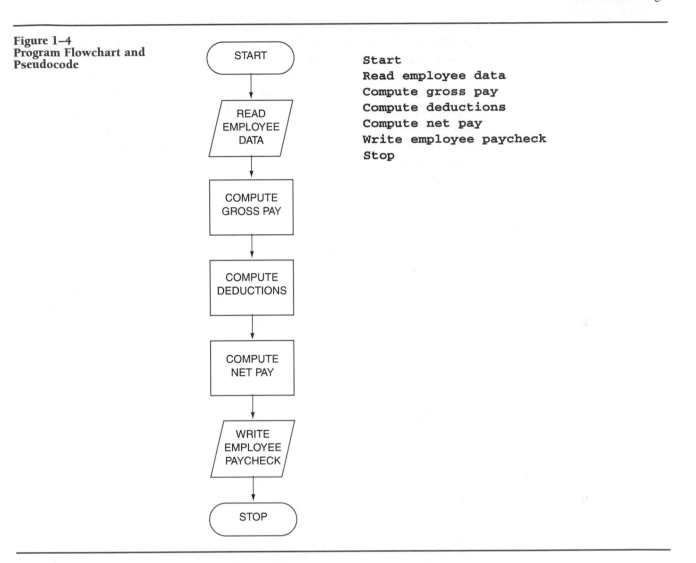

```
Start
Read employee data
Compute gross pay
Compute deductions
Compute net pay
Write employee paycheck
Stop
```

Another important program design tool is the **hierarchy chart**, or **structure chart**. Structure charts are used in program design to show the relationships among parts of a program. Each program part is called a **module**, and the process of breaking down a program into parts or modules is called **modularization**. Figure 1–5 illustrates a structure chart for a payroll program.

In this example, the payroll program is subdivided into three major modules. Each module is identified by a name and a number. Module B010 (compute pay) is further subdivided into three modules. Notice that descending levels on the structure chart represent modules of greater detail or refinement.

There are many ways to divide a program into modules. Some are better than others. We will explore modularization in more detail in Chapter 6.

Program flowcharts, pseudocode, and structure charts are very useful tools during program design. They also provide handy references during program modification. So do **class diagrams** expressed in the **Unified Modeling Language**, or **UML** (see Figure 1–6). Notice that the names of

Figure 1–5
Structure Chart

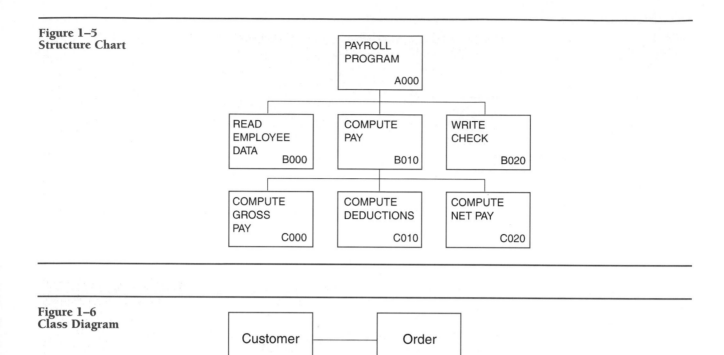

Figure 1–6
Class Diagram

the modules at the second and third levels on the structure chart in Figure 1–5 contain verbs (Read, Compute, Write). The names of the classes represented by the rectangles in Figure 1–6 do not contain verbs; they are nouns. That is one clue to a major difference between procedure-oriented design and object-oriented design. **Procedure-oriented design**, discussed in Parts 1 and 3 of this text, emphasizes processes. **Object-oriented design**, discussed in Part 2, emphasizes data.

Once the design of a program has been completed, it should be reviewed by a small group of programmers and analysts. In this way, potential problems that might have gone unnoticed by the person who designed the program surface early in the program development cycle. (It is always easier to find someone else's error than to locate your own.) The goal of the review is to ensure that high-quality software is produced. A good program is one that is reliable, producing what it is expected to produce. In other words, it works!

Another characteristic of high-quality software is ease of use. As a user, you want programs to be **user-friendly**. For example, assume you are about to use a program to compute the average of a series of numbers. You load the program into the computer and execute it. You watch the screen,

Figure 1–7
Sample Input Screen (Not User-Friendly)

Figure 1–8
Sample Input Screen (User-Friendly)

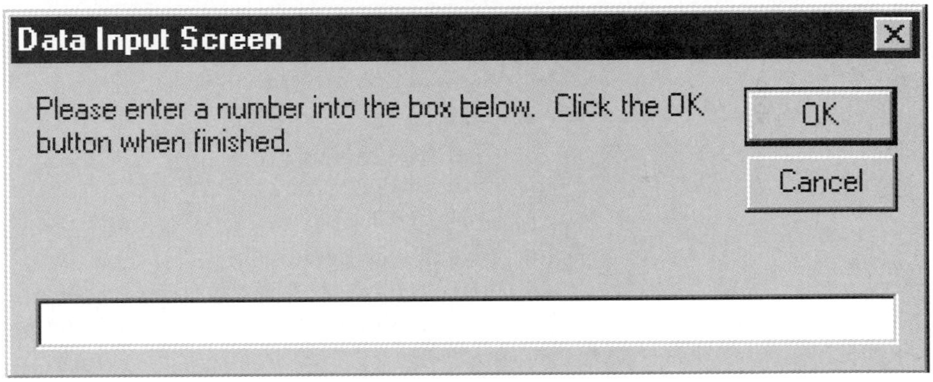

and a question mark (?) appears (see Figure 1–7). What should you do? It's not clear whether you should type in all the numbers or only one number at a time. If you type in all the numbers, should they be separated by spaces or commas? You cannot tell from the screen. Now, suppose you load and execute another program to compute an average. This time a message on the screen tells you exactly what to do (see Figure 1–8). Which program would you rather use? Which one is user-friendly?

Both Figures 1–7 and 1–8 are examples of what is called a **graphical user interface (GUI)**. Graphical user interfaces make it easier for the user to interact with the computer. They present information in an easy-to-use point-and-click fashion. For example, once the user enters a number in the box in Figure 1–8, the user can simply click the OK button using some type of pointing device. Common pointing devices include mice and track balls. Thus, a GUI minimizes the amount of typing required by presenting the user with choices from which desired options can be selected by point-and-click actions.

What about ease of use from a programmer's point of view? Although a programmer does not use a program in the same sense that a user does, he or she must be able to understand the steps in the design if modifications are required. Consider a program that handles income tax calculations based on current tax laws. What happens when some of these laws change? The program will have to be modified to reflect the changes. In all probability, the person modifying the program will not be the one who wrote it originally, so the program logic must be clear and easy to follow. This textbook discusses methodologies that are used to design reliable, well-written programs.

Write the Program   After the program logic has been developed, it must be expressed in a programming-language form. The selection of the language depends on the type of application and on the software development tools available. In the past, most business applications were written in COBOL because it is oriented to business use. Then structured languages such as Pascal and C were used more frequently because they contain language elements that closely parallel the structures used in program design. A programmer who uses these **procedure-oriented languages** codes the specific steps in the problem-solving process. Programming examples in Basic, a simple-to-use procedural language, are illustrated in subsequent chapters of Part 1 in this book.

In addition to this traditional procedural approach, another approach to program development using **fourth-generation languages (4GLs)** has become common. Most 4GLs are **nonprocedural languages**; a programmer using a 4GL defines *what* needs to be done, not *how*. For example, many popular software packages (word processors, electronic spreadsheets, and database managers, to name a few) contain programming statements that can be used alone or in combination (macros and programs) to accomplish a task more efficiently. Even personal computer (PC) users will find that a strong base in programming logic can be a tremendous help when using a particular package to implement decision-making activities.

With the increased use of applications containing graphical user interfaces, visual languages such as Visual Basic and Visual C++ have become popular. These languages support the concept of **event-driven programming**. An event-driven program is one that is designed to respond to actions that occur when the program is executing. The actions can be initiated by the computer or by the user. For example, when the user clicks the OK button illustrated in Figure 1–8, a click event is triggered. The programmer must design the logic associated with that click event. We will see many examples of event-driven programs in the Visual Basic examples later in this book.

Finally, widespread use of the Internet and the World Wide Web have been accompanied by rapid growth in the use of the Java programming language. Java is often described as similar to C++, but without some of C++'s complex programming constructs. Java and C++ are **object-oriented languages**. Java and C++ programmers deal with classes and objects, which are instances of those classes, as discussed in Part 2. More and more web users are creating their own display screen designs, or web pages. Web pages are becoming increasingly more sophisticated as users learn to use web authoring tools. Small segments of Java code, called Java applets, are commonly embedded in web pages to create exciting and dynamic Internet-based applications.

Thus, programming logic is an essential skill in computer-related jobs. Some users write programs using conventional or event-driven

**Figure 1–9
Online Help Screen**

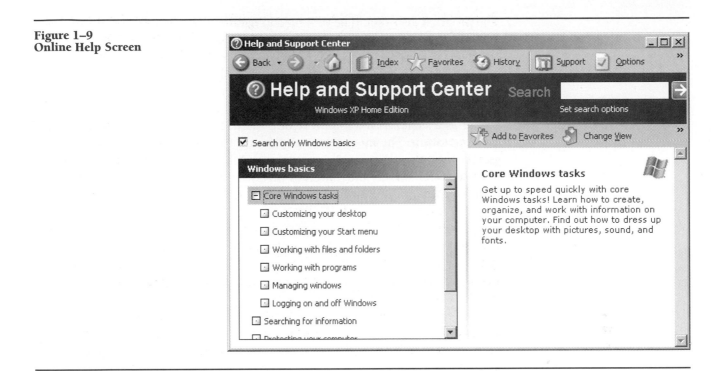

programming languages; some use popular packages as tools; some use previously written programs as applications. Many users write programs designed for the Web. This is not to say that anyone can program. Training in program design and the use of the tools is necessary. The amount of detailed logic that users need differs according to what they do. Many of the CASE tools include basic tutorials and **online help systems** designed to help users, analysts, designers, and programmers use the applications and tools effectively (see Figure 1–9).

Test and Debug the Program   In this step the program is checked for errors **(bugs)** and tested with sample data to see if actual results produced by the program match expected results. There are two main categories of errors that may exist in a program: syntax errors and logic errors. **Syntax errors** occur when the programmer does not follow the rules of the language he or she is using. For example, many statements in C must end with semicolons. If the programmer forgets to enter a semicolon, he or she makes a syntax error. Such errors are usually easy to find and eliminate. A **logic error** occurs when a step in the program logic is incorrect. For example, an averaging program will produce the wrong answer if the sum of the numbers to be averaged is divided by a number other than the total count of numbers. Such errors are not always easy to detect. The programmer may need to go back and review the logic created in the second step of the program development cycle in order to locate the error.

The program also must be tested thoroughly with computer help using simple data. Every possible condition should be tested, if feasible. For example, if a payroll program computes an employee's pay at time and a half for all hours worked over 40, the program should be tested using data values that are less than 40 hours, exactly 40 hours, and greater than 40 hours.

The program also should be designed to handle input data that is invalid. For example, an hours value that is less than 0 in the payroll example should be detected and noted in an error message or on an input exception report.

The programs, modules, or classes that make up a system should first be tested individually to make sure that each works correctly. This is called **unit testing**. Then the programs, modules, or classes should be tested together as a system, either entirely or in part. This is called **system testing**, or **integration**. The more thorough the unit testing is, the more likely that the integration will proceed successfully. All too often, the unit testing is not adequate and the integration of the system components becomes a tedious, time-consuming sequence of dos and redos.

Complete the Program Documentation   Documentation is an ongoing process that occurs throughout the program development cycle. A technical reference needs to be created for programmers who may need to modify one or more of the programs. User guides need to be created to tell users how to use and operate the system. Some of the documentation may be in hard copy (printed) form; other documentation may be provided as online help for convenient reference by users. We cannot overemphasize the importance of good documentation. It is a tremendous help during program development and a vital aid during program modification.

Implement the New System

After the program development cycle is completed, the new system is implemented. Users are trained, and operating procedures are defined. The system documentation is reviewed, revised as necessary, and prepared in its final form. Any of several implementation strategies may be employed: The new system may be run concurrently with the existing (old) system; the new system may completely replace the old system; or the new system may be phased in gradually. The specifics of each strategy are beyond the scope of this book, but you should be aware that they exist.

Evaluate the New System   In the sixth step of the SDLC, the new system is evaluated to determine if it is meeting the required objectives. If some of the objectives are not being met, parts of the system may have to be modified. In any case, a system evaluation report should be prepared for management. It is likely to serve as one input to a subsequent system development life cycle.

Structured Programming

As stated earlier, this text will focus on the second step of the program development cycle, developing the program logic. There are many ways to solve a problem, particularly a complex one. There are usually many correct solutions; however, some are more desirable than others. In this text we will adhere to the principles of a methodology called **structured programming**, a technique that has proven to be very effective in solving problems as well as in modifying solutions. The principles of structured programming are fundamental to procedure-oriented programming (Parts 1 and 3 of this book). They are also relevant to object-oriented programming

(Part 2). That is why we need to look at them now, at the outset of our discussion of program design.

Structured programming is the ability to express a problem solution using only three basic patterns of logic. These patterns are referred to as **control structures**. The patterns are based on the computer's ability to execute instructions in a step-by-step, sequential manner; its ability to make decisions; and its ability to repeat instructions. The theoretical framework for this approach is usually traced to a paper by C. Bohm and G. Jacopini, initially published in Italian in 1965, then republished in English in 1966.[1] Their "structure theorem," which appears in that paper, is generally accepted as a proof of the claim that the three structures are sufficient for programming. In addition, as early as 1965, Professor E.W. Dijkstra of The Netherlands insisted that programs using definite structuring were easier to write, read, and verify.[2] There is a vast amount of literature documenting numerous program development projects in which this does, indeed, appear to be the case. This information is as relevant today as it was more than 35 years ago.

The following sections will use familiar examples to illustrate the three patterns of structured programming. They should give you a sense of how each structure works. The next several chapters will teach you how to use each structure to solve specific, well-defined problems.

Basic Control Structures

The three patterns referred to in the previous sections are called the SIMPLE SEQUENCE control structure, the IFTHENELSE control structure, and the DOWHILE control structure.

SIMPLE SEQUENCE Control Structure   The **SIMPLE SEQUENCE control structure** represents the computer's ability to execute instructions in a step-by-step, sequential manner. A simple example not involving the computer can be used to illustrate this concept. The sequential steps given in Figure 1–10 represent directions you might give a friend to get to your house. The steps must be followed in a sequential manner. If not, your friend might never find your house.

Figure 1–10
SIMPLE SEQUENCE:
An Example

1. Proceed down Main Street for two miles.
2. Turn left on Ocean Drive.
3. Proceed on Ocean Drive for three blocks to the fork.
4. At the fork, take Swan Street to the left.
5. Proceed two blocks.
6. House is second on the left (246 Swan Street).

---

[1]C. Bohm and G. Jacopini, "Flow Diagrams, Turing Machines and Languages with Only Two Formation Rules," *Communications of the ACM* 9,5 (May 1966): 366–71.

[2]Among Dijkstra's writings on the subject are "GOTO Statement Considered Harmful," Letter to the Editor, *Communications of the ACM* 11,3 (March 1968): 147–48; "The Structure of the Multiprogramming System," *Communications of the ACM* 11,5 (May 1968): 341–46; and "Structured Programming," in J.N. Buxton and B. Randell, eds., *Software Engineering Techniques* (NATO Scientific Affairs Division: Brussels 39, Belgium, April 1970): 84–88.

Figure 1–11
IFTHENELSE: An Example

1. Proceed down Main Street for two miles.
2. Turn left on Ocean Drive.
3. Proceed on Ocean Drive for three blocks to the fork.
4. IF left turn at fork is blocked THEN
   Take right turn at fork onto Eagle Street.
   Proceed five blocks.
   Turn left at Clifton Avenue.
   Proceed three blocks.
   Turn left at Circle Drive.
   Proceed two blocks.
   Turn left at Swan Street.
   House is fourth on the right (246 Swan Street).
   ELSE
   At the fork, take Swan Street to the left.
   Proceed two blocks.
   House is second on the left (246 Swan Street).
   ENDIF

IFTHENELSE Control Structure   The **IFTHENELSE control structure** represents the computer's ability to make a decision. Let us modify the SIMPLE SEQUENCE example to illustrate this concept. Suppose the left turn at the fork is blocked at certain times. You will need to provide alternative directions to your friend. You will also need to include the original directions (because they are shorter) to use if the left turn at the fork is not blocked. Notice how the directions in Figure 1–11 incorporate decision-making logic.

Notice that, if the road is blocked, there are eight separate instructions to follow. If the road is not blocked, there are three instructions. Both sequences of instructions are included because you don't know if the road will be blocked. However, your friend will follow only one of the sequences. Although this example is nontechnical, we have included the keywords and indentation that identify an IFTHENELSE control structure. You will learn more about this control structure in Chapter 3.

DOWHILE Control Structure   The **DOWHILE control structure** represents the computer's ability to repeat a series of instructions. A series of repeated instructions is called a **loop**. Let us look at a different example to illustrate this concept. Consider the simple instructions on the back of a shampoo bottle (listed in Figure 1–12). Although these instructions are simple enough for us to understand, a computer would get confused. First of all, the second instruction says to "Rinse," but "Rinse" what? Also, the third instruction says to "Repeat." Clearly, we know to repeat the process from the beginning. However, we need to specify to the computer which steps it should repeat. Figure 1–13 gives a new version of these instructions.

These new instructions clarify some unknowns, but the computer would still have trouble. Can you see why? Every time the computer got to Step 3, it

Figure 1–12
DOWHILE: Example 1

Wash hair.
Rinse.
Repeat.

Figure 1–13
DOWHILE: Example 2

1. Wash hair.
2. Rinse hair.
3. Repeat from Step 1.

Figure 1–14
DOWHILE: Example 3

```
DOWHILE hair is not clean
    Wash hair.
    Rinse hair.
ENDDO
```

would begin again at Step 1, just as the instructions say. This is an example of an **infinite loop**: There is no way out. The computer would theoretically execute forever. In common practice, this would not actually happen. A special program would intercept the program execution and issue an error message indicating that too much time had passed since the program began executing.

Every loop must include a statement that defines how many times to execute the loop steps or under what condition to continue or to stop the looping process. In this example, we could indicate that the steps be performed two times, or that the steps be performed as long as the hair is not clean. Consider our example one more time (see Figure 1–14). We have included the keywords and indentation that identify a DOWHILE control structure. The two indented steps represent what needs to be done, and the first statement (DOWHILE) specifies the condition necessary to continue processing. We will explore DOWHILE loops in more detail in Chapters 4 and 5.

Key Terms

system
computer-based
  information system
system development
  life cycle (SDLC)
system analyst
user
system designer
system flowchart
computer-assisted
  software engineering
  (CASE)
program development
  cycle (PDC)
programmer
program flowchart
pseudocode
hierarchy (structure)
  chart

module
modularization
class diagram
Unified Modeling
  Language (UML)
procedure-oriented
  design
object-oriented design
user-friendly
graphical user interface
  (GUI)
procedure-oriented
  language
fourth-generation
  languages (4GLs)
nonprocedural language
event-driven
  programming
object-oriented
  language

online help system
bug
syntax error
logic error
unit testing
system testing
  (integration)
structured
  programming
control structure
SIMPLE SEQUENCE
  control structure
IFTHENELSE control
  structure
DOWHILE control
  structure
loop
infinite loop

Exercises

1. Name the steps in the system development life cycle.

2. Name the steps in the program development cycle.

3. (a) What does the acronym *CASE* stand for?
   (b) List four major objectives of CASE technologies.

4. What are the characteristics of a high-quality program?

5. (a) What does it mean to say that a program is user-friendly?
   (b) Give an example of a user-friendly program.
   (c) Give an example of a program that is not user-friendly.

6. What is a graphical user interface?

7. What does it mean to say that a language is procedure-oriented?

8. (a) Give an example of a procedure-oriented language.
   (b) Give an example of a visual language.
   (c) Give an example of an object-oriented language.

9. What is meant by event-driven programming?

10. (a) What is the difference between a syntax error and a logic error?
    (b) Give an example of each type of error.

11. What is the difference between unit testing and system testing?

12. Give some examples of documentation.

13. Name the three basic control structures of structured programming.

SIMPLE SEQUENCE Control Structure

Objectives

Upon completion of this chapter you should be able to

- Define an algorithm.
- Name the characteristics of an algorithm.
- Name the four parts, in order, of a data hierarchy.
- Identify, and use in program design, the SIMPLE SEQUENCE control structure.
- Define the terms *information processing* and *information-processing system*.
- Distinguish between a program flowchart and a system flowchart.
- Identify, and use in program design, the terminal interrupt, general I/O, and process program flowcharting symbols.
- Distinguish between a variable and a constant.
- Identify, and use in program design, variables, constants, and assignment statements.
- Distinguish between input and output, and explain why input and output steps are needed in program design.
- Perform a procedure execution of a simple algorithm.

Introduction

A computer is an extremely powerful, fast machine. In less than a second, it can perform difficult calculations that otherwise would take days, months, or years to perform. Yet a computer has no magical power; it is only a tool. It cannot devise a plan or decide to act. It can do only what it is specifically told to do. We can direct the computer to do what we want by specifying our needs in a discrete step-by-step manner. Specifically, we must develop an **algorithm**, which is a step-by-step procedure to solve a problem.

A computer cannot act independently; it has no intelligence of its own. For this reason, any algorithm that we use to direct it must be set up to identify all aspects of a processing situation and to present, in detail, all steps to be performed.

The algorithm must meet the following requirements:

- Use operations from only a given set of basic operations
- Produce the problem solution, or answer, in a finite number of such operations

The concept of *a given set of basic operations* is important because the computer can perform only certain operations; its capabilities are planned very carefully by hardware designers who lay out specifications that direct subsequent construction of the machine. The concept of *a finite number of such operations* is important because each operation performed by a computer takes a certain amount of time (say, from one-millionth to one-billionth of a second, depending on the machine architecture). If an unlimited number of steps is required, it is impossible, even using the fastest computer available, to obtain the solution in a finite amount of time.

The term *algorithm* may be new to you, but most of us use numerous algorithms daily. For example, we adopt routine procedures, or algorithms, for getting up in the morning, fixing meals, going to work, and so on. A typical wake-up algorithm is shown in Figure 2–1.

Figure 2–1 is an example of a **program flowchart**. It may also be called a **block diagram**, or **logic diagram**. Very simply, a program

Figure 2–1
Wake-Up Algorithm

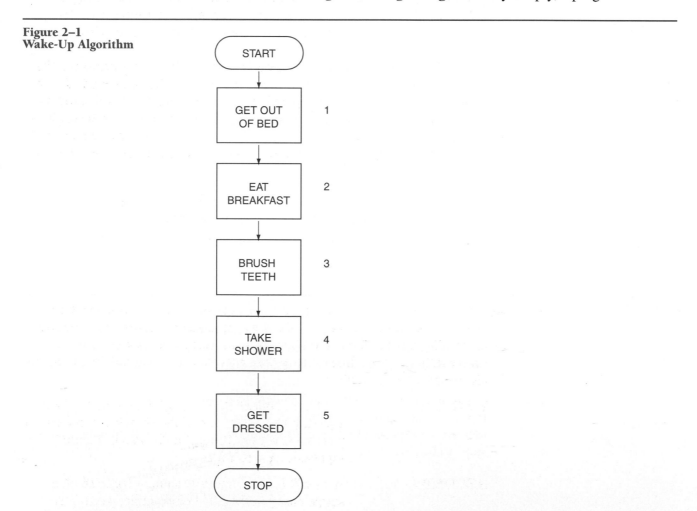

flowchart is a picture of an algorithm. Each step in this algorithm is defined in a rectangle, the symbol used to indicate a specific process. The rectangle represents the **process symbol**. Every algorithm must have one entry point (START) and one exit point (STOP). These are indicated by the ellipsis symbols, called **terminal interrupt symbols**.

All symbols on a program flowchart are connected by flowlines, which may or may not have arrowheads. The arrowheads show the direction of flow. They can be omitted when the flow is from top to bottom or left to right, but are otherwise required. Program flowcharts are usually easier to read when arrowheads are used.

Notice that the flowchart in Figure 2–1 has five steps, all processed one after another. The flowchart illustrates the first of the three basic control structures used to develop algorithms. As you may recall from Chapter 1, this structure is called the **SIMPLE SEQUENCE control structure**. It is the simplest and most frequently used of the three basic control structures.

In common practice, the algorithm shown in Figure 2–1 is carried out somewhat informally. The person who executes the algorithm may even be unaware that he or she is following an algorithm. In other situations, however, algorithms are more formally defined. For example, business operations within a company are firmly established—uniform accounting procedures must be followed; inventory must be tightly controlled; manufacturing volumes must be correlated with both distribution and sales; and so on.

All of these business applications involve some form of paperwork. The term **information processing** is really another name for paperwork. It is a series of planned actions and operations upon data to achieve a desired result. The methods and devices that achieve the result form an **information-processing system**. Regardless of the kind of data processed or the methods and devices used, all information processing systems involve at least three basic elements:

- The source data, or input, entering the system
- The orderly, planned processing within the system
- The end result, or output, from the system

Data Hierarchy

Before we look at our first formal algorithm, we will discuss some terms associated with the structure of the data processed by algorithms. The four terms, **file**, **record**, **field**, and **character**, make up what is known as the **data hierarchy**. This hierarchy defines how data is structured for use in the algorithms that we will explore.

FILE   A file is a collection of related data or facts. For example, payroll facts for *all* the employees in a company form a payroll file. All the data in Figure 2–2 represent a payroll file.

RECORD   A record is a collection of data, or facts, about a single entity in the file. For example, payroll facts for a *single* employee in a company form a payroll

Figure 2–2
Data Hierarchy

EMPLOYEE NUMBER	EMPLOYEE LAST NAME	EMPLOYEE FIRST NAME	PAY RATE	CITY	STATE
06337	Jones	Fred	10.50	New York	NY
09155	Smith	Susan	7.25	Atlanta	GA
16840	Adams	John	8.35	Atlanta	GA
31719	Baker	Jane	12.60	Detroit	MI
⋮					

Character — *Field*
Record
File

record. Each line in Figure 2–2 represents one payroll record.

FIELD A field is any single piece of data, or fact, about a single entity (record) in a file. For example, an employee number might be one of the fields in a payroll record. The employee number 06337 represents a field in the first payroll record in Figure 2–2

CHARACTER A character is a letter (A–Z, a–z), number (0–9), or special character (for example, . or ? or %, and so on). For example, in Figure 2–2 the employee number contains all numeric characters; the employee last name contains all alphabetic characters; and the pay rate contains both numeric characters and a special character.

Sales Application Example

Consider the sales operations of a large department store. A sales manager delegates responsibilities to supervisors in various departments. Each supervisor submits a weekly sales report that is checked against inventory changes and then used as the basis for figuring commissions. Each supervisor is confronted with the task of preparing the weekly sales report for his or her department. In some stores, each supervisor may be asked to develop his or her own procedure. In other stores, the sales manager develops a detailed set of instructions for each supervisor. In either case, the supervisor starts with the sales data available (the input), thinks about the weekly sales report that is needed (the output), and plans how to prepare the report on the basis of the sales data (the process).

Sales Application System Flowchart

Before a program flowchart is developed to solve a problem, a **system flowchart** is often created to show more general information about the application. A system flowchart illustrating the basic inputs, processes,

**Figure 2–3
Sales Problem (System
Flowchart)**

and outputs for the sales program needed for this application is shown in
Figure 2–3.

The leftmost symbol on the system flowchart is the **general input/
output (I/O) symbol**; that is, the parallelogram. It always represents either input (data available for processing) or output (processed data, or
information, available to users or for additional processing). Because
this symbol can be used for both input and output, it appears twice on
the flowchart.

The rectangular symbol appears again in Figure 2–3. Here it is a general-purpose symbol indicating, collectively, all the processing steps within
the sales program. When this symbol appears on a system flowchart, it represents an action or a series of actions performed with computer help—the
carrying out of an algorithm expressed in a computer-program form.

One important function of a flowchart is to aid in problem analysis and
solution planning; another is to aid the problem solver in communicating
ideas to others. To help standardize such communication, the American National Standards Institute (ANSI) coordinated the development of a standard set of flowcharting symbols and associated meanings. We shall use
many of these symbols in the flowcharts in this book. Their shapes and
meanings are summarized in Appendix A.

Thus, a system flowchart shows the data, flow of work, and workstations
within an information-processing system. In this book we shall deal with
simple problems requiring only one computer program for their solution. A
system flowchart may show the flow of work through several programs,
each having several inputs and outputs, but such systems are beyond the
scope of this book.

Sales Application Program Flowchart

Although a system flowchart is very helpful in showing the inputs, major
processing functions, and outputs of an information-processing system, it
gives only a limited amount of detail about how the computer performs the
specific processing steps. The system flowchart in Figure 2–3 shows that a
program for computing a weekly sales report is to be written and executed.
It does not show which mathematical operations are needed or the order in
which the operations must be performed. To provide this detailed information, we construct a program flowchart.

In the program flowchart, the detailed steps needed to process the data
about one person's sales must be specified. It is sometimes helpful to work
backward. In this flowchart, for example, we should first decide what we
need to know to compute a salesperson's pay. Usually, a salesperson's pay
is determined in part by how much he or she has sold in a given week. In
this example, the sales will be divided into two categories: those items sold

at regular price, and those sold at a reduced or sale price. Each person will be paid a base amount as well as a percentage of his or her weekly sales. This percentage (commission) will be 6 percent on the sales of regular-priced items and 3 percent on the sales of sale-priced items.

Here is a representative sequence of instructions to solve this problem:

1. Input the two weekly sales totals for any employee.
2. Compute the regular commission by multiplying the regular sales amount by 6 percent.
3. Compute the sales commission by multiplying the reduced sales amount by 3 percent.
4. Compute the total pay due: $200.00 (base pay) + regular commission + sales commission.
5. Output the total pay on the payroll report.

These instructions also can be expressed pictorially, whereby the steps to be followed can be seen at a glance. A program flowchart of the required steps is shown in Figure 2–4.

Figure 2–4
Sales Problem (Program Flowchart)

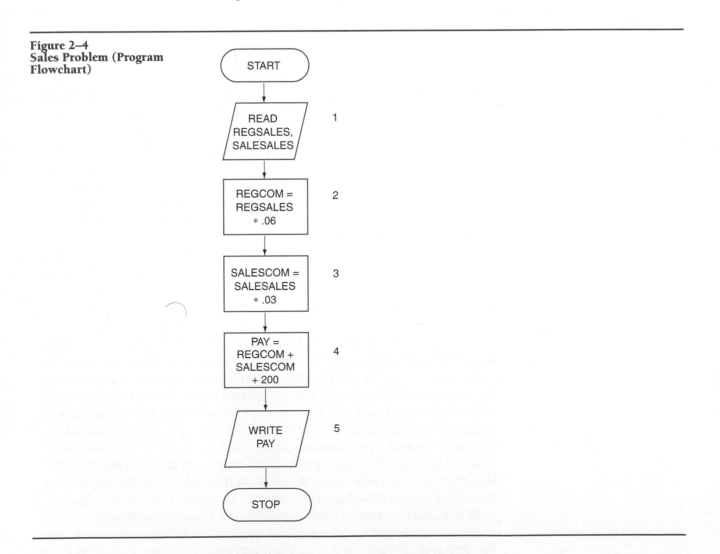

We have just traced the first two major parts of any problem-solving task:

- Define the problem to be solved.
- Develop a solution algorithm—steps to be taken to solve the problem.

The five steps on the flowchart in Figure 2–4 correspond to Steps 1 through 5 of the written procedure. Notice, however, the use of the general I/O symbol in Steps 1 and 5. It is important to understand that, before the pay can be computed, the sales data must be known to the computer. We get this data into the computer in an **input** step (notice the I/O symbol). This is almost always one of the first steps in an algorithm. Typically, the word *READ* is used to indicate input, but a word such as *ENTER*, *GET*, or *INPUT* can also be used.

In Step 1 the word *READ* is followed by the words *REGSALES* and *SALESALES*. These words are names for variables. **Variables** are data items whose values may change, or vary, during processing. We create **variable names** to represent, or refer to, these data items. One common mistake is to put the specific data right in the program, as follows:

```
READ
$1000,
$3000
```

This step may lead the reader to believe that the salesperson sold $1000 worth of regular items and $3000 worth of sale items. These numbers, although correct for one employee, are likely to need changing each time the program is run because each employee's sales data is likely to differ. This problem is handled by the use of variables. The variable names are place-holders for the specific data values. The value of a variable can change each time a program is run, or even within one run of the same program. Think of a mailbox. The address (variable name) is always the same, but the mail (content) is different every day.

When a variable is used in conjunction with a READ statement, we can assume that the value of the variable is now known to the computer; this is called a **defined value**. The use of variables helps to ensure that the program is **data independent**, which means that the program will perform the required processing steps on any set of input data. Figure 2–5 shows a snapshot of the computer's memory at this point. We will refer to the figure as a **memory diagram**. On the left is a list of all the variable names referenced in the algorithm. The boxes on the right represent the actual values that these variable names refer to in the computer's memory. At this point REGSALES and SALESALES represent values that were input in Step 1. Actually, these two amounts can be any numbers equal to the values of whatever was sold. In our example, we will assume that the regular sales amount (REGSALES) was $1000.00 and the reduced sales amount (SALE SALES) was $3000.00. Only the values of REGSALES and SALESALES are defined to the computer at this point.

In Step 2, the regular commission is computed. Because this step involves arithmetic, it is represented by a process symbol, not by an I/O symbol. In this step we are illustrating the use of an important fundamental

Figure 2–5
Sales Problem—Memory
Diagram 1

REGSALES	1000
SALESALES	3000
REGCOM	
SALESCOM	
PAY	

statement—the **assignment statement**. This statement is commonly used when describing computer processing. An assignment statement must adhere to the following rules of formation:

1. Only a single variable name may appear to the left of the assignment symbol, which in our example is =.

2. Only a single variable name, constant, or expression may appear to the right of the =.

3. Everything to the right of the = must be known (defined) to the computer.

Step 2 uses an assignment statement to compute the commission on the regular sales. Because the percentage is always 6 percent, the .06 can be used in the statement. The .06 is an example of a **constant**, a value that never changes. Because .06 is a constant, it can be placed in computer storage when the program itself is placed in computer storage. Therefore it does not have to be input during processing. The times sign (*) is defined because it refers to a basic arithmetic operation the computer can perform. The asterisk symbol is used to show multiplication because the traditional times sign, ×, may be confused with the variable name X. The only other item to the right of the = is the variable name REGSALES. Because the value of this variable was previously input (Step 1), it is also known at this time. Because everything to the right of the = is defined, the computer can now do the calculation and assign the result to the variable on the left of the =. When carrying out an assignment statement, the computer operates on everything defined to the right of the = first. In this example, $1000 is multiplied by .06, and the result, $60, is assigned to the variable name REGCOM. REGCOM is now known or defined to the computer (see Figure 2–6).

Step 3 computes the commission on the sales items (remember 3 percent). Can you see that everything to the right of the = is defined to the computer? SALESALES is multiplied by .03, and the result is assigned to SALESCOM (see Figure 2–7).

Step 4 then adds the two computed commissions to the base pay of $200 to come up with the salesperson's total pay. Notice that the base pay is a constant—all salespeople get at least $200. In this assignment statement, values for the two variables on the right were not input but are still defined to the computer. Any variable that is the result of a computation (left side)

Figure 2–6
Sales Problem—Memory
Diagram 2

REGSALES	1000
SALESALES	3000
REGCOM	60
SALESCOM	
PAY	

Figure 2–7
Sales Problem—Memory
Diagram 3

REGSALES	1000
SALESALES	3000
REGCOM	60
SALESCOM	90
PAY	

in a previous step is defined and can be used in subsequent steps. As a result of Step 4, the variable name PAY is now defined and has a value of $350 (see Figure 2–8).

At this point, the pay for a single employee has been computed, but it is still inside the computer. Often we want to see that pay on a piece of paper. If so, it must be output from the computer to a printer. We show this operation in Step 5, again using an I/O symbol, with the word *WRITE* indicating **output**. Words such as *OUTPUT*, *PRINT*, or *DISPLAY* also can be used. The current value of a variable can be output, provided it has been defined, either by being input or by being computed. We see here that PAY was computed in Step 4.

Throughout this process, several variable names have been introduced. Usually, variable names are chosen by the programmer. It is a wise idea to choose descriptive names, such as SALES, instead of S. Descriptive variable names make the algorithm (in flowchart form and later in programming-language form) much more self-documenting and easier to read.

Figure 2–8
Sales Problem—Memory
Diagram 4

REGSALES	1000
SALESALES	3000
REGCOM	60
SALESCOM	90
PAY	350

Design Verification

When the programmer is satisfied that all processing steps have been identified and provided for, the solution algorithm should be verified. The objective is to prevent errors from occurring, or, if some have already occurred, to detect and eliminate them as soon as possible. In the past, a major portion of a programmer's time was spent, not in program design and coding, but rather in debugging and testing. Today, many computer professionals insist that this need not be the case: A program can be written correctly, so that it executes properly the first time it is run. A careful, early verification of the program design, or solution algorithm, is an essential step in achieving this objective.

Under one approach to verification of design, the design documentation is distributed to selected reviewers, who are asked to study it and respond within a set time. Each reviewer is directed to note, individually, any required changes, additions, and deletions. This approach is known as an **informal design review**.

Another approach is the use of **structured walkthroughs**. At this point in algorithm development, the walkthrough is a **formal design review**. Here, the design documentation is made available to several people selected to serve as members of a review team. After the reviewers have had time to prepare, they meet together with the program designer and a moderator for an established period, usually about two hours. Each reviewer is expected to have studied the design documentation and is asked to comment on its completeness, accuracy, and general quality. Then the moderator "walks" the group through each step of the documentation, covering any points raised by the review team.

How does one start when reviewing a solution algorithm set forth in a program flowchart or other form of design documentation? Some reviewers, individually or in groups, find that one effective approach is to pretend to be the computer. Representative values for all types of input are selected: (1) data that is normally expected; (2) valid but slightly abnormal data (e.g., minimum and maximum values allowable); and (3) invalid data. The individual or group follows the problem-solving logic step by step to process the input and determine what output will be produced. If the output matches predetermined correct results, the logic within the solution algorithm is upheld. Pretending to be the computer in this way is called **simulation**, **procedure execution**, or **desk checking**. While tracing the problem-solving logic, each reviewer must be careful not to make any assumptions. The computer can do only what it is told; it is unable to make assumptions.

Sample Problem 2.1 (Temperature Conversion Problem)

Problem:

The International Broadcasting Company wants a computer program that will accept a temperature reading expressed in Fahrenheit degrees as input, convert the value to Celsius degrees, and provide both the Fahrenheit value and the Celsius value as output for its hourly weather report.

Solution:

The system flowchart for this application is shown at the left in Figure 2–9. A program flowchart representation of the solution algorithm, showing how the problem is to be solved, is given at the right.

The terminal interrupt symbol containing START identifies the beginning of the program. First, the Fahrenheit value (FARENHT) is read as input. Figure 2–10 shows a memory diagram assuming an input value of 77 degrees.

In the next step, a familiar mathematical formula is executed to convert this value from Fahrenheit to Celsius. This formula contains three constants (32, 5, and 9) and several mathematical operators. The computer evaluates mathematical expressions in a specific order. All multiplication and division operations are evaluated first, in a left-to-right order. Next, all addition and subtraction operations are evaluated, again from left to right. We can place parentheses around part of an expression if we want the computer to evaluate that part first. If there are several sets of parentheses, the innermost parentheses are evaluated first. Figure 2–11 shows how two expressions, one with parentheses and one without parentheses, are evaluated. Notice that the results are different.

**Figure 2–9
Temperature Conversion
Problem**

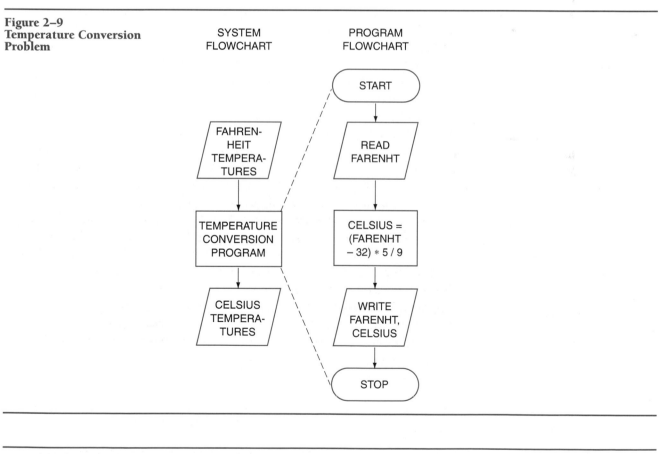

SYSTEM
FLOWCHART

PROGRAM
FLOWCHART

**Figure 2–10
Temperature Conversion
Problem—Memory Diagram 1**

FARENHT	77
CELSIUS	

Figure 2–11
Expression Evaluation

Expression—no parentheses	Expression—with parentheses
3 + 5 * 2 – 8 / 4	(3 + 5) * 2 – 8 / 4
3 + 10 – 8 / 4	8 * 2 – 8 / 4
3 + 10 – 2	16 – 8 / 4
13 – 2	16 – 2
11	14

Figure 2–12
Temperature Conversion
Problem—Memory Diagram 2

FARENHT	77
CELSIUS	25

In Sample Problem 2.1, parentheses are placed around the expression FARENHT–32 because the subtraction needs to be done before the multiplication. Without the parentheses, first the 32 would have been multiplied by 5, then that result would have been divided by 9, and then that result would have been subtracted from the current value of the variable FARENHT. This sequence of operations would have produced an incorrect result.

Figure 2–12 shows a memory diagram after the value of CELSIUS has been computed correctly.

The two temperatures, FARENHT and CELSIUS, are then written as output. Notice that the value of the variable FARENHT is not computed in the algorithm. However, because it is input, it is known to the computer (look again at Figure 2–10). At this point the computer provides two numbers as output without identifying them. In Chapter 3 we will discuss how to label output values. Finally, program execution is terminated.

When this simple sequence of basic operations is executed by the computer, it will provide the solution to this temperature conversion problem.

Sample Problem 2.2 (Billing Problem)

Problem:

A major department store needs a program to prepare a monthly bill for each customer. For simplicity, let us assume that each customer purchases (at most) one type of item each month. For each purchase, there will be four input values: customer name, item, quantity purchased, and price. The output will be the customer's monthly bill after a 10 percent discount is taken before taxes, and a 5 percent sales tax is added.

Solution:

The system flowchart to represent this billing problem appears in Figure 2–13. The program flowchart for the billing program to provide the solution appears in Figure 2–14.

**Figure 2–13
Billing Problem (System
Flowchart)**

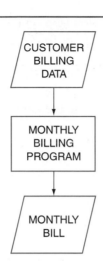

**Figure 2–14
Billing Problem (Program
Flowchart)**

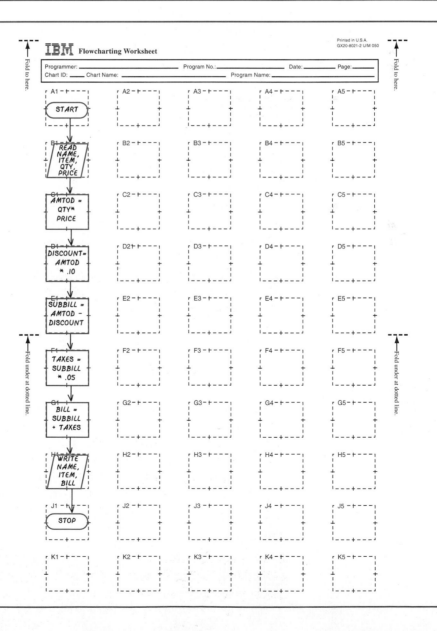

Four values are input in the first step on the program flowchart. Figure 2–15 shows a memory diagram after four specific values have been read during processing.

Notice that both AMTOD and DISCOUNT must be defined before they can be used on the right-hand side of an assignment statement. The next two steps on the flowchart compute these values. The results are illustrated in Figure 2–16.

What other variables must be defined? Desk-check the solution algorithm to verify that each assignment statement operates only on values that are well defined to the computer. What are the names of the variables that are input? What are the names of the variables that are output? What constants are used in this algorithm? (See Exercise 11 at the end of the chapter.)

Before you go on to the exercises, give some thought to the following questions: Think about why we used constants to represent the discount percent (.10) and the tax percent (.05). Could we have represented this data by variables instead? (See Exercise 12.) Also, why do you think the billing data was not constant? Can you think of some general criteria that might

Figure 2–15
Billing Problem—Memory Diagram 1

NAME	MR. JOHN LEE
ITEM	SCARF
QTY	2
PRICE	19.50
AMTOD	
DISCOUNT	
SUBBILL	
TAXES	
BILL	

Figure 2–16
Billing Problem—Memory Diagram 2

NAME	MR. JOHN LEE
ITEM	SCARF
QTY	2
PRICE	19.50
AMTOD	39.00
DISCOUNT	3.90
SUBBILL	
TAXES	
BILL	

determine when data used in a program should be constant and when it should be allowed to vary, that is, to be represented by variables?

Finally, think about the following questions: What happens within the computer during the stages of input? Processing? Output? Do you think each stage is necessary in every algorithm? We'll answer many of these questions, and more, in future chapters.

Flowcharting Tools

The program flowchart in Figure 2–14 is superimposed on a **flowcharting worksheet**. Forms of this type are designed to assist programmers in placing symbols on flowcharts. In full size, the 11-by-16 $1/2$-inch worksheet provides an arrangement of 50 blocks with alphabetic and numeric coordinates: The 10 horizontal rows are lettered from top to bottom—*A* to *K*; the 5 vertical rows are numbered from left to right—1 to 5. The blocks are aids for squaring up flowlines, maintaining uniform spacing between symbols, and providing coordinates (for example, *A1* and *K3*) that can be referred to elsewhere on the flowchart. The worksheet itself is usually printed in light-blue ink so that its guidelines do not appear on photographic copies of the flowchart.

Another tool that may be provided for the programmer's use is a **flowcharting template**. The template is a plastic or metallic card containing flowcharting symbols as cutout forms. The programmer can easily trace the outlines of the symbols needed for both system and program flowcharts. For example, a flowcharting template made available by IBM is shown in Figure 2–17. The flowcharting symbols on this template generally comply with the American National Standards Institute (ANSI) and International Standards Organization (ISO) recommendations summarized in Appendix A. Use of such templates is not only convenient, but also encourages uniformity in flowcharting, which in turn provides for better communication between the programmer and others who refer to the programmer's flowcharts.

Figure 2–17
Flowcharting Template

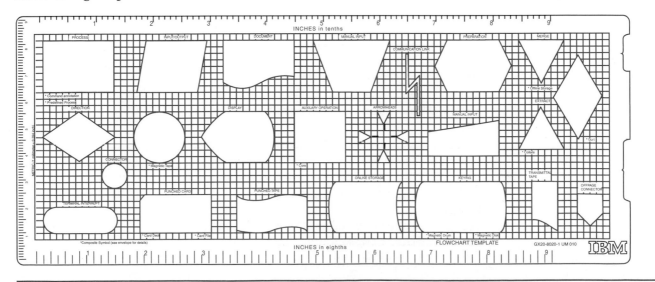

In recent years, software vendors have developed numerous CASE tools that can be used to construct and modify system and program flowcharts. These CASE tools can save the analyst and programmer many hours of work and, consequently, have become quite popular.

Enrichment (Basic)

Basic is a procedural programming language that is relatively simple to use. This section, as well as the other Basic sections in subsequent chapters, does not attempt to teach you how to program. These sections are simply illustrations of some of the program design examples in this book. As you read through these Basic examples, you should get a sense of what actually is involved in the creation and execution of a Basic program.

Figure 2–18 illustrates a listing of the program that solves the temperature conversion problem (see Figure 2–9). Each step on the program flowchart is shown as a separate line in the listing. In Basic there are several ways in which data can be input. In this example, we use the *Input* statement, which will request a Fahrenheit temperature from the user when the program is executed. Note that this Input statement is preceded by a *Print* statement. The Print statement outputs directions to the user that specify what input is needed. Remember our discussion of user-friendly programs in Chapter 1. We have also included two additional Print statements, one for each temperature value. We have added identifying labels to make the output clearer. Text enclosed in double quotation marks will be printed out exactly as entered. The text is a *character-string constant*. The Basic language requires that character-string constants be enclosed in double quotation marks. Some programming languages require that character-string constants be enclosed in single quotation marks, and some programming languages accept either double quotation marks or single quotation marks. (We'll learn more about character-string constants in Chapter 3.) The *End* statement is used in Basic to stop the execution of the program.

Figure 2–19 illustrates the output that will be produced when the program is executed. When the Input statement is executed, a question mark is presented to the user, indicating that the computer is waiting for input.

Figure 2–18
Temperature Conversion Problem (Basic List)

```
PRINT "Enter a Fahrenheit temperature"
INPUT FARENHT
CELSIUS = (FARENHT-32)* 5 / 9
PRINT "The Fahrenheit temperature is "; FARENHT
PRINT "The Celsius temperature is "; CELSIUS
END
```

Figure 2–19
Temperature Conversion Problem (Basic Run)

```
Enter a Fahrenheit temperature
? 212
The Fahrenheit temperature is   212
The Celsius temperature is   100
```

After the user enters a temperature, program execution continues and two lines of output are written, one for each temperature value. Note that in Figure 2–18 a blank space is included in each character-string constant prior to the ending quotation mark. We include a blank space in each character-string constant to ensure that at least one blank space will be output between the label and the temperature value. If the temperature value is positive, the computer outputs an additional blank space instead of a plus (+) sign. However, if the temperature value is negative, the computer outputs a minus (−) sign to the left of the temperature value. Thus, positive temperature values will be preceded by two blank spaces, while negative temperature values will be preceded by one blank space as shown below:

The Fahrenheit temperature is 45   (positive value)

The Fahrenheit temperature is –45   (negative value)

Enrichment (Visual Basic)

Windows is a program that creates a graphical user interface on the screen while managing all the applications running on a computer. Visual Basic is used to create applications that will run in the Windows programming environment. Visual Basic is an event-driven programming language that incorporates graphical user interface design and programming that is similar to Basic. This section, as well as the other Visual Basic sections in subsequent chapters, does not attempt to teach you how to program. These sections are simply illustrations of some of the program design examples in this book. As you read through these Visual Basic examples, you should get a sense of what is actually involved in the creation and execution of a Visual Basic program.

Figure 2–20 illustrates the graphical interface for the temperature conversion problem (see Figure 2–9). In Visual Basic, the interface is created by the programmer by placing *objects*, or *controls*, on the screen. In this example, five objects are created—three *labels*, one *text box*, and one *command button*. The two labels "Fahrenheit Temperature" and "Celsius Temperature" are used to identify other controls. A text box is created to accept user input. In this example, the empty rectangle below the "Fahrenheit Temperature" label is the text box. This is the area where the user will enter the temperature. Next to the text box is another empty rectangle. Although this rectangle looks exactly like a text box, it is actually another label control. Label controls should be used for items that are output only. Because the Celsius temperature is not an input, using a text box would be inappropriate for holding the Celsius value. Text boxes give the user the ability to enter data; labels do not. Thus, labels are used either to identify other controls or to display items that are output only. The other control on the screen is a command button called Convert. When the user clicks the Convert button, a *click event* is generated and the Fahrenheit temperature in the text box is converted to a Celsius amount and displayed in the label. The programmer must create the program that will be executed when the click event occurs.

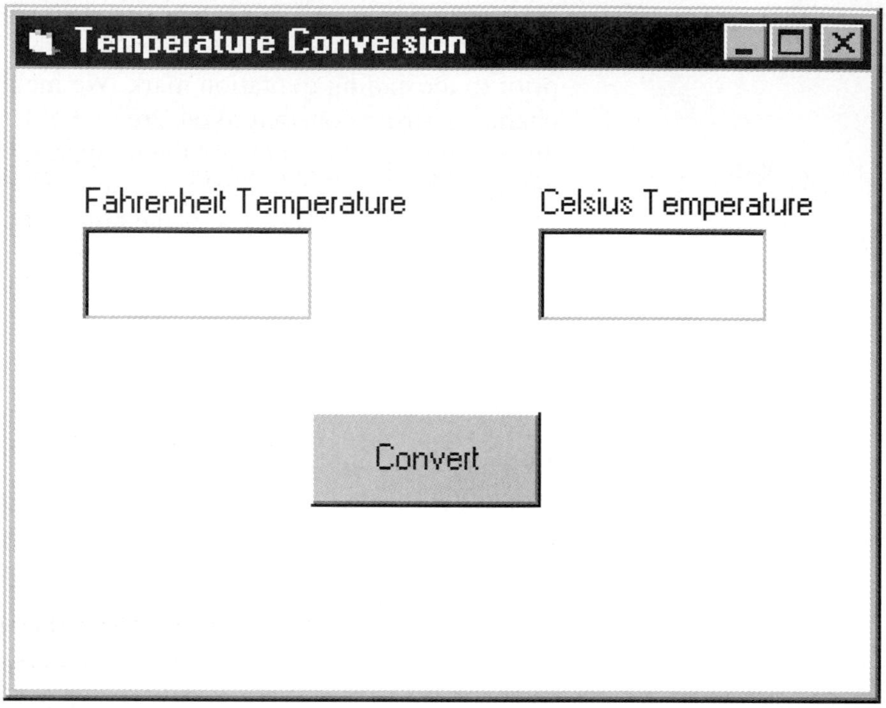

Figure 2–20
Temperature Conversion
Problem (Visual Basic—
Screen 1)

Figure 2–21 illustrates the screen after the user has entered a Fahrenheit value, and Figure 2–22 illustrates the screen after the user has clicked the Convert button.

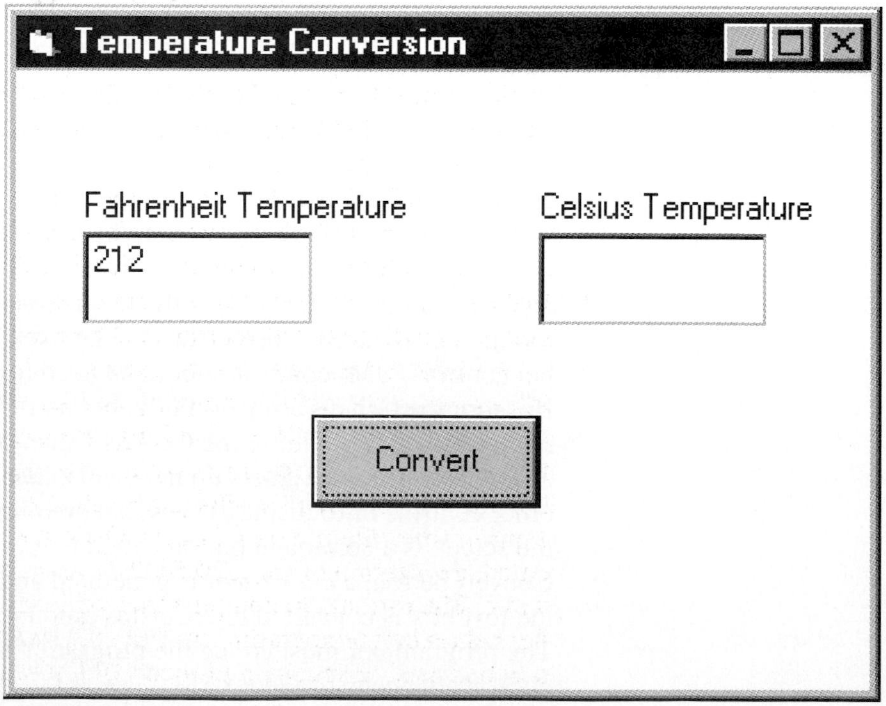

Figure 2–21
Temperature Conversion
Problem (Visual Basic—
Screen 2)

Figure 2–22
Temperature Conversion
Problem (Visual Basic—
Screen 3)

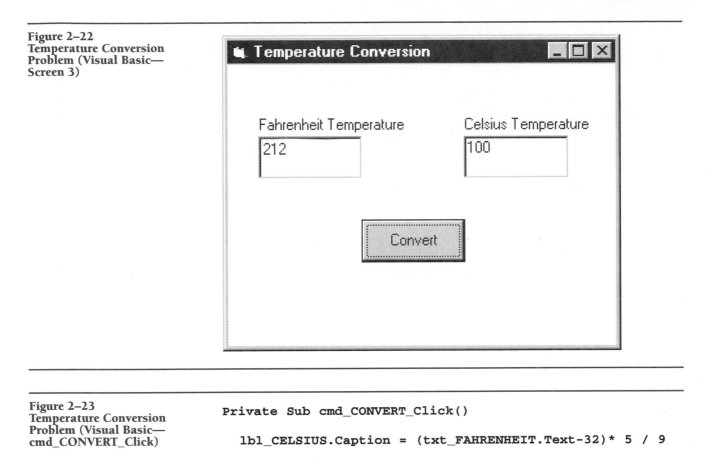

Figure 2–23
Temperature Conversion
Problem (Visual Basic—
cmd_CONVERT_Click)

```
Private Sub cmd_CONVERT_Click()

   lbl_CELSIUS.Caption = (txt_FAHRENHEIT.Text-32)* 5 / 9

End Sub
```

Figure 2–23 illustrates the program that is associated with the click event. In Visual Basic each event is written as a separate small program called a *Sub* (Subprogram). The name of the Sub is identified by the name of the control (cmd_CONVERT) followed by an underscore and the name of the event (Click). Each control has a unique name defined by the programmer. Standard naming conventions dictate that command buttons begin with *cmd*, labels with *lbl*, and text boxes with *txt*. The rest of the name can be chosen by the programmer, but should be descriptive of the control's function. Note that this segment of code contains one assignment statement that parallels the assignment statement in the design. The only difference is that the variable names are replaced by the control names.

In Visual Basic each control has a set of characteristics called *properties*. For example, the *caption* property of a label control determines what will be displayed in that label. Similarly, the *text* property of the text box control determines what will be displayed in the text box. For example, the label that will hold the Celsius temperature (lbl_CELSIUS) contains a blank caption when the program begins. Once the program in Figure 2–23 is executed, the caption of lbl_CELSIUS is replaced with the computed Celsius value. The formula to compute this value must access the text property of the Fahrenheit temperature control (txt_FAHRENHEIT). The following notation is used to specify a property of a given control:

controlname.property

It is important to note that in Visual Basic programs, variable names in design are often replaced by control names and properties in code.

Key Terms

algorithm
program flowchart
block diagram
logic diagram
process symbol
terminal interrupt
  symbol
SIMPLE SEQUENCE
  control structure
information processing
information-processing
  system
file
record

field
character
data hierarchy
system flowchart
general input/output
  (I/O) symbol
input
variable
variable name
defined value
data independent
memory diagram
assignment statement
constant

output
informal design
  review
structured
  walkthrough
formal design review
simulation
procedure execution
desk checking
flowcharting
  worksheet
flowcharting template

Exercises

1. What is an algorithm?

2. State in your own words the two required characteristics of an algorithm. Explain why each is necessary.

3. (a) What is an information-processing system?
   (b) Name three basic elements involved in an information-processing system of any type.

4. (a) Name the four terms (highest to lowest) associated with the data hierarchy.
   (b) Give an example to illustrate each of these terms.

5. Name the fields in the payroll records shown in Figure 2–2.

6. Distinguish between system flowcharts and program flowcharts, discussing how they are similar and how they differ.

7. (a) What is the normal, or assumed, direction of flow on both system and program flowcharts?
   (b) What must the designer do to indicate when the normal direction of flow is not adhered to on a particular path on a flowchart?

8. (a) How has the work of the American National Standards Institute (ANSI) affected flowcharting?
   (b) Why is this work important?

9. Evaluate the following expressions:
   (a) $7 + 4 - 6 * 5 * 2$
   (b) $10 / 2 - 8 + 41 * 3$

(c) $(7 + 2) - 6 / 3 * 2 / 2$
(d) $8 * (6 + (4 + 2) / 3) - 2 * 2 - 1$

10. Look at the flowchart in Figure 2–4. On the basis of this flowchart, what are the total wages for an employee if the employee's sales of regular items are $1,430.00 and his sales of sale items are $820.00?

11. Look at the flowchart in Figure 2–14. On the basis of this flowchart, answer the following questions:
    (a) For what variables are values read as input?
    (b) What variables' values are output?
    (c) What constants are used?
    (d) Simulate the execution of this algorithm, assuming the following values are read as input for the first four variables named.

NAME	MRS. A.B. WALLACE
ITEM	BLOUSE
QTY	3
PRICE	49.95
AMTOD	
DISCOUNT	
SUBBILL	
TAXES	
BILL	

What values should the computer provide as output?

12. Redo Sample Problem 2.2, allowing the discount rate and the tax rate to vary. Provide the two rates as input at the beginning of the algorithm.

13. The computer is to read values for regular hours, overtime hours, and hourly wage rate for one employee from an employee time sheet. Payment for regular hours is to be computed as rate times hours. Payment for overtime hours is to be computed at time and a half, or 1.5 times rate times hours. The computer is to compute and output the total pay for the employee for the week. Construct a program flowchart for this application.

14. Draw a program flowchart for a program that will accept one number as input. Assume this number represents some amount of yards. Compute the corresponding values of feet and inches, and output all three values—that is, yards, feet, and inches.

15. Draw a program flowchart for a program that will compute the average of five grades. Input the five grades and output the average.

16. Draw a program flowchart for a program that will compute the area and perimeter of a rectangle. The input will contain the length and the

width of the rectangle. The output will contain the length, width, area, and perimeter of the rectangle.

17. Repeat Exercise 16, but now find the area and circumference of a circle rather than the area and perimeter of a rectangle. Assume that one value, the diameter of the circle, is provided as input. Output the diameter, area, and circumference of the circle.

IFTHENELSE Control Structure

3

Objectives

Upon completion of this chapter you should be able to

- Identify, and use in program design, the IFTHENELSE control structure.
- Identify, and use in program design, the decision and connector program flowcharting symbols.
- Use pseudocode as a tool in program design.
- Identify, and use in program design, the null ELSE pattern.
- Identify, and use in program design, the sequential IFTHENELSE pattern.
- Identify, and use in program design, the nested IFTHENELSE pattern.
- Distinguish between the logic represented by a sequential IFTHENELSE pattern and the logic represented by a nested IFTHENELSE pattern.

Introduction

The programs discussed in Chapter 2 were straightforward. Each program directed the computer to accept some input, perform one or more calculations, and then provide written output. The computer was directed to follow a SIMPLE SEQUENCE control structure and, accordingly, the steps were carried out in a sequential manner.

In many cases, however, it may be desirable to vary the sequence of processing steps carried out within a solution algorithm. For example, the computer may need to handle different kinds of input data or respond to different situations that arise during processing. Event-driven programs often have to deal with a wide range of possibilities. To accomplish this, we want to take advantage of the computer's logical decision-making capabilities.

Billing Example

Refer again to the task of preparing monthly statements to be mailed to the customers of a large department store (see Sample Problem 2.2). Let us modify the problem slightly by stating that only customers whose bills exceed $200 (before taxes) are to receive a discount. A flowchart of an algorithm to solve this problem is shown in Figure 3–1. The diamond-shaped symbol in the flowchart is a **decision symbol**. It indicates that, at a particular point in processing, a choice between two alternative paths, or sequences of instructions, is to be made.

Figure 3–1
Billing Problem (Flowchart)

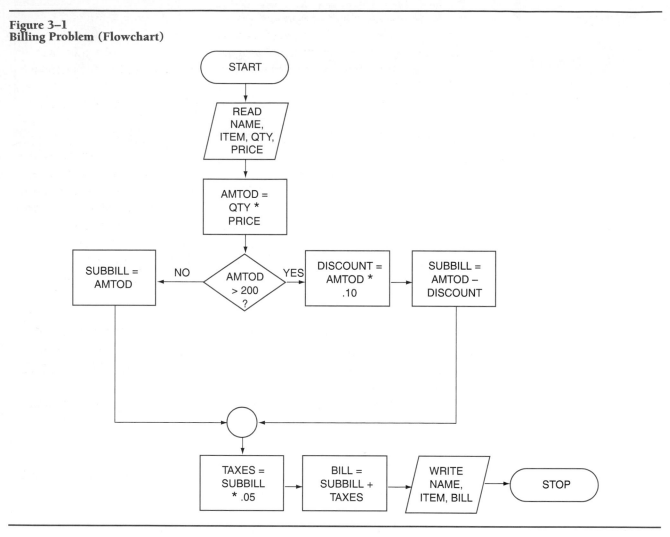

The computer reads the data and multiplies quantity by price. Then the result of the multiply operation is tested: Is the amount owed (AMTOD) greater than (>) $200? Obviously, the question can be answered in either of two ways—yes or no. If the amount owed is not greater than $200, then the NO path is taken and the computer sets the variable named SUBBILL to the original amount owed. No discount is computed in this case. On the other hand, if the amount owed is greater than $200, the YES path is taken. It is composed of two steps: First the actual discount (DISCOUNT) is computed (in this example the discount multiplier is 10 percent). DISCOUNT is then deducted from the amount owed, and the result is placed in the variable named SUBBILL. Notice that there are two assignment statements beginning with SUBBILL =; however, only one of them will actually be executed by the computer. This decision point in the algorithm is called a **conditional branch**.

Notice that the test responses (YES and NO) are clearly indicated on the flowlines from the decision symbol. This documentation is necessary to make the flowchart easier to understand. Programming errors are less likely to occur when the solution algorithm is expressed in computer-program form.

Pseudocode

A flowchart is one way of expressing the decision-making logic in this solution algorithm, but there are other ways as well. One very common technique is the use of an informal language known as **pseudocode**. We can use pseudocode to express the same decision-making logic, as shown in Figure 3–2.

Whereas flowcharts express algorithms pictorially, pseudocode is a text form of representation. Pseudocode is similar to some high-level programming languages (e.g., Visual Basic and Visual C++), but it does not require that we follow strict rules as we would if actually writing a program. When using pseudocode, it is acceptable to use English words and mathematical symbols as we have in the flowchart. The pseudocode presents the solution algorithm in an easy-to-read, top-to-bottom fashion. For emphasis and clarity, the keywords IF, THEN, ELSE, and ENDIF are written in uppercase letters. The THEN and ELSE clauses are indented a few positions. The keywords ELSE and ENDIF are aligned at the left margin with the keyword IF to show that they are part of the same decision-making step.

IFTHENELSE Control Structure

Recall that in Chapter 2 we stated that the SIMPLE SEQUENCE pattern was basic to the design of an algorithm and one of the three basic control structures. Here we see another basic control structure, the **IFTHENELSE control structure**. Its general form is shown in Figure 3–3.

First, condition *p* is tested. If *p* is true, then statement *c* is executed and statement *d* is skipped. Otherwise (else), statement *d* is executed and *c* is skipped. Control then passes to the next processing step.

Figure 3–2
Pseudocode—Introduction

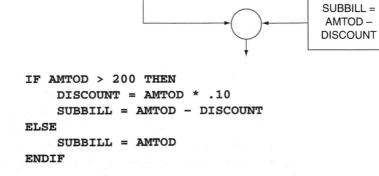

```
IF AMTOD > 200 THEN
     DISCOUNT = AMTOD * .10
     SUBBILL = AMTOD - DISCOUNT
ELSE
     SUBBILL = AMTOD
ENDIF
```

Figure 3–3
IFTHENELSE—Generic

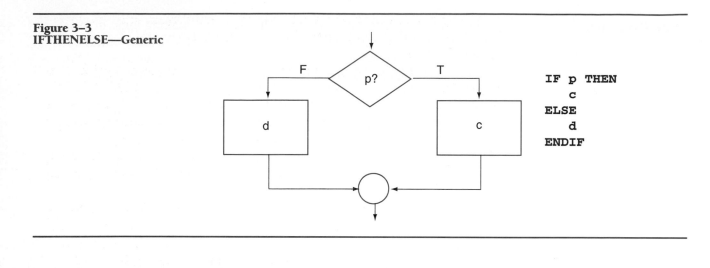

```
IF p THEN
      c
ELSE
      d
ENDIF
```

Figure 3–4
Billing Problem (Pseudocode)

```
Start
Read NAME, ITEM, QTY, PRICE
AMTOD = QTY * PRICE
IF AMTOD > 200 THEN
     DISCOUNT = AMTOD * .10
     SUBBILL = AMTOD - DISCOUNT
ELSE
     SUBBILL = AMTOD
ENDIF
TAXES = SUBBILL * .05
BILL = SUBBILL + TAXES
Write NAME, ITEM, BILL
Stop
```

Before leaving this figure we should note that a small, circular symbol, called a **connector symbol**, is used to represent the decision-making logic within the IFTHENELSE pattern. It acts as a collector, emphasizing that the IFTHENELSE pattern has only one entry point and only one exit point. When a connector symbol is used in this manner, it always has two flowlines entering and one exiting. A well-structured program requires that both decision paths join together in a common exit point. Figure 3–4 shows the pseudocode for the entire algorithm shown in flowchart form in Figure 3–1.

Time Card Example

Now let's consider another variation of the IFTHENELSE structure. Assume we have been given the following problem statement: An employee time card containing employee number, name, and hours worked is to be read as input. If the employee has worked more than 40 hours, his or her number, name, and hours worked are to be printed on a weekly overtime report provided as output. If the employee has not worked more than 40 hours, no print action is required.

A program flowchart representation of the algorithm to solve this problem is given at the left in Figure 3–5. The same algorithm is expressed in pseudocode form at the right.

This algorithm introduces one new idea: the *no-function condition*, usually called a **null ELSE**. When the tested condition of the IFTHENELSE pattern (HOURS > 40, in this case) is true, we follow the YES path. When the tested condition is not true, no special alternative action is required. Thus the NO path goes directly to the connector symbol closing the IFTHENELSE. In pseudocode, the no-function condition is represented by enclosing the keyword ELSE in parentheses. It is followed immediately by the keyword ENDIF.

As a system designer or programmer, you may choose to use either program flowcharts or pseudocode, or both, in developing an algorithm. Flowcharts are a good learning tool and are usually easier for the beginner; remember the old saying—"A picture is worth a thousand words." Pseudocode, however, is faster to write, and may be a more suitable tool when working on complex problems.

Another advantage of pseudocode is that it can be created using any of several available CASE tools. It can also be included as comments in program coding or as part of the prologue for a program. Once entered, pseudocode can be updated easily with computer help whenever the logic of the program is changed. We shall continue to use both program flowcharts and pseudocode as we study the basic patterns of well-structured algorithms; both techniques offer certain advantages to system designers and programmers.

Figure 3–5
Time Card Problem

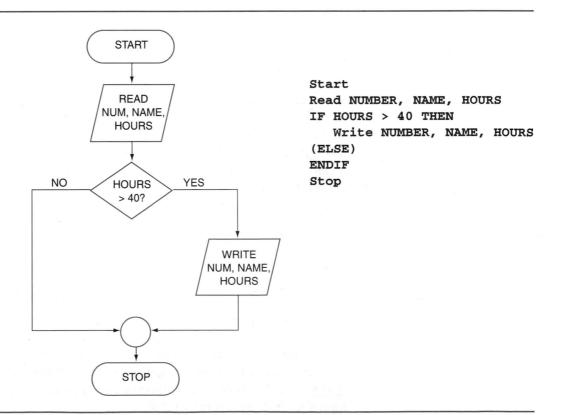

```
Start
Read NUMBER, NAME, HOURS
IF HOURS > 40 THEN
    Write NUMBER, NAME, HOURS
(ELSE)
ENDIF
Stop
```

Sample Problem 3.1 (Payroll Problem)

Problem:

Compute the pay for an employee, assuming that name, Social Security number, hours worked, and hourly rate are input. The output will be the name, Social Security number, and pay for the employee. Regular pay will be computed as hours (up through 40) times rate, and overtime pay will be computed at time and a half (1.5 times hours times rate) for all the hours worked over 40.

Solution:

A solution to this problem is shown in both flowchart form (see Figure 3–6) and pseudocode form (see Figure 3–7).

We see here numerous examples of the SIMPLE SEQUENCE control structure and one example of the IFTHENELSE control structure. Once the employee data is input, the value of the variable HOURS, which is now defined, can be tested to determine if the employee did, in fact, work overtime

Figure 3–6
Payroll Problem (Flowchart)

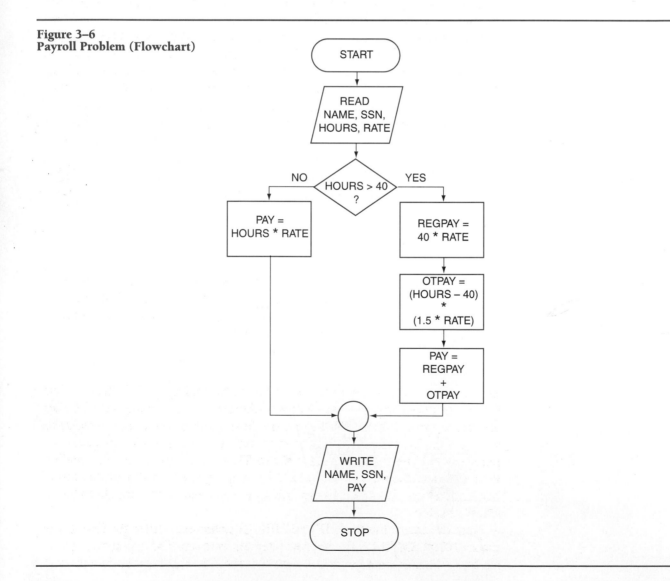

**Figure 3–7
Payroll Problem
(Pseudocode)**

```
Start
Read NAME, SSN, HOURS, RATE
IF HOURS > 40 THEN
    REGPAY = 40 * RATE
    OTPAY = (HOURS - 40) * (1.5 * RATE)
    PAY = REGPAY + OTPAY
ELSE
    PAY = HOURS * RATE
ENDIF
Write NAME, SSN, PAY
Stop
```

(i.e., is HOURS greater than 40?). If not, the pay is computed normally (HOURS * RATE). However, if the employee did work overtime, we break the computation for pay down into three parts. First we compute the pay received for the first 40 hours (REGPAY). In this path, we know the employee worked at least 40 hours. Next, we compute the overtime pay (OTPAY), that is, the pay received for all hours over 40 (HOURS–40), remembering to multiply the rate by 1.5 to get the time-and-a-half rate. In the last step along the YES path of the IFTHENELSE, the two previously computed amounts are added to determine the total pay. The WRITE statement that follows the connector will be executed, regardless of the way the pay was computed. This statement follows the IFTHENELSE structure; therefore, it is not contained within the IFTHENELSE logic. This is indicated in the pseudocode by placing the WRITE statement on the line below the keyword ENDIF. The ENDIF in pseudocode is equivalent to the connector on the flowchart.

Sample Problem 3.2 (Finding the Smallest Number)

Problem:

Now that we have seen how a single IFTHENELSE structure works let us look at a problem that combines several IF statements. Figure 3–8 shows the flowchart representation and Figure 3–9 shows the pseudocode representation of an algorithm to find the smallest of four numbers.

Solution:

The four numbers are input as values for the variables N1, N2, N3, and N4. One additional variable is needed; it is identified by the name SMALL. We assume the first number (N1) is the smallest number and assign its value to the variable SMALL. All the other numbers, N2, N3, and N4, are compared one by one to the current value of SMALL. If any is, in fact, smaller than SMALL, then the variable SMALL is assigned the value of that number. In this way, SMALL always holds the value currently thought to be the smallest value.

Note the use of the three IFTHENELSE statements. After the first test is made (N2 < SMALL), the appropriate path is taken. The second test is made, and again, an appropriate path is taken. The third test is then made.

Figure 3–8
Finding the Smallest Number
(Flowchart)

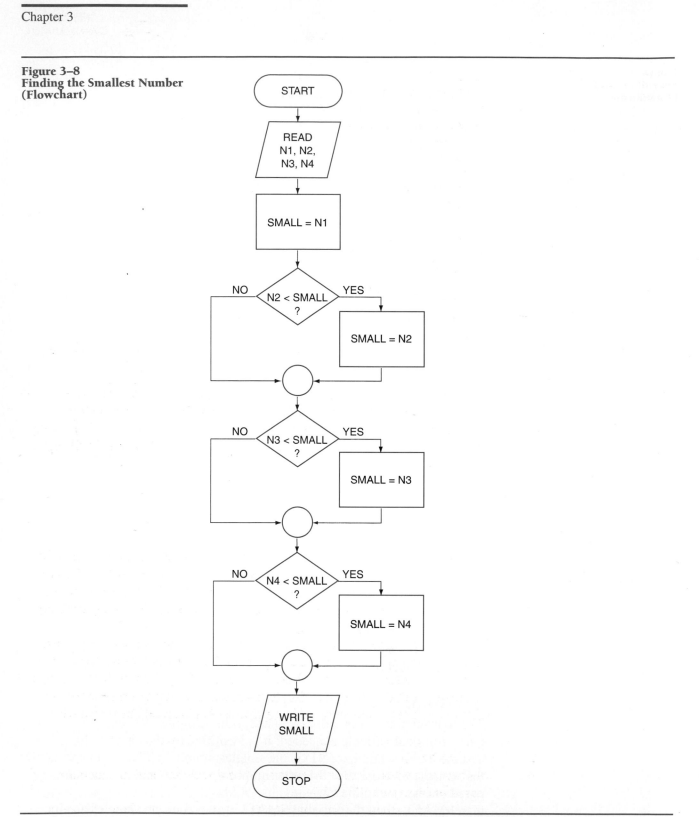

After its YES or NO path is executed, the smallest value is output. It is important to see that all three IFTHENELSE statements are executed, regardless of the outcomes of previous tests. This makes sense because each number in the group must be checked. This is an example of a **sequential IFTHENELSE pattern**; that is, all the tests are always made one

Figure 3–9
Finding the Smallest Number
(Pseudocode)

```
Start
Read N1, N2, N3, N4
SMALL = N1
IF N2 < SMALL THEN
    SMALL = N2
(ELSE)
ENDIF
IF N3 < SMALL THEN
    SMALL = N3
(ELSE)
ENDIF
IF N4 < SMALL THEN
    SMALL = N4
(ELSE)
ENDIF
Write SMALL
Stop
```

after another. Note that in the pseudocode in Figure 3–9 we see three ENDIFs. The second IF test is placed immediately following the first ENDIF, and the third IF test immediately follows the second ENDIF. Note also that, in this example, all three IF statements contain a null ELSE path. Do you see why?

Sample Problem 3.3 (Bank Problem)

Problem:

Let us look at another problem using several IFTHENELSE structures. In this example the **nested IFTHENELSE pattern** will be illustrated. Like the sequential IF pattern, the nested IF pattern utilizes several IFs but the logic behind the pattern is quite different. Consider the following problem: Assume we want to compute the new balance in a customer's bank account. The input will be the customer's name, account number, previous balance, transaction amount, and a code indicating whether the transaction was a deposit (code of 1) or withdrawal (code of 2). For simplicity, each input record will describe only one transaction. The output will contain the customer's name, account number, previous balance, and new balance.

Solution:

A solution to this problem is shown in both flowchart and pseudocode forms in Figures 3–10 and 3–11.

After the customer data is input, the code must be checked for a value of 1 or 2 to determine whether to add or subtract the transaction amount from the old balance. The first IF statement checks if the code is 1, which would indicate a deposit. If it is 1, the new balance is computed by adding the transaction amount; in other words, a deposit is credited. The second IF statement determines if the code is 2, which would indicate a withdrawal. If it is 2, the transaction amount is subtracted from the old balance. The

Figure 3–10
Bank Problem (Flowchart)

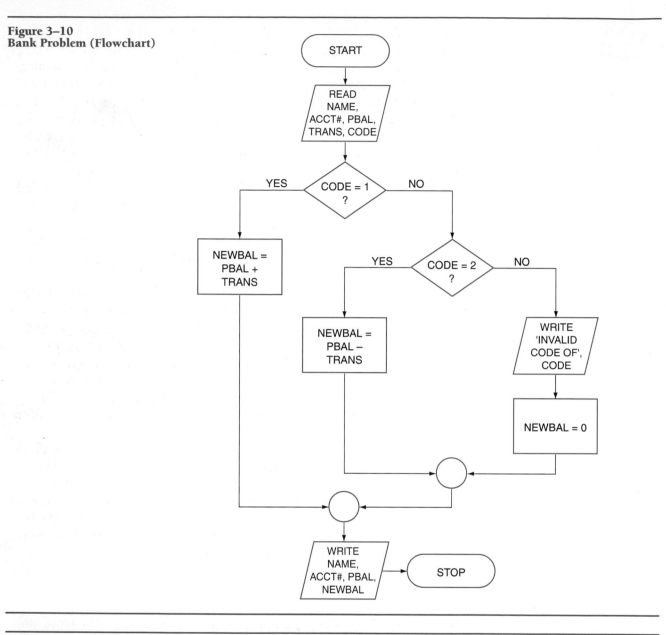

Figure 3–11
Bank Problem (Pseudocode)

```
Start
Read NAME, ACCT#, PBAL, TRANS, CODE
IF CODE = 1 THEN
   NEWBAL = PBAL + TRANS
ELSE
   IF CODE = 2 THEN
      NEWBAL = PBAL - TRANS
   ELSE
      Write 'Invalid code of', CODE
      NEWBAL = 0
   ENDIF
ENDIF
Write NAME, ACCT#, PBAL, NEWBAL
Stop
```

second test is made only if the first test is not true. (It is not necessary to test for a code of 2 if we already know the code is 1.)

According to the logic of this nested IFTHENELSE pattern, each test is made only if the preceding test is not true. (On the other hand, in a sequential IFTHENELSE pattern, all tests are made in all cases.) The entire second IFTHENELSE is totally contained within the first IFTHENELSE; that is, the CODE = 2 test is actually a statement in the false path of the CODE = 1 test. Notice also that we indicate the exit point of each IFTHENELSE with a separate connector, thus giving us two connectors on the flowchart. This is shown by two consecutive ENDIFs within the pseudocode.

The nesting of each additional nested IFTHENELSE is indicated by additional indentation within the pseudocode. It is important to ensure that every IFTHENELSE has its own ending point (ENDIF), and that the words IF, ELSE, and ENDIF are aligned at the left at a uniform indentation. The statements within the IF statement's true and false paths are further indented.

Finally, in our current example, notice what will happen if both tests fail; that is, if the code is neither 1 nor 2. In this case, we assume that the code was keyed incorrectly, and we write a statement that instructs the computer to output an error message. The information within the single quotation marks is a **character-string constant** that will be output. The current value of the variable named CODE will also be output. (What do you think would have happened if we had also included the variable name CODE within the quotation marks?) We will discuss this concept in more detail in the next section.

The second WRITE statement in this example is not contained within either of the IF statements. It is placed after both connectors—so it will be executed regardless of the value of the code. Therefore, the new balance (NEWBAL) will be printed even if the code is invalid. We assign a value of 0 to NEWBAL after writing the error message, so that this value of 0 will be written as output. (What would have happened if we had not?) Note that two lines of output will be printed if the code is invalid.

Character-String Constants

In the previous example we included a statement to output an invalid code message as well as the value of the variable CODE. It is important to note that we are actually outputting two different entities—the character-string constant 'Invalid code of' and the value of the variable CODE. For example, if a code of 3 had been input, then the following information would be output as a result of the WRITE statement:

```
Invalid code of 3
```

We usually designate the value of a character-string constant by enclosing it in single or double quotation marks, thereby distinguishing it from a variable name. Because a character-string constant is, in fact, a type of constant, the value of the character-string constant does not change during processing. Remember our discussion of variables, variable names, and constants in Chapter 2. In our example, the text between the quotation marks (but not including them) is actually output. If we included the variable name CODE within the quotation marks ('Invalid code of CODE') in the WRITE statement, then the word CODE would be considered part of the character-string constant and the following information would be output:

```
Invalid code of CODE
```

Character-string constants can be found in other statements as well. For example, you could assign the value of a character-string constant to a variable by means of a statement such as NAME = 'Mary' or CODE = 'A'. Notice the difference between the assignment statements CODE = 'A' and CODE = A. In the first one, CODE is assigned the actual letter A. In the second assignment statement, CODE is assigned the value of the variable A, which should have been defined previously. Note the effect of each of the following statements:

CODE = 'A'	Memory location CODE contains the letter A.
A = 6	
CODE = A	Memory location CODE contains the number 6.
CODE = A	Memory location CODE is undefined because A is undefined and this will likely produce an error message.

You could also use a character-string constant in a decision statement. For example, consider the statement

```
IF TRANSCODE = 'D' THEN
```

The value of the variable TRANSCODE will be compared to the letter D and appropriate action will be taken. It is assumed that the variable TRANSCODE has been defined previously. Now consider the statement

```
IF TRANSCODE = D THEN
```

In this example, the value of the variable TRANSCODE will be compared to the value of the variable D. Again, both variables should have been defined previously.

Sample Problem 3.4 (Sales Problem)

Problem:

Let us now consider a more complex problem illustrating the nested IFTHENELSE pattern. Assume we want to determine the commission received by a salesperson. The commission rate is based on two factors, the amount of sales and the class to which the salesperson belongs. The input will be the salesperson's name, number, sales amount, and class. The output will be the salesperson's name, number, and commission. The commission rate will be based on the following criteria:

CLASS = 1	If sales is equal to or less than $1000, the rate is 6 percent.
	If sales is greater than $1000 but less than $2000, the rate is 7 percent.
	If sales is $2000 or greater, the rate is 10 percent.

CLASS = 2	If sales is less than $1000, the rate is 4 percent.
	If sales is $1000 or greater, the rate is 6 percent.
CLASS = 3	The rate is 4.5 percent for all sales amounts.
CLASS = 4	The rate is 5 percent for all sales amounts.
CLASS = any other value	Output an appropriate error message.

Solution:

A solution to this problem is shown in both flowchart and pseudocode forms in Figures 3–12 and 3–13.

After the sales data is input, the class must be checked for a value of 1, 2, 3, or 4 to determine the appropriate commission rate. The first IF statement determines if the class is 1. If so, another test is made to determine the value of sales. If SALES is less than or equal to 1000, 6 percent is assigned to the rate. If SALES is not less than or equal to 1000, it is checked again to determine if it is less than 2000. If so, 7 percent is assigned to the rate. Note that we did not have to check whether SALES was greater than 1000. Do you see why? Finally, if SALES is not less than 2000 (meaning that SALES is greater than or equal to 2000), 10 percent is assigned to the rate.

If the class is not 1, we check for 2. If so, the sales must also be tested. If SALES is less than 1000, 4 percent is assigned to the rate; otherwise, 6 percent is assigned to the rate. If the class is not 2, we check for 3. In this case 4.5 percent is assigned to the rate regardless of the value of SALES. Therefore, the sales amount does not have to be tested. If the class is not 3, we check for 4. In this case 5 percent is assigned to the rate, regardless of the value of SALES. Note again that the sales amount does not have to be tested. Finally, if the class is not 4, we output an error message. We also assign a value of 0 to the rate (for the same reason we assigned a value of 0 to NEWBAL in Sample Problem 3.3). At this point, the rate has been given the proper value for all possible values of SALES and CLASS. We can now compute the commission based on this rate and output the salesperson information.

Notice on the flowchart that each IFTHENELSE structure has an associated connector. Similarly, the pseudocode indicates the keyword ENDIF for each IFTHENELSE. There are several levels of nested IFTHENELSE structures in this example. The pseudocode shows the nesting very clearly by the indentation. Make sure you understand what statements belong to each IFTHENELSE and how each IFTHENELSE structure begins and ends.

It is important to understand how a nested IFTHENELSE pattern differs from a sequential IFTHENELSE pattern. In the latter, all tests are always made in all cases. In the nested pattern, subsequent tests are made depending on the outcome of a previous test. In the nested pattern it is probable that many of the tests will be skipped. Do you see why the nested IFTHENELSE structure was more appropriate for this problem? What would happen if we had tested CLASS using a series of sequential IF statements? (See Exercise 17) We will see additional examples of the nested IFTHENELSE pattern in later chapters.

Figure 3–12
Sales Problem (Flowchart)

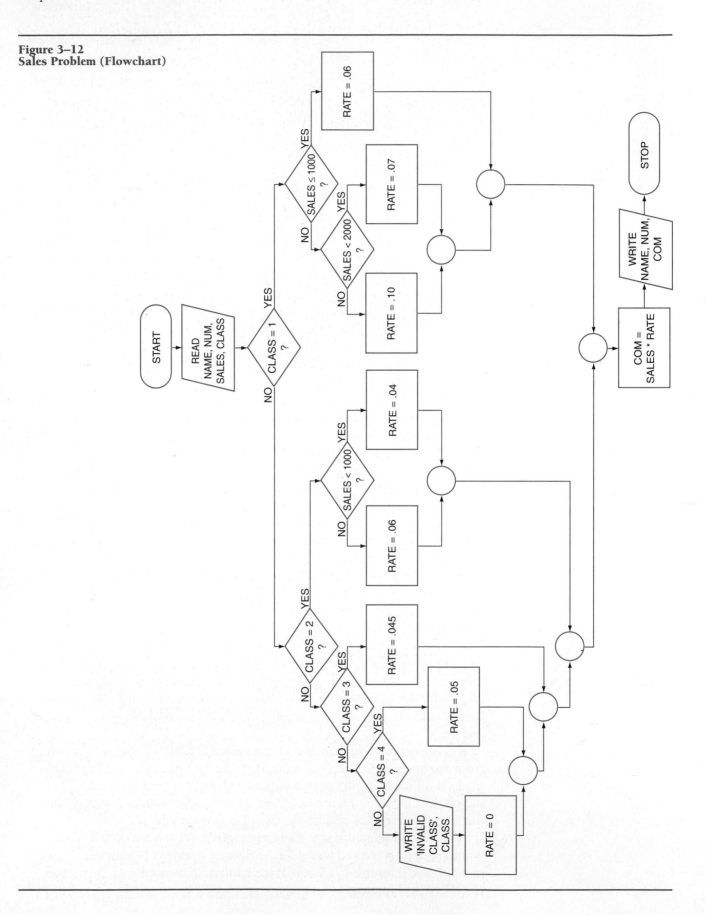

Figure 3–13
Sales Problem (Pseudocode)

```
Start
Read NAME, NUM, SALES, CLASS
IF CLASS = 1 THEN
    IF SALES ≤ 1000 THEN
        RATE = .06
    ELSE
        IF SALES < 2000 THEN
            RATE = .07
        ELSE
            RATE = .10
        ENDIF
    ENDIF
ELSE
    IF CLASS = 2 THEN
        IF SALES < 1000 THEN
            RATE = .04
        ELSE
            RATE = .06
        ENDIF
    ELSE
        IF CLASS = 3 THEN
            RATE = .045
        ELSE
            IF CLASS = 4 THEN
                RATE = .05
            ELSE
                Write 'INVALID CLASS', CLASS
                RATE = 0
            ENDIF
        ENDIF
    ENDIF
ENDIF
COM = SALES * RATE
Write NAME, NUM, COM
Stop
```

Enrichment (Basic)

Figure 3–14 illustrates a listing of a program that solves the sales problem (see Figures 3–12 and 3–13). In this example we use the Input statement, which will request a name, number, sales, and class from the user when the program is executed. Note that this Input statement is preceded by a Print statement. The Print statement outputs directions to the user that specify what input is needed. Note that the variable name for the salesperson name is NAME$. In Basic, variables that hold non-numeric data must be represented by names ending with a $. As you can see, the Basic program is very similar to the pseudocode in Figure 3–13. Identifying text was added in the last Print statement to make the output more readable.

Figure 3–15 illustrates the output that will be produced when the program is executed. There are eight possible paths that the computer can follow, depending on the input values of the variables SALES and CLASS. This program was executed eight separate times to show what output will

Figure 3–14
Sales Problem (Basic List)

```
PRINT "Enter the salesperson name, number, amount of sales, and class"
INPUT NAME$, NUMBER, SALES, CLASS
IF CLASS = 1 THEN
        IF SALES <= 1000 THEN
                RATE = .06
        ELSE
                IF SALES < 2000 THEN
                        RATE = .07
                ELSE
                        RATE = .1
                END IF
        END IF
ELSE
        IF CLASS = 2 THEN
                IF SALES < 1000 THEN
                        RATE = .04
                ELSE
                        RATE = .06
                END IF
        ELSE
                IF CLASS = 3 THEN
                        RATE = .045
                ELSE
                        IF CLASS = 4 THEN
                                RATE = .05
                        ELSE
                                PRINT "Invalid Class",CLASS
                                RATE = 0
                        END IF
                END IF
        END IF
END IF
COMMISSION = SALES * RATE
PRINT "Name is "; NAME$; "  Number is "; NUMBER; " Commission is "; COMMISSION
END
```

be produced with varying values for the variables SALES and CLASS. Note that the last example prints two lines of output when CLASS is invalid. You should always test a program with data that corresponds to every possible path within the program. This type of testing will ensure that the program will be reliable. Note that blank spaces were positioned in the character-string constants shown in Figure 3–14 to ensure that the output fields are spaced appropriately.

Enrichment (Visual Basic)

Figure 3–16 illustrates the graphical user interface for the sales problem (see Figures 3–12 and 3–13). In this example, three text boxes are created

Figure 3-15
Sales Problem (Basic Run)

```
Enter the salesperson name, number, amount of sales, and class
?James,555,1000,1
Name is James   Number is   555   Commission is   60

Enter the salesperson name, number, amount of sales, and class
?John,321,1500,1
Name is John   Number is   321   Commission is   105

Enter the salesperson name, number, amount of sales, and class
?Steve,246,3000,1
Name is Steve   Number is   246   Commission is   300

Enter the salesperson name, number, amount of sales, and class
?Kate,642,500,2
Name is Kate   Number is   642   Commission is   20

Enter the salesperson name, number, amount of sales, and class
?Tom,678,1000,2
Name is Tom   Number is   678   Commission is   60

Enter the salesperson name, number, amount of sales, and class
?Joan,111,1000,3
Name is Joan   Number is   111   Commission is   45

Enter the salesperson name, number, amount of sales, and class
?Ted,222,5000,4
Name is Ted   Number is   222   Commission is   250

Enter the salesperson name, number, amount of sales, and class
?Harry,333,1200,5
Invalid Class 5
Name is Harry   Number is   333   Commission is   0
```

to accept user input. A label is created to hold the computed commission. A new control, called an *option button*, will be used to accept the user input for the class. Four option buttons are created with captions representing the names of the classes. Two command buttons are created, one to compute the commission and one to end execution of the program.

Figure 3-17 illustrates the screen after the user has entered the input values. Note that in this example the user clicked the first option button to indicate that the salesperson belongs to class 1. The small circle becomes darkened when the option button is selected. It is important to note, at this point, that option buttons should only be used to represent values that are mutually exclusive. Only one option can be selected, or darkened, at once. If the user selects the second button, then the second small circle will be darkened, rather than the first. Because each salesperson can be associated with only one class, option buttons are a reasonable choice in this example.

Figure 3-18 illustrates the screen after the user has clicked the Compute Commission button. The value for the commission is displayed in the label.

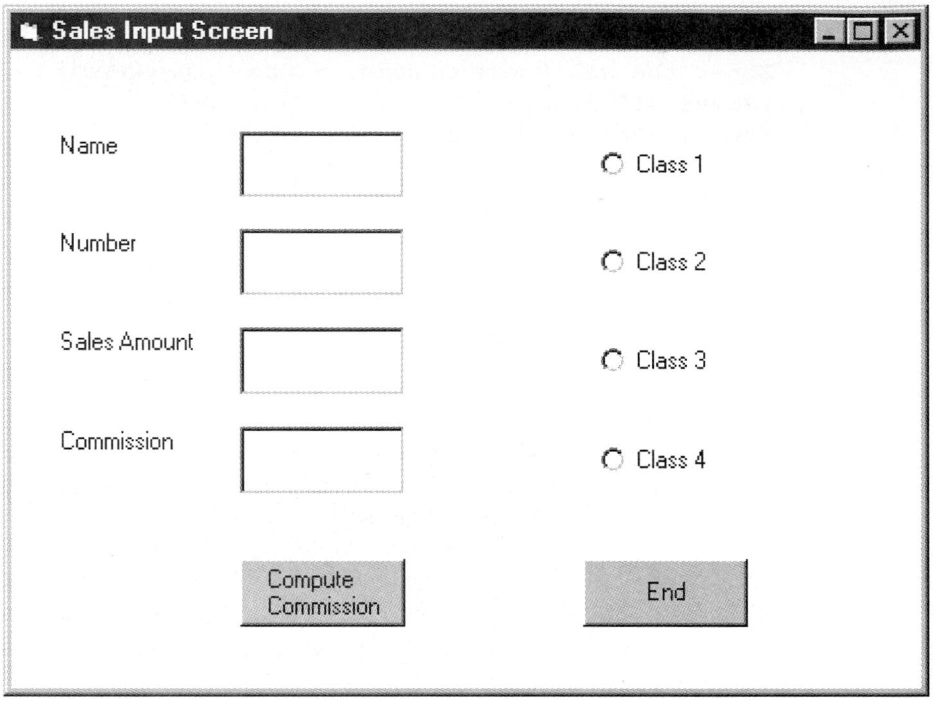

Figure 3–16
Sales Problem (Visual Basic—Screen 1)

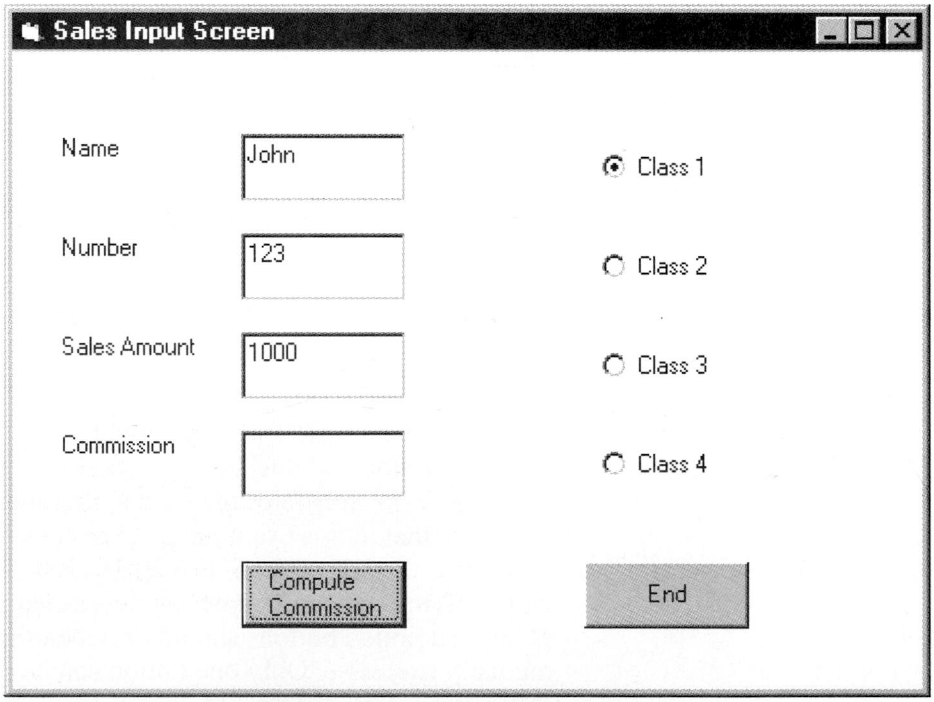

◼ Sales Input Screen	_ ☐ ✕

Name   [ ]   ○ Class 1

Number   [ ]   ○ Class 2

Sales Amount   [ ]   ○ Class 3

Commission   [ ]   ○ Class 4

[Compute Commission]   [End]

Figure 3–17
Sales Problem (Visual Basic—Screen 2)

◼ Sales Input Screen	_ ☐ ✕

Name   [John]   ◉ Class 1

Number   [123]   ○ Class 2

Sales Amount   [1000]   ○ Class 3

Commission   [ ]   ○ Class 4

[Compute Commission]   [End]

**Figure 3–18
Sales Problem (Visual
Basic—Screen 3)**

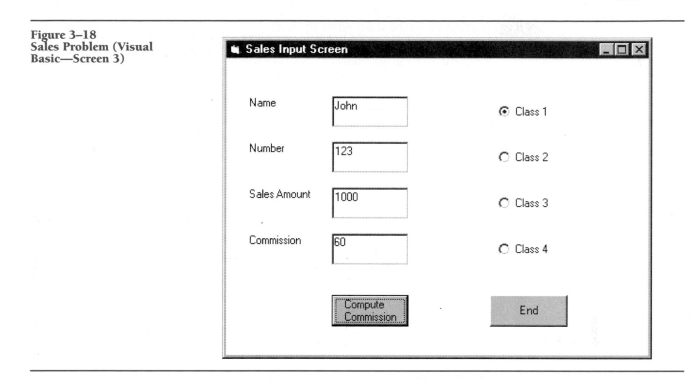

Because option buttons are used, the user cannot enter an invalid class. The user simply chooses one of the buttons. Consider, however, what would happen if the user clicked the Compute Commission button prior to selecting one of the options for class. In this case, since no class was selected, an error message needs to be output. This situation is illustrated in Figure 3–19. In Visual Basic we can use a *message box* to display

**Figure 3–19
Sales Problem (Visual
Basic—Screen 4)**

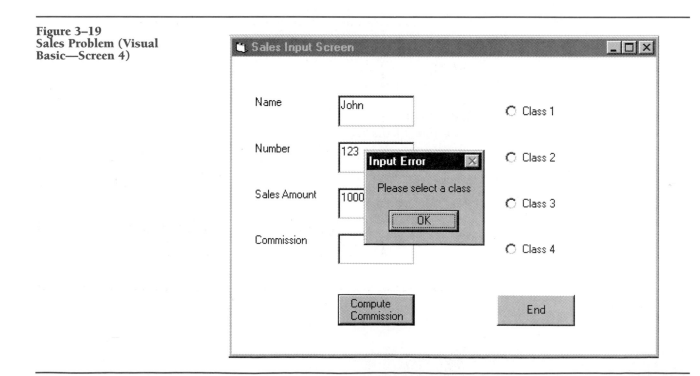

information to the user. After the user reads the message, the user clicks the OK button and the program execution continues.

Figure 3–20 illustrates the program that is associated with the click event of the Compute Commission button. Note that these statements very closely parallel the pseudocode in Figure 3–13. Only the IF statements that test the value of CLASS are different. Option buttons have a property called *value*, which can be either true or false. If an option button is selected, the value of the value property is true. If an option button is not selected, the value of the value property is false. Only one option button can have its value property set to true at any given time. Thus, each IF test checks the value of one of the option button's value property. Note that option button control names begin with *opt*. Note also that the Invalid Class error message is replaced by the MsgBox statement. This statement causes a message box to be displayed. The programmer determines both the message to be displayed and the title of the message box window.

Figure 3–20
Sales Problem (Visual Basic—cmd_COMPUTECOMMISSION_Click)

```
Private Sub cmd_COMPUTECOMMISSION_Click()

If opt_CLASS1.Value = True Then
    If txt_SALES.Text <= 1000 Then
        RatePercent = 0.06
    Else
        If txt_SALES.Text < 2000 Then
            RatePercent = 0.07
        Else
            RatePercent = 0.1
        End If
    End If
Else
    If opt_CLASS2.Value = True Then
        If txt_SALES.Text < 1000 Then
            RatePercent = 0.04
        Else
            RatePercent = 0.06
        End If
    Else
        If opt_CLASS3.Value = True Then
            RatePercent = 0.045
        Else
            If opt_CLASS4.Value = True Then
                RatePercent = 0.05
            Else
                MsgBox "Please select a class",, "Input Error"
                RatePercent = 0
            End If
        End If
    End If
End If
lbl_COMMISSION.Caption = txt_SALES.Text * RatePercent

End Sub
```

**Figure 3–21
Sales Problem (Visual
Basic—cmd_END_Click)**

```
Private Sub cmd_END_Click()

End

End Sub
```

In our example, only an OK button is included in the message box. The programmer can display other buttons in the message box by including special numeric codes in the MsgBox statement between the two commas. Because no special code was included in the MsgBox statement in this example, only the OK button was displayed.

In this example an additional button was included with the caption *End*. Figure 3–21 illustrates the program that is associated with the click event of the End button. When the user clicks the End button, the Visual Basic End statement is executed and the program execution terminates. It is usually a good idea to include an End button as part of the graphical user interface for a program.

Key Terms

decision symbol	connector symbol	nested IFTHENELSE
conditional branch	null ELSE	pattern
pseudocode	sequential	character-string
IFTHENELSE control	IFTHENELSE	constant
structure	pattern	

Exercises

1. State in your own words the purpose of a decision-making step in a solution algorithm.

2. Which of the symbols below represents the question: Is DUE greater than or equal to CREDIT?

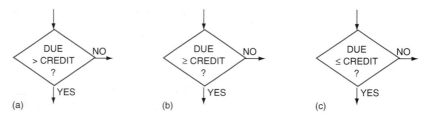

3. **(a)** What is pseudocode?
   (b) What are some of the advantages that the use of pseudocode offers?

4. **(a)** Using ANSI-approved flowcharting symbols, sketch the logic of an IFTHENELSE control structure.
   (b) State the logic of an IFTHENELSE control structure in pseudocode form.

(c) What pseudocode keywords did you use in your response to Exercise 4(b)?

(d) What indentations did you use in your response to Exercise 4(b)? Why?

5. (a) Explain what a null ELSE indicates.
   (b) Describe a problem in which a null ELSE occurs.
   (c) How is a null ELSE condition indicated on a flowchart?
   (d) How is a null ELSE condition indicated when using pseudocode?

6. How is a connector symbol used in flowcharting an IFTHENELSE control structure? Why is it used?

7. Explain how the decision-making logic represented by a sequential IFTHENELSE pattern differs from that represented by a nested IFTHENELSE pattern.

8. Write a program flowchart and corresponding pseudocode to solve the following problem: Assume the input for a student is name, student number, and three grades. Output the student's name and an S if the average of the three grades is 65 or more. Otherwise (else), output the student's name, a U, and the number of additional points needed for an S.

9. Write a program flowchart and corresponding pseudocode to prepare a monthly credit card billing report. The input will contain the name of the person who has purchased on credit, the person's previous balance, total purchases, and total payments. The output lists the person's name, previous balance, total purchases, total payments, the amount subject to a finance charge, the interest, and the new balance.
   The amount subject to a finance charge is obtained by adding the total purchases to the previous balance and subtracting the total payments. If the amount subject to a finance charge is $250.00 or more, interest must be calculated by multiplying this amount by 1.5 percent. If the amount subject to a finance charge is less than $250.00, interest is calculated by multiplying this amount by 1 percent. The new balance is obtained by adding the interest to the amount subject to a finance charge.

10. Write a program flowchart and corresponding pseudocode to prepare an inventory report. The input will contain the item number, quantity, and unit price of a particular product. The inventory value will be computed by multiplying the quantity by the unit price. If the value of the inventory is more than $1000.00, then output the item number and the amount by which that item exceeds $1000.00. If the inventory value is $1000.00 or less, then output the item number and the computed inventory value. Include appropriate labels to identify the output values.

11. Write a program flowchart and corresponding pseudocode to prepare a tuition bill. The input will contain the student name, Social Security number, and total number of credits for which the student has enrolled. The bill will contain the student name, Social Security number, and computed tuition. Total credits of 10 or more indicate that the student is full-time. Full-time students pay a flat rate of $1000.00 for tuition. Total credits of less than 10 indicate that the student is part-time. Part-time students pay $100.00 per credit for tuition.

12. Write a program flowchart and corresponding pseudocode to prepare a monthly bank statement. The input will contain the customer name, account number, old balance, total monthly deposit amount, and total monthly withdrawal amount. The statement will contain the customer name, account number, old balance, and new balance. The new balance is computed by adding the total monthly deposit amount to, and subtracting the total monthly withdrawal amount from, the old balance. If the new balance is less than $100.00, then a $5.00 service fee will be subtracted from the new balance to determine the correct new balance.

13. Write a program flowchart and corresponding pseudocode to solve the following problem: Assuming that one number is input, output the number and a message indicating whether it is positive or negative. Output nothing if the number is 0.

14. Write a program flowchart and corresponding pseudocode to solve the following problem: Input a person's name, height (in inches), and weight (in pounds). If the person's height exceeds 5 feet (60 inches), and the person's weight exceeds 100 pounds, output the person's name. If the height and weight do not meet these criteria, output nothing.

15. Write a program flowchart and corresponding pseudocode to prepare a contract labor report for heavy equipment operators. The input will contain the employee name, job performed, hours worked per day, and a code. Journeyman employees have a code of J, apprentices a code of A, and casual labor a code of C. The output consists of the employee name, job performed, hours worked, and pay. Journeyman employees receive $20.00 per hour. Apprentices receive $15.00 per hour. Casual labor receives $10.00 per hour. Write two solutions, one using a nested IF pattern and another using a sequential IF pattern. (a) What output is produced by each solution? (b) Discuss which approach is more efficient and why.

16. Write a program flowchart and corresponding pseudocode for the following problem: You are given an input record that contains a student's name and three exam scores. Compute the average of the three exam scores and output a grade corresponding to the average as follows:

Average	Grade
90-100	A
80-89	B
70-79	C
60-69	D
under 60	F

Your output should look like this:

_____ received a grade of _____.

17. Redo Sample Problem 3.4 using a series of sequential IFTHENELSE statements to test the class.
   (a) Determine whether this approach works and why or why not.
   (b) Discuss which approach (nested IFs or sequential IFs) is more efficient and why.

18. Write a program flowchart and corresponding pseudocode to solve the following problem: Input name, age, and code for an applicant for a job in ABC Company. A code of 1 indicates the applicant is a U.S. citizen, and a code of 2 indicates the applicant is not a U.S. citizen. The specifications for the job require that the applicant be a U.S. citizen 21 years of age or older. Your algorithm should check that these conditions are met and output an appropriate message(s) if one or both tests fail. If the applicant passes both tests, output the applicant's name only. Place only one test in each decision symbol (flowchart) or IF statement (pseudocode). Solve this problem using the nested IF approach. Solve it again using the sequential IF approach.

DOWHILE Control Structure—Counter-Controlled Loops

Objectives

Upon completion of this chapter you should be able to

- Identify, and use in program design, counters, accumulators, and program loops.
- Construct and identify the characteristics of a simple counter-controlled program loop.
- Identify, and use in program design, the DOWHILE control structure.
- Identify, and use in program design, the preparation flowcharting symbol.
- Design programs using counter-controlled loops.
- Design programs using header record logic.
- Define the term *proper program*.

Introduction

The solution algorithms discussed so far have at least one characteristic in common: They show the program logic required to process only one set of input values. Generally, however, a computer program must be designed and coded to process many sets of input values. To provide for this, a **program loop** must be included in the program flowchart. A program loop is a sequence of processing steps that may be repeated. The computer may reexecute the instructions. Consider, for example, the following problem: Construct an algorithm that will accept six numbers as input and will compute and output the sum of these six numbers. Figure 4–1 illustrates a solution to this problem without the use of a program loop.

First, six numbers are read into the memory locations reserved for the variables represented by the variable names N1, N2, N3, N4, N5, and N6. An assignment statement is then used to add the numbers, and the result is placed in memory under the variable named SUM. SUM is then output and the program execution stops. This solution is quite simple and straightforward. You should be familiar with all of the steps in this algorithm from previous chapters. Consider, however, the limitations of this approach. What if you were now asked to revise this algorithm to compute and output the sum of 100 numbers? You would need 100 separate variable names to represent the data being input. In addition, these 100 variable names would need to be included in the

Figure 4–1
Adding Six Numbers (No Loop)

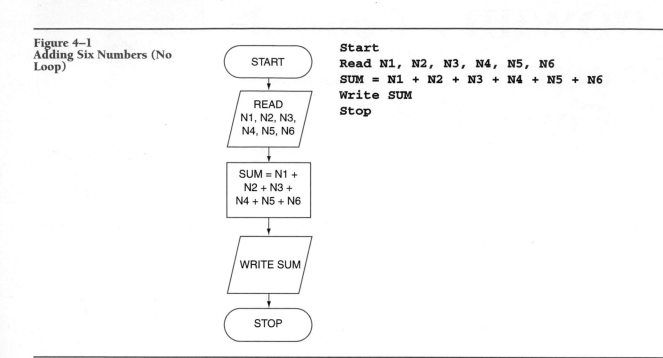

```
Start
Read N1, N2, N3, N4, N5, N6
SUM = N1 + N2 + N3 + N4 + N5 + N6
Write SUM
Stop
```

assignment statement to compute the sum. Clearly, this is not a desirable approach to take.

Let us look for a moment at how we would perform this task manually, using a calculator. First, we would either turn the calculator on or clear it if we had already been using it for another task. We would then enter the first number, say 5. The number 5 would appear on the display and we could then hit the + sign. The number 5 would again appear on the display. At this point we could enter a second number, say 7. A 7 would now appear on the display. When we hit the + sign this time the two numbers would be added and 12 would appear on the display. Consider the sequence of steps that occurs as we enter four more numbers, say 3, 4, 8, and 1, and add them together. Figure 4–2 illustrates this process.

Figure 4–2
Calculator Simulation (No Loop)

	ACTION	SCREEN DISPLAY
1.	Clear calculator memory	
2.	Enter a number (5)	5
3.	Hit the + sign	5
4.	Enter a number (7)	7
5.	Hit the + sign	12
6.	Enter a number (3)	3
7.	Hit the + sign	15
8.	Enter a number (4)	4
9.	Hit the + sign	19
10.	Enter a number (8)	8
11.	Hit the + sign	27
12.	Enter a number (1)	1
13.	Hit the + sign	28

Figure 4–3
Calculator Simulation (Loop)

1. Clear calculator memory

2. Enter a number

3. Hit the + sign

} Perform these steps
    six times.

Do you notice any repetition? Steps 2, 4, 6, 8, 10, and 12 all perform the exact same task; only the entered number is different. Similarly, Steps 3, 5, 7, 9, 11, and 13 perform the same task. The screen, meanwhile, displays sums that get increasingly closer to the final sum of 28.

Now look at Figure 4–3. It shows how this procedure can be reorganized into a simpler one using a loop.

Problem (Adding Six Numbers)

Now look at the flowchart and pseudocode (see Figure 4–4) of an algorithm that illustrates this type of logic.

Figure 4–4
Adding Six Numbers

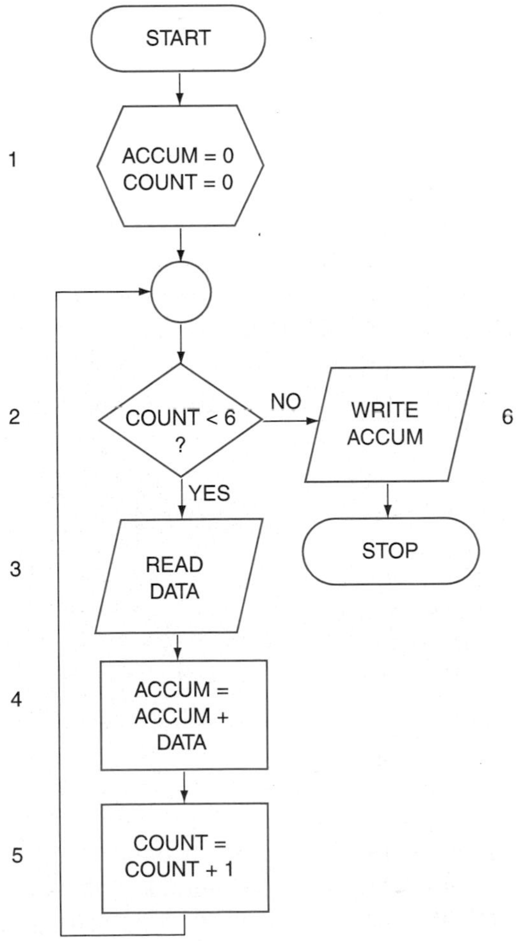

```
Start
ACCUM = 0
COUNT = 0
DOWHILE COUNT < 6
    Read DATA
    ACCUM = ACCUM + DATA
    COUNT = COUNT + 1
ENDDO
Write ACCUM
Stop
```

A new flowcharting symbol is introduced in Figure 4–4—the **preparation symbol**. It represents an operation performed on data in preparation for a major sequence of operations. Often this operation is performed before a loop is entered (as in this flowchart); in these cases, the operation is referred to as an **initialization step**. In Step 1 of this algorithm, two variables are set to 0. One is used as an accumulator (ACCUM), and the other is used as a counter (COUNT). An **accumulator** is a variable that is used to hold the sum of a group of values. This sum is computed by gradually adding (accumulating) each value to the variable each time the loop is executed. A **counter** is a special type of accumulator. A counter adds or accumulates by a constant amount, usually 1. The value of such a counter increases by 1 each time the loop is executed.

ACCUM plays the role of our calculator's memory. It is initialized to 0 just as the calculator's memory is cleared at the start of an activity. It will represent what we see on the calculator's screen when we hit the + sign. Simply put, ACCUM is used to hold the sum of the numbers so far, often called a **partial sum**. COUNT, on the other hand, is used to keep track of how many numbers have been added so far. When we add a series of numbers using a calculator, we keep track of how many we have already added and how many remain to be added. In our algorithm, we must name a variable that the computer can use for this task. The computer increases the value of COUNT (and then checks its value) to know when to stop adding numbers.

Simulation (Adding Six Numbers)

Now look at what happens during one cycle of the loop in Figure 4–4. First, a test is made (Step 2): Is COUNT less than 6? If so, a number is read (let's say 5) and stored in the memory location represented by the variable called DATA (Step 3). This step is equivalent to the calculator step: Enter a number. Now look at Step 4. Do you recognize an assignment statement? Notice that the variable ACCUM appears on both sides of the assignment statement. The computer will add the value of DATA (5, for example) to the current value of ACCUM (0 at the moment) and place that result back into the storage location used for the variable ACCUM. ACCUM now contains the value 5. We have lost the previous value of ACCUM, because this statement assigned a new value to it (overlaying the previous value in the storage location). Step 4 is thus equivalent to the calculator step: Hit the + sign. The same kind of process affects the variable COUNT in Step 5. The computer adds 1 (a constant) to the current value of COUNT (0 at the moment), and places the result back into the storage location represented by the variable COUNT. COUNT now holds the value 1. Look again at ACCUM and COUNT. Can you see that ACCUM does in fact hold the sum of the numbers so far? Does COUNT indicate how many numbers have been added so far? The flowline then takes us back to the test that determines if COUNT is less than 6. Since this is still true, the YES path is again taken and the READ statement is executed. Continue this process with five more numbers—7, 3, 4, 8, and 1. Keep track of the values for DATA, ACCUM, and COUNT each time through the loop. Check your results with Figure 4–5 before you go on. It shows a simulation of the steps in this algorithm.

Figure 4–5
Adding Six Numbers—
Simulation

INPUT: 5
7
3
4
8
1

☐ VALUE CHANGES
IN THIS STEP

STEP	ACCUM	COUNT	DATA
1	☐0☐	☐0☐	UNDEFINED
3	0	0	☐5☐
4	☐5☐	0	5
5	5	☐1☐	5
3	5	1	☐7☐
4	☐12☐	1	7
5	12	☐2☐	7
3	12	2	☐3☐
4	☐15☐	2	3
5	15	☐3☐	3
3	15	3	☐4☐
4	☐19☐	3	4
5	19	☐4☐	4
3	19	4	☐8☐
4	☐27☐	4	8
5	27	☐5☐	8
3	27	5	☐1☐
4	☐28☐	5	1
5	28	☐6☐	1

When COUNT finally reaches 6, the answer to the question asked at the beginning of the loop will be NO and the WRITE statement will be executed to output the sum of the six numbers. In this example, the sum should be 28, which is the most recent value of ACCUM. The program execution will then stop. Notice that the value of ACCUM is not output until the loop is exited; that is, until all six numbers have been added. This sequence of operations makes sense, since the computer can't output the sum until it finishes computing the sum, and the computer can't finish computing the sum until all numbers have been read and processed.

The DOWHILE Loop

The previous example was a simple but informative one. You can see that the YES flowline extending from the decision symbol in Figure 4–4 causes three steps to be executed and then flows back toward the beginning of the flowchart to the small circular symbol. Recall from Chapter 3 that this symbol is a **connector symbol**. It serves as a collector at the beginning of the loop. As before, it has two flowlines entering and only one exiting. It performs no action of its own. Its presence on the flowchart serves only to emphasize that there is only one entrance point to the loop (at the connector symbol) and only one exit from it (at the decision symbol). If (and only if) the tested condition is true (COUNT < 6), the processing steps along the true (YES) path are executed. If the tested condition is not true, the false (NO) path is followed. If the false path is taken, we say that the program loop is exited.

This loop pattern, called the **DOWHILE loop**, is the third basic pattern of structured program logic. Note that the DOWHILE pattern is a **leading-decision program loop**. The test to determine whether the loop should be executed or exited is encountered immediately upon entering the loop. In other words, the test immediately follows the connector symbol. Whether or not the additional steps inside the loop are executed depends on the outcome of the test. Notice the significance of the fact that, if the tested condition is not true, the loop is exited immediately thereafter. Thus, if the tested condition is not true the first time it is tested, the remaining steps in the loop are not executed at all.

As outlined above, the DOWHILE loop has three basic structural characteristics that must be adhered to. To create a properly formed DOWHILE loop you must

1. Place the loop test before any other step within the loop (leading decision).

2. Place the loop steps in the YES path of the loop test.

3. Indicate that the loop will exit in the NO path of the loop test.

Make sure that you can identify these characteristics in the DOWHILE loop in the previous example and in any algorithm you write using the DOWHILE loop.

The general form of the DOWHILE pattern is shown in Figure 4–6. To help you become familiar with this pattern, it is flowcharted in two ways.

Figure 4–6
DOWHILE Loop—Generic

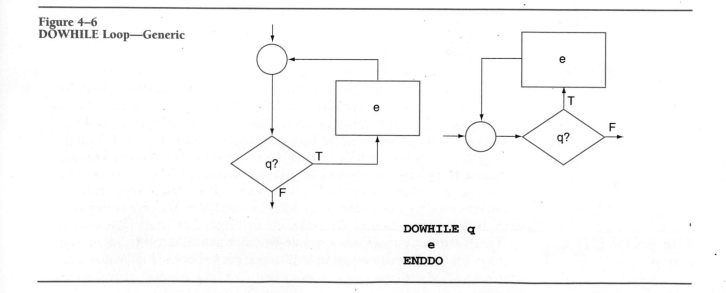

```
DOWHILE q
    e
ENDDO
```

The same control structure appears in each representation; you should learn to recognize and use either one. First, condition q is tested. If q is true, statement e is executed and control returns to the test of q. If q is false, control passes to the next processing step. The DOWHILE pattern is always set up this way—the steps in the loop are executed while the outcome of the test is true, and the loop is exited when the outcome is false.

DOWHILE Pseudocode

Now look at the pseudocode representation in Figure 4–4. Note the keywords DOWHILE and ENDDO. They are aligned at the left and mark the beginning and ending of the program loop, respectively. The initialization steps are placed before the loop, that is, on the line above the DOWHILE statement. The actual loop steps that occur in the YES path on the program flowchart are placed on the lines below the DOWHILE statement and directly above the ENDDO statement. The WRITE instruction that occurs in the NO path outside the loop on the program flowchart is placed directly after ENDDO. In pseudocode, the steps within the loop (YES path in the flowchart) are always placed between DOWHILE and ENDDO. Similarly, the steps after the loop (NO path in the flowchart) are always placed on the line after ENDDO. All processing steps within the loop (the YES path on the flowchart) are indented a few positions for clarity.

Counter-Controlled Loops

It is also important to look at the technique used to control the loop. In this example a counter (COUNT) is used to determine whether the loop steps will be repeated yet another time. This is an example of a **counter-controlled loop**. The algorithm clearly shows the number of times the loop steps will be done (six times in this case). A counter-controlled loop has two basic characteristics:

1. The loop is controlled by a counter.

2. The number of times the loop will be executed is known or preset.

These characteristics are shown pictorially in Figure 4–7. Each loop is executed a predetermined number of times. In the flowchart to the left, the COUNT is initialized to 0 before the loop is entered, and then it is tested in the first step of the loop. The number used in this test always corresponds to the number of times the loop is to be executed. After the processing steps within the loop are completed, the count is incremented by one and the test is made again. COUNT is often referred to as the **loop control variable**. In the flowchart to the right in Figure 4–7, a slightly different approach is used. The count is initialized to a number representing how many times the loop is to be done. Then the loop test determines if the count is still greater than 0. If so, the count will be decremented by one (rather than incremented) after the processing steps in the loop are executed. Do you see why? Both flowcharts accomplish exactly the same thing; they just approach it in two different ways. Both ways are satisfactory, and you will see examples of each throughout this book. Make sure you can identify each of these steps in the previous

Figure 4–7
Simple Counter-Controlled
Loop—Generic

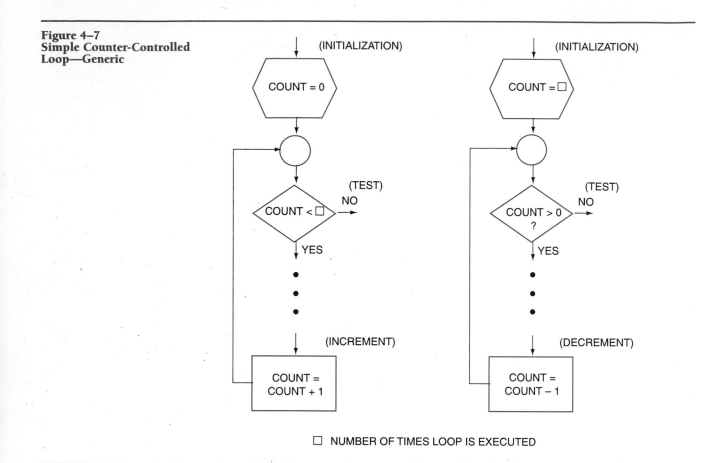

☐ NUMBER OF TIMES LOOP IS EXECUTED

example as well as in any algorithm you write that uses a counter-controlled loop.

Sample Problem 4.1 (Payroll with Counter Loop)

Problem:

Redo the payroll problem from Chapter 3 to process the payroll data for exactly 10 employees.

Solution:

Because we want to repeat the same processing steps for each employee, we can code the steps just once and use a counter-controlled DOWHILE loop structure to repeat the processing steps 10 times. A solution is shown in program flowchart form in Figure 4–8 and in pseudocode form in Figure 4–9.

A counter (COUNT) is used to control the loop. It is initialized to 0 at the beginning of the program. The DOWHILE loop is structured with the test first, as required: COUNT is checked for a value less than 10. If the YES path is taken, then the processing steps in the loop will be executed for one employee.

Figure 4–8
Payroll Problem—10
Employees (Flowchart)

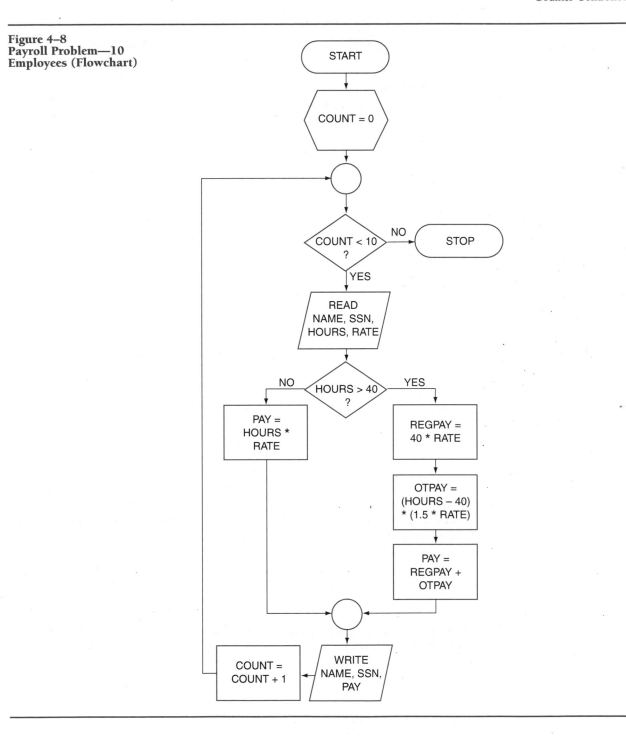

Look at these steps closely. The sequence of operations is the same as the sequence of operations we used in Figure 3–6. Remember, however, that that solution processed data for only one employee. In this example, after the WRITE statement is executed, the processing for one employee is complete, the COUNT is incremented by 1, and control is returned to the beginning of the loop. This series of actions is repeated 10 times (until the data for all 10 employees has been processed). At that point the value of COUNT is 10, the NO path is taken, and program execution stops.

<table>
<tr><td>Figure 4–9
Payroll Problem—10
Employees (Pseudocode)</td><td></td></tr>
</table>

Figure 4–9
Payroll Problem—10
Employees (Pseudocode)

```
Start
COUNT = 0
DOWHILE COUNT < 10
   Read NAME, SSN, HOURS, RATE
   IF HOURS > 40 THEN
      REGPAY = 40 * RATE
      OTPAY = (HOURS - 40) * (1.5 * RATE)
      PAY = REGPAY + OTPAY
   ELSE
      PAY = HOURS * RATE
   ENDIF
   Write NAME, SSN, PAY
   COUNT = COUNT + 1
ENDDO
Stop
```

Sample Problem 4.2 (Averaging Problem with Counter Loop)

Problem:

An instructor at Cloverdale School wants a computer program that will compute and print a student's term average. Each student has five scores. Each score will be input, one at a time, and will need to be accumulated. To determine the student's average, the scores' sum will be divided by five.

Solution:

A solution to this problem is shown in Figure 4–10 (flowchart) and Figure 4–11 (pseudocode). This algorithm accumulates the sum of the scores so that the average can be computed. This processing can be done using the same technique for adding as was used in Figure 4–4. Note that we are decrementing the count in this example. Consider what changes you would have to make to increment the count instead. Both approaches are satisfactory.

Notice that the average is not computed until we exit the loop. One common mistake of the beginner is to include the computation of the average (AVG = SUM / 5) inside the loop. This is inefficient because the average depends on the final value (last partial sum) of SUM. Remember that the variable SUM is being used as an accumulator of the partial sum of the student scores. The complete sum isn't known until all the scores have been accumulated. It is thus appropriate to compute the average only after exiting the loop.

Header Record Logic

Let us now turn our attention back to the mechanisms by which a loop can be controlled. This chapter has introduced DOWHILE loops and the counter approach for controlling them. Recall that a simple counter-controlled controlled loop is always executed a preset number of times. In our previous examples, the limits were 6 (adding problem), 10 (payroll problem), and 5 (averaging problem). We now look at a more general and somewhat

**Figure 4–10
Averaging Problem
(Flowchart)**

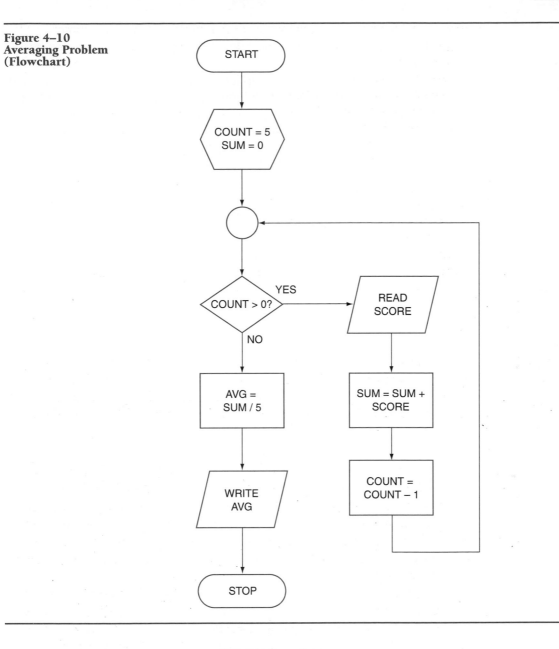

**Figure 4–11
Averaging Problem
(Pseudocode)**

```
Start
COUNT = 5
SUM = 0
DOWHILE   COUNT > 0
    Read SCORE
    SUM = SUM + SCORE
    COUNT = COUNT - 1
ENDDO
AVG = SUM / 5
Write AVG
Stop
```

more flexible approach to loop control that builds on simple counter-controlled loops. This approach is called **header record logic**. In the header record logic approach, a counter is still used to control the loop as before. However, the number of loop iterations can vary with each program execution. This flexibility is provided by inputting a special record before any of the regular input records. This special record is called a **header record**, and it specifies how many additional input records will follow. The additional input records contain the regular data that will be processed by the program. It is important to understand that the header record is a separate record and not part of the first regular input record.

For example, we could have included a header record containing a 10, followed by 10 employee records in our payroll problem. If we set the logic up properly, the same algorithm should work at a later time if the header record is changed to 50, followed by 50 employee records. The general logic required to accomplish this processing is shown in Figure 4–12.

Figure 4–12
Header Record Logic—
Generic

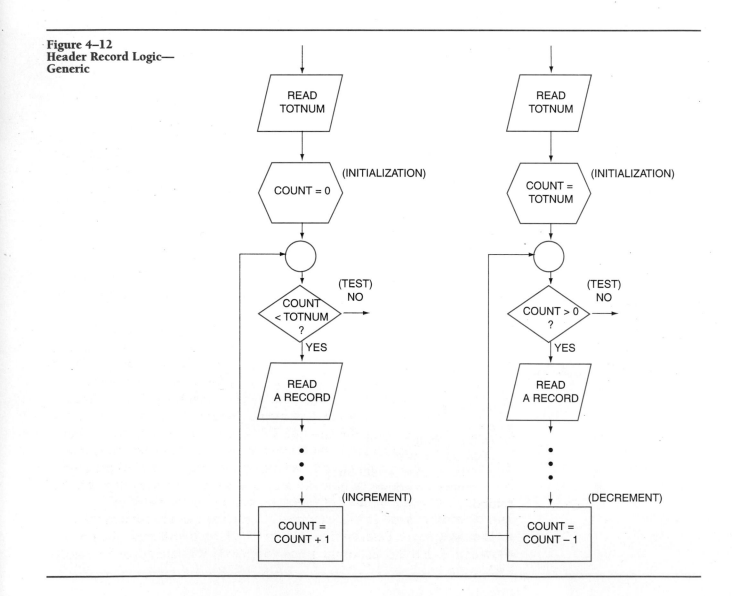

These flowcharts expand the notion of simple counter-controlled loops, shown in Figure 4–7. Both flowcharts include the addition of an initial input step to read the header record information. In this example, we have used the variable named TOTNUM to represent the total number of records to be processed. The flowchart on the left (incrementing the count) refers to TOTNUM in the loop test instead of to a specific value such as 6, 10, or 5. The flowchart on the right (decrementing the count) indicates that COUNT will initially be set to the value of TOTNUM instead of to a specific value such as 6, 10, or 5. In this way the algorithm becomes more flexible. No mention is made of any specific number on the program flowchart. Also note the second READ statement in each of these flowcharts. This READ step is positioned as the first step within the DOWHILE loop; it inputs the regular data to be processed, one record at a time. This is the step that would actually input the employee records or the individual numbers to compute a sum. Make sure you understand the purpose of the two READ steps. The first READ is done only once—it inputs the header record. The second READ is done every time the loop steps are executed—each time reading one of the regular records containing data to be processed.

Sample Problem 4.3 (Payroll with Header Record)

Problem:

Redo Sample Problem 4.1 to include a header record. A special input record containing the number of employees will be provided as the first input to the program. The individual employee records will then follow that record. The output will still contain the employee name, number, and pay for each employee as in Sample Problem 4.1.

Solution:

A solution to this problem is shown in Figure 4–13 (flowchart) and Figure 4–14 (pseudocode). This algorithm computes the pay for each employee in exactly the same manner as Sample Problem 4.1. What sets this problem apart from Sample Problem 4.1 is that we do not just process data for 10 employees; that is, we cannot compare COUNT to 10 or any other known number in the loop test. How, then, can we know how many times to execute the loop steps? Remember that the first input record will contain the number of employees for whom data is to be processed. This number is represented by the variable TOTEMP. Since this data needs to be input, the first step in Figures 4–13 and 4–14 shows a READ statement with the variable TOTEMP representing the number of employees. This record is the header record. After this record is input, TOTEMP is defined and can be used in the loop test instead of the number 10. In this way, every time the program is used, the loop steps can be executed a variable number of times, depending on the value of TOTEMP. Note that, in this example, the regular records contain the individual employee data. Notice also that the loop is still controlled by a counter and that the rest of the steps are exactly the same as in Sample Problem 4.1.

Figure 4–13
Payroll Problem with Header
Record (Flowchart)

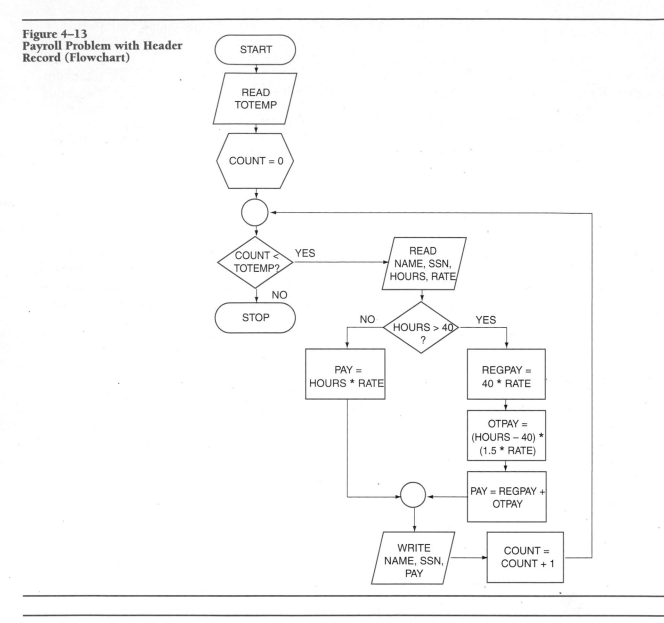

Figure 4–14
Payroll Problem with Header
Record (Pseudocode)

```
Start
Read TOTEMP
COUNT = 0
DOWHILE  COUNT < TOTEMP
     Read NAME, SSN, HOURS, RATE
     IF HOURS > 40 THEN
          REGPAY = 40 * RATE
          OTPAY = (HOURS - 40) * (1.5 * RATE)
          PAY = REGPAY + OTPAY
     ELSE
          PAY = HOURS * RATE
     ENDIF
     Write NAME, SSN, PAY
     COUNT = COUNT + 1
ENDDO
Stop
```

Sample Problem 4.4 (Averaging Problem with Header Record)

Problem:

Redo Sample Program 4.2 to include a header record. This time the number of individual scores that must be added will depend on the number of assignments completed. A special input record containing the student's name and number of scores to be added will be provided as the first input to the program whenever it is used. The individual scores will then follow that record. The output is to contain the student's name, number of assignments completed, and term average computed during processing.

Solution:

A solution to this problem is shown in Figure 4–15 (flowchart) and Figure 4–16 (pseudocode).

This algorithm accumulates the value of the scores so that the average can be computed. The processing can be done using the same technique for adding as was used in Figure 4–10. What sets this problem apart from the previous one is that we cannot just add five numbers—that is, we cannot

Figure 4–15
Averaging Problem with Header Record (Flowchart)

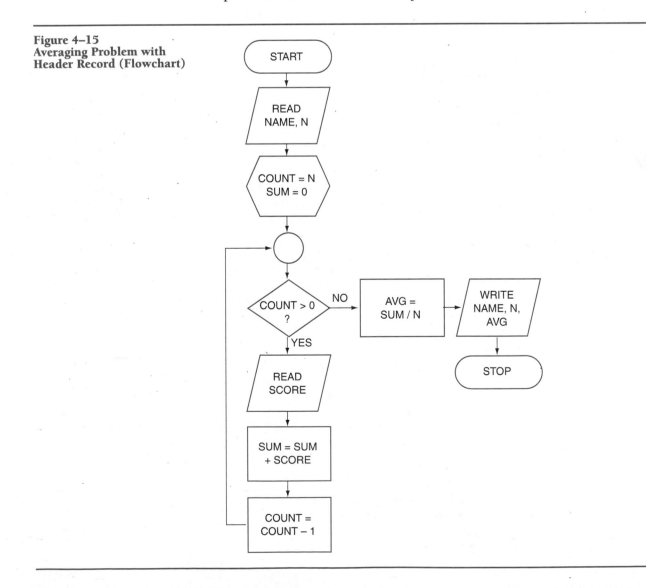

**Figure 4–16
Averaging Problem with
Header Record (Pseudocode)**

```
Start
Read NAME, N
COUNT = N
SUM = 0
DOWHILE COUNT > 0
   Read SCORE
   SUM = SUM + SCORE
   COUNT = COUNT - 1
ENDDO
AVG = SUM / N
Write NAME, N, AVG
Stop
```

assign a 5 or any other known number to count. How, then, can we know how many times to execute the loop steps? Remember that the first input record will contain the name of the student and the number of assignments completed. This number will be used in the loop test. Since this data needs to be input, the first step in Figures 4–15 and 4–16 shows a READ statement with the variable N representing the number of assignments completed. This record is the header record—it contains not only the number of assignments completed but also the name of the student. The regular records contain the individual scores.

Notice that the loop is still controlled by a counter. However, the initial value of the counter (COUNT) is set to N (just input), not to a fixed number (i.e., a constant such as 5). When COUNT finally reaches 0, we exit the loop and compute the student's average. This is accomplished by dividing the accumulated sum by the total number of scores, which is N, the value input on the header record. Notice that the value of N never changes throughout program execution. N is still equal to its initial input value.

The No-Data Condition

Before we leave this problem, an important question needs to be addressed: What if the value read in for N is less than or equal to 0? In this case, COUNT will be set to either 0 or a negative value after the name record is input. When the test is then made the first time to check if COUNT is greater than 0, the NO path will be taken immediately, since the condition is not true even the first time. The step immediately following the loop is the computation of the average, which involves a division by N. If N is negative, the average computed will be 0. (Do you see why?) If N is 0, a division by 0 will be attempted, and an error will result. We can handle these two cases by adding an IFTHENELSE statement right after N is input, as shown in Figures 4–17 and 4–18. Under this test the computer will check the current value of N and output a special message if the current value of N is 0 or negative. The program execution will then stop. If the NO path is taken as a result of that first test, we know that N must be positive, and we can proceed to compute the average as was shown in the previous example.

Figure 4–17
Averaging Problem with No-Data Test (Flowchart)

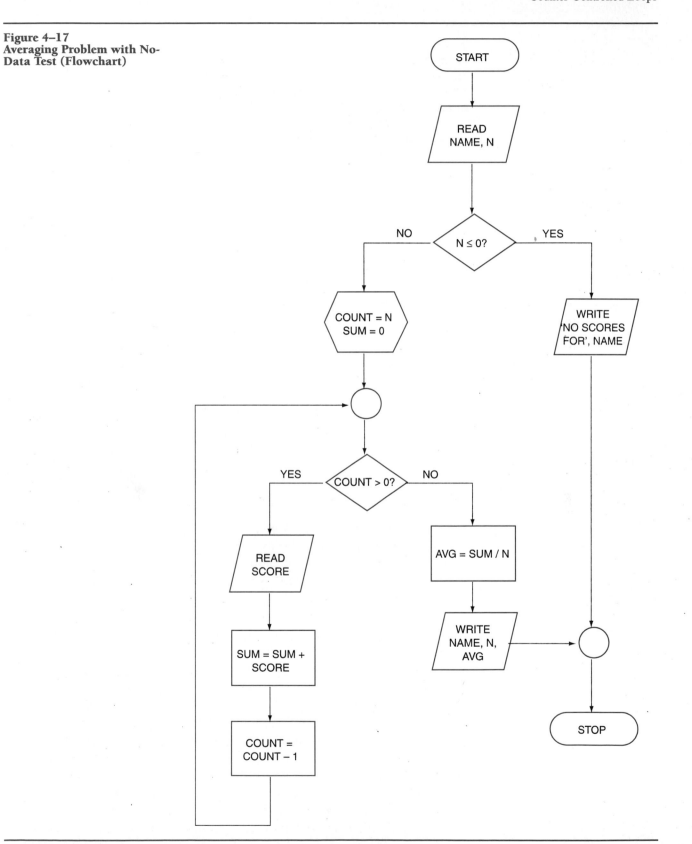

Figure 4–18
Averaging Problem with No-Data Test (Pseudocode)

```
Start
Read NAME, N
IF N ≤ 0 THEN
      Write 'No scores for ', NAME
ELSE
      COUNT = N
      SUM = 0
      DOWHILE COUNT > 0
            Read SCORE
            SUM = SUM + SCORE
            COUNT = COUNT - 1
      ENDDO
      AVG = SUM / N
      Write NAME, N, AVG
ENDIF
Stop
```

Notice the structures that are now part of our program. There is a DOWHILE loop completely nested in the NO path of the IFTHENELSE statement. Do you see it? Do you also see the role that each connector plays?

Proper Programs

Although you may find it hard to realize at this time, it is important to note that any solution algorithm can be expressed using only the three basic patterns of logic we have learned thus far: SIMPLE SEQUENCE, IFTHENELSE, and DOWHILE. As mentioned earlier, this approach is known as **structured programming**. When we construct an algorithm using only the three basic patterns of structured programming, we are at the same time taking advantage of another idea set forth by Bohm and Jacopini: the **building-block concept**. We have noted already that each basic pattern is characterized by a single point of entrance and a single point of exit. A SIMPLE SEQUENCE may be only a single statement, or it may be a series of single statements. It may also include IFTHENELSE patterns and DOWHILE patterns, and these may in turn include other SIMPLE SEQUENCEs comprising single statements, IFTHENELSEs, and DOWHILEs. We say that the contained patterns are *nested*.

A solution algorithm should have only one entry point and only one exit point. At the same time, every basic pattern in the algorithm should also have one entry and one exit point. A computer program representation of such an algorithm can be viewed conceptually as a single statement. A program that can be viewed as a single statement is called a **proper program**. Look again at some of the algorithms presented thus far in this book—they are all examples of proper programs.

Enrichment (Basic)

Figure 4–19 illustrates a listing of the program that solves the adding six numbers problem (see Figure 4–4). In this example we implement the

Figure 4–19
Adding Six Numbers Problem
(Basic List)

```
CLS
COUNT = 0
ACCUM = 0
DO WHILE COUNT < 6
    PRINT "Input a number";
    INPUT NUMBER
    ACCUM = ACCUM + NUMBER
    COUNT = COUNT + 1
LOOP
PRINT "The sum of the numbers is "; ACCUM
END
```

Figure 4–20
Adding Six Numbers Problem
(Basic Run)

```
Input a number? 1
Input a number? 2
Input a number? 3
Input a number? 4
Input a number? 5
Input a number? 6
The sum of the numbers is   21
```

DOWHILE loop with a construct in Basic that is very similar to DOWHILE pseudocode. The keyword LOOP is used in Basic instead of the keyword ENDDO. The same type of indentation is used in the Basic program, and the logic of the Basic DOWHILE loop is identical to the logic of the DOWHILE loop in pseudocode. In addition, most of the statements in the Basic program parallel the pseudocode. Again, an Input statement is used to request a number from the user. Because the Input statement is included within the loop, it will be executed six times and the user will be prompted for six numbers. The Print statement is used again to output directions to the user. Note that these directions will be printed six times, since the Print statement is also included within the loop. Identifying text was added in the last Print statement to make the output more readable.

Figure 4–20 illustrates the output that will be produced when the program is executed. After the user enters the first number (1 in this case), the user is prompted to enter another number. As the user enters each number, the number is accumulated and stored in the variable ACCUM. After six numbers have been entered by the user, the sum (21 in this case) is output with an identifying label.

Enrichment (Visual Basic)

Figure 4–21 illustrates the graphical user interface for the adding six numbers problem (see Figure 4–4).

In this example, only one control is created—a command button. When the user clicks this command button, a special window called an *input box* is presented to the user, as illustrated in Figure 4–22. The input box is another way (in addition to the text box control) to accept input from the user.

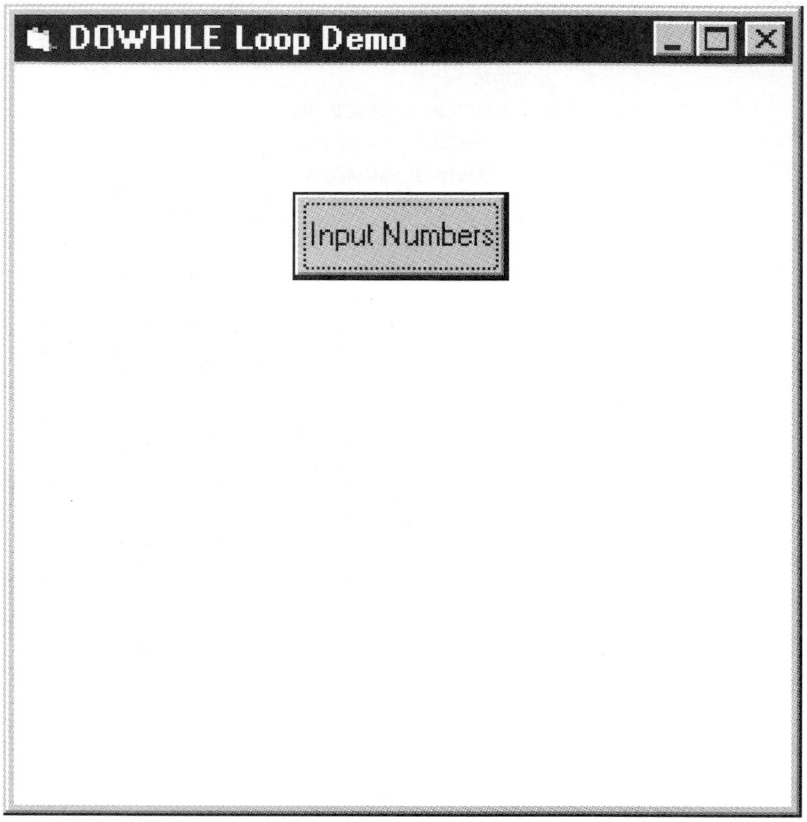

Figure 4–21
Adding Six Numbers Problem
(Visual Basic—Screen 1)

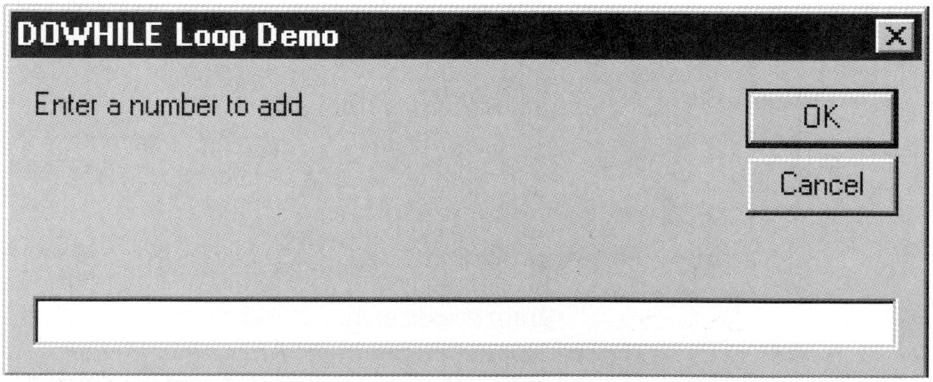

Figure 4–22
Adding Six Numbers Problem
(Visual Basic—Screen 2)

At this point the user should enter a number and click the OK button. This same input box will be presented five more times as the user enters five additional numbers. After the last number is input and the user clicks the OK button, a message box is presented displaying the sum of the six numbers, as shown in Figure 4–23.

Figure 4–24 illustrates the program that is associated with the click event of the Input Numbers button. In this example we implement the DOWHILE loop with a construct in Visual Basic that is very similar to DOWHILE pseudocode. The keyword LOOP is used in Visual Basic instead of the

Figure 4–23
Adding Six Numbers Problem
(Visual Basic—Screen 3)

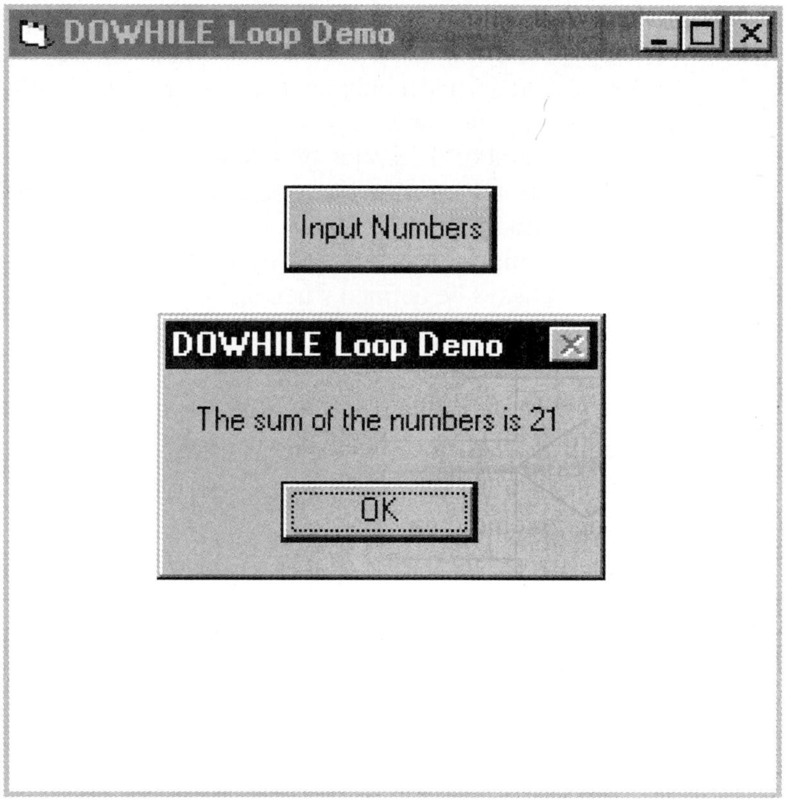

Figure 4–24
Adding Six Numbers Problem
(Visual Basic—
cmd_INPUT_Click)

```
Private Sub cmd_INPUT_Click()

Cnt = 0
Accum = 0
Do While Cnt < 6
    Number = InputBox$("Enter a number to add", "DOWHILE Loop Demo")
    Accum = Accum + Number
    Cnt = Cnt + 1
Loop
MsgBox "The sum of the numbers is " & Accum,,"DOWHILE Loop Demo"

End Sub
```

keyword ENDDO. The same type of indentation is used in the Visual Basic program, and the logic of the Visual Basic DOWHILE loop is identical to the logic of the DOWHILE loop in pseudocode. In addition, most of the statements in the Visual Basic program parallel the pseudocode. An InputBox$ statement is used to request a number from the user. This statement causes the input box window to be displayed. Because the InputBox$ statement is included within the loop, it will be executed six times and the user will be prompted for six numbers. The programmer determines both the message to be displayed and the title of the input box window. The programmer has no control over the buttons that will be displayed in the input box. An OK button and a Cancel button are always displayed. When a user enters a value into the input box and clicks the OK button, that value is assigned to the variable Number. The value of Number is then accumulated, the counter is incremented, and the loop test is executed. If the user clicks the Cancel button instead of the OK button, no value will be assigned to Number. This situation could cause an error to occur since the value of the variable Number may not always be defined when the assignment statement following the InputBox$ statement is executed. In addition, if the user enters a non-numeric value in the input box, another type of error may occur since the variable Number is meant to hold only numeric data. Clearly, this program is not complete. Potential errors are likely. We will discuss some of these issues in later chapters. At this point, we simply need to realize this program's limitations.

Key Terms

program loop	connector symbol	header record logic
preparation symbol	DOWHILE loop	header record
initialization step	leading-decision	structured
accumulator	program loop	programming
counter	counter-controlled loop	building-block concept
partial sum	loop control variable	proper program

Exercises

1. State in your own words the purpose of a program loop.

2. What does it mean to say that a loop is "exited"?

3. (a) Using ANSI-approved flowcharting symbols, sketch the logic of a DOWHILE control structure.
   (b) State the logic of a DOWHILE control structure in pseudocode form.
   (c) What pseudocode keywords did you use in your response to Exercise 3(b)?
   (d) What indentations did you use in your response to Exercise 3(b)? Why?

4. (a) Why do we call the DOWHILE pattern a leading-decision loop?
   (b) What is particularly significant about the potential effects of this execution sequence?

5. Explain how the building-block concept can be used in developing a solution algorithm.

6. What is a proper program?

7. Name the three basic patterns of structured programming.

8. (a) What is a counter-controlled program loop?
   (b) Draw a generic program flowchart to explain header record logic.
   (c) How are the loop constructions in Exercises 8(a) and (b) similar?
   (d) How do they differ?

9. Use Figure 4–4 to answer the following questions:
   (a) What would happen if the step COUNT = 0 were placed inside the loop?
   (b) What would happen if the step ACCUM = 0 were placed inside the loop?
   (c) What would happen if the step Write ACCUM were placed inside the loop?
   (d) What would happen if only three numbers were input?
   (e) What would happen if 10 numbers were input?
   (f) How could you change the algorithm to work for 10 numbers?

10. (a) Using ANSI-approved flowcharting symbols, sketch the logic showing the processing steps required to output the first 10 numbers (the numbers 1 to 10). Use a counter-controlled program loop.
    (b) State the logic of Exercise 10(a) in pseudocode form.

11. (a) Using ANSI-approved flowcharting symbols, sketch the logic showing the processing steps required to compute and output the square (number times number) and cube (number times number times number) for each of the numbers 1 through 100. Use a counter-controlled program loop.
    (b) State the logic of Exercise 11(a) in pseudocode form.

12. Construct a program flowchart and corresponding pseudocode describing the processing steps needed to solve the following problem: Initial values of 5.00 and 3.00 are to be assigned to A and B, respectively. The value for C, which is 95 percent of A, is to be computed. A, B, and C are to be printed. A is to be increased by twice the value of B. B is to be increased by 10 percent. The steps beginning with the computation of a value for C are to be repeated five times, and then program execution should terminate. Be sure to plan a well-structured program.

13. Assume the problem statement in Exercise 12 is modified as follows: A new value is to be computed for C on the first, third, and fifth passes through the loop only; otherwise, C is not to be changed. Construct a program flowchart and corresponding pseudocode for this revised algorithm.

14. Do a procedure execution of your solutions to Exercises 12 and 13. Be sure to list all the variables you used in the design, and to show how their values change with each iteration of the loop. Do not make any assumptions.

15. Construct a program flowchart and corresponding pseudocode to develop a solution algorithm for a program that will compute and print the sum of the numbers 1, 3, 5, 7, . . . , 99. Be sure to plan a well-structured program.

16. The Fibonacci series is defined to be the following numbers:

    0, 1, 1, 2, 3, 5, 8, . . .

    where each number is the sum of the previous two numbers. Construct a program flowchart and corresponding pseudocode that will compute and output the first number greater than 100 in the above series. No input is required.

17. Construct a program flowchart and corresponding pseudocode to compute and print a sum as shown below:

$$1^1 + 2^2 + 3^3 + 4^4 + \ldots + N^N$$

N will be input.

18. N factorial (N!) is defined as the following series of numbers:

$$N! = N \times (N - 1) \times (N - 2) \times (N - 3) \times \ldots \times 1$$

Construct a program flowchart and corresponding pseudocode to compute N! and output both N and N!. N will be input, and N! should only be computed if N is greater than 0. If it is not, output an appropriate message.

19. Construct a program flowchart and corresponding pseudocode to solve the following problem: ABC Company needs a weekly payroll report for its salespeople. Input to the program is a salesperson's name, number, and weekly sales. Output is the salesperson's name, number, and pay. Each salesperson receives a base pay of $300.00 as well as a 10 percent commission on his or her total sales up to and including $500.00. Any sales over $500.00 merit a 15 percent commission for the employee. (For example, if sales = $600.00, then pay = $300.00 + $50.00 [or 10 percent * 500] + $15.00 [or 15 percent* 100] = $365.00.) Use a DOWHILE loop and a counter to compute the weekly payroll for exactly 20 employees. Be sure to plan a well-structured solution.

20. Redo the solution to Exercise 19 to process sales data for any number of employees. Assume that the total number of employees for whom data is to be processed will be indicated on the first input record. This record will be followed by all the employee records. Be sure to check that the header record contains a positive number. Output an appropriate message if the header is not positive.

21. Redo the solution to Sample Problem 4.3 to include an initial IF statement to check that the header record contains a positive number. Output an appropriate message if the header is not positive.

22. Redo the solution to Exercise 21 to output the total amount paid at the regular rate for all employees, the total amount paid at the overtime rate for all employees, and the total pay for all employees.

23. Construct a program flowchart and corresponding pseudocode to solve the following problem: You are in a pumpkin patch looking for the great pumpkin. The first input record indicates how many pumpkin weight records follow. Several input records, each containing the weight of a pumpkin, follow the first input record. Assume that all the pumpkin weights are greater than 0. You are to find the largest weight and the average weight of all the pumpkins. You are to output each weight, as well as the largest weight and the average weight. Include an initial IF in your design that will output a descriptive message if the header record indicates that there are no input records to process.

DOWHILE Control Structure—Trailer Record Logic

5

Objectives

Upon completion of this chapter you should be able to

- Distinguish between header record logic and trailer record logic.
- Design programs using trailer record logic.
- Distinguish between heading, detail, and total lines.
- Design programs that require heading, detail, and total lines.
- Design the logic needed to handle invalid input data.
- Define and distinguish between the priming read and the loop read.
- Design the logic required for automatic end-of-file processing.
- Design a program that outputs headings on every page of a report.

Introduction

In Chapter 4, the concept of a loop was introduced. Specifically, the DOWHILE loop was used in the solutions for all of the problems. In this chapter, we will continue to use the DOWHILE loop. Remember what characterizes this type of loop. The test to determine if the processing steps within the loop should, in fact, be executed always comes first within the loop structure. Also, the test is worded so that the loop steps are always executed in the YES (true) path and exited in the NO (false) path. We also looked at the important issue of loop control, that is, which factor determines when the loop should be exited. In Chapter 4 we began with the concept of the simple counter-controlled program loop, and then expanded that idea to a specific technique called header record logic. Both approaches use a counter to control the loop. In header record logic, the first input record (the header record) contains a number that specifies how many times to execute the loop. In this chapter we will explore a different procedure for loop control called trailer record logic.

As the name **trailer record logic** implies, this approach involves the use of a **trailer record** (last input record) to control the loop. This last record serves as a signal to the computer that no more records are to be processed. Because this record serves as an indicator that no more records are to follow, it must contain some special information that differentiates it from the rest of the input, and that can be used in the loop test to determine when all input has been processed. The question is: What type of signal should we place in a trailer record? The answer depends on the usual contents of the input records. The signal used may vary from problem to problem. For example, if the input consists of employee records, and each record contains a Social Security number, then a record with a Social Security number of 0 could serve as a trailer record.

The key requirement is that a trailer record (also called a **dummy data value** or **sentinel value**) contain a value in the particular field chosen for the loop test that is different from any possible value for the field. (No one, for example, has a Social Security number of 0.) That is, we must be sure that our choice of the trailer value will not occur as a valid value for the field in other input records. If this were to happen, none of the records following the record containing the trailer value as a valid value would be input or processed. In addition, a trailer record may need to contain values for all fields even though only one of these values will be tested. It is also important to understand that a DOWHILE loop controlled by a trailer record does not require a counter for loop control. Let us explore this concept with a specific example.

Sample Problem 5.1 (Defective Parts Problem)

Problem:

Car/Go Manufacturing Company produces a large number of automotive parts each year in its two plants. Some of these parts are returned to the main sales office because of defects in manufacture. An input record is prepared for each defective part, containing a code, part number, type of part, and date returned. The code is either 1 or 2, indicating whether plant 1 or plant 2 made the defective part.

Car/Go needs a read-and-print program to process these records and produce a listing of their contents. In addition, the number of defective parts manufactured by each plant is to be totaled for future reference. A trailer record containing a plant code of 9 is to signal the end of the input (see Figure 5–1). The output is to consist of a printed listing of the contents of the input records, followed by the total computed for each plant.

Solution:

A solution to this problem is shown in both flowchart and pseudocode forms in Figures 5–2 and 5–3.

Figure 5–1
Defective Parts Problem—
Input Format

Plant Code	Part #	Part Type	Date Returned
2	L45603	TAILLIGHT	04/16/2004
1	W07722	WIPER BLADE	04/17/2004
1	C19654	CLOCK	04/16/2003
2	D33045	DOMELIGHT	04/20/2003
⋮	⋮	⋮	⋮
9	999999	END OF FILE	00/00/0000

Figure 5–2
Defective Parts Problem
(Flowchart)

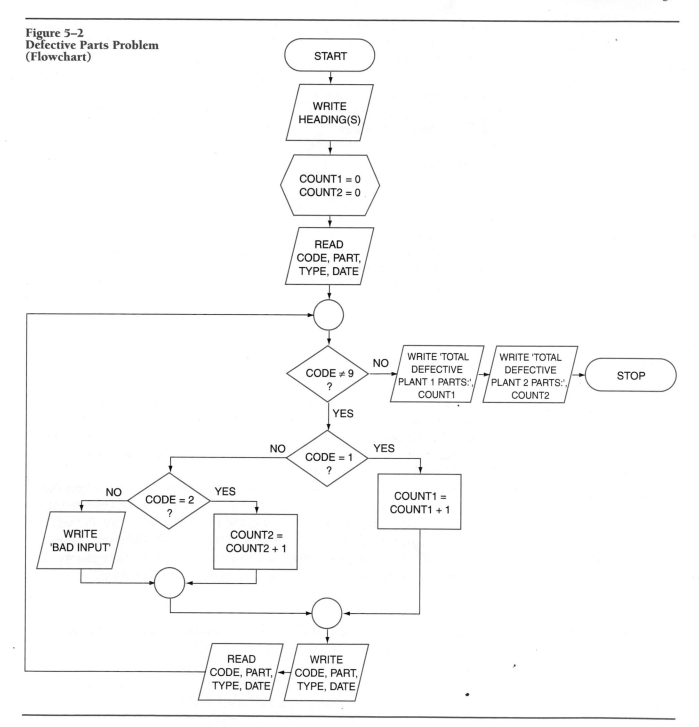

Heading Lines

The algorithm begins with an output statement that produces one or more headings. A **heading line** serves as a title to a report that is output. Headings are almost always used on business reports to identify them to readers. A report might include a report heading such as PAYROLL REPORT or DEFECTIVE PARTS REPORT. It might also include one or more column

Figure 5–3
Defective Parts Problem
(Pseudocode)

```
Start
Write heading(s)
COUNT1 = 0
COUNT2 = 0
Read CODE, PART, TYPE, DATE
DOWHILE CODE ≠ 9
   IF CODE = 1 THEN
      COUNT1 = COUNT1 + 1
   ELSE
      IF CODE = 2 THEN
         COUNT2 = COUNT2 + 1
      ELSE
         Write 'Bad Input'
      ENDIF
   ENDIF
   Write CODE, PART, TYPE, DATE
   Read CODE, PART, TYPE, DATE
ENDDO
Write 'Total defective plant 1 parts:', COUNT1
Write 'Total defective plant 2 parts:', COUNT2
Stop
```

headings such as NAME, HOURS, RATE, and so on. It is not usually necessary to define the headings in the early stages of algorithm development. The detailed text of the headings can be worked out later, but an indication that one or more headings will be output should be included in the design at this point. Notice that headings are output at the beginning of processing; therefore, this output step is not included in the loop. This makes sense because a title is usually shown only once—at the start of a report, not preceding every line within the report. Later in this chapter we will look at the logic to output headings on every page of the report. At this point, however, our design indicates that the headings will be output only once—at the beginning of the report.

Detail Lines

In the next step, two counters are initialized to 0 to keep a tally of the number of defective parts produced by each plant. These counters are not used to control the loop, as was the case with header record logic. At this point, we are ready to process each input record, so the first record is input. Notice that the READ statement refers to four variables, each representing one of the input fields. Notice also that the READ statement in Figure 5–2 is followed by a connector indicating the beginning of the DOWHILE loop.

Now notice the test at the beginning of the loop. We need to know if the loop steps should be executed; that is, is the record we've just read a normal input, or is it the trailer record? If CODE (just input) is not 9, we assume that we have a normal input record and proceed with the "true" path—the actual processing steps within the DOWHILE loop. A nested IFTHENELSE structure checks to see if the code is 1 or 2. The appropriate counter is then increased because we need to keep track of the number of defective parts from each plant. If both tests fail, the code must be invalid because the company has only two plants, 1 and 2. In this case we output an error message indicating that the input is invalid.

Both IFTHENELSE structures are then closed with their respective connectors on the flowchart and ENDIFs in the pseudocode. The contents of the input record are output. Note that if the code is in fact invalid, two lines of output are generated: an error message ("Bad Input") and the contents of the input record itself. We can then see the contents of all invalid records, with an error message preceding each one.

Note that the last WRITE statement within the loop is executed each time the loop steps are processed. It causes one line of output to be printed for each input record. This output is an example of a **detail line**, that is, a line of output associated with one input record. Detail lines are always output in a step within the main processing loop. Remember, though, that heading lines are output before the loop is entered.

The last step in the DOWHILE loop is a READ statement that is identical to the first READ statement. This statement causes the next available record to be input; then the loop processing is started again, and the initial test is made to determine whether the code just input is 9. If this second record is not the trailer record (CODE is not 9), the loop steps are repeated for the record and a third record is read. This processing continues until all the input records have been tested and output. When the trailer record is finally input, the loop test is not true (because the code is 9). Notice the double negative; that is, a code of 9 means that the test for a code not equal to 9 is false. Why this confusing verbiage? Remember, the test must be worded in such a way as to cause the loop steps to be done in the true path and the loop exit to be the false path.

Total Lines

When the loop is finally exited, the values for the two counters, as well as identifying labels, are output. Each of these two lines is an example of a **total line**; that is, a line that represents cumulative information from one or more input records. Total lines are typically output after the loop is exited. It is important to remember the differences between a heading line, a detail line, and a total line; similarly, you must remember where the statement for each type of line should be written relative to the loop.

Priming Read and Loop Read

In trailer record logic there are always two READ statements in the algorithm that look exactly the same. The first READ statement, placed before the loop, causes the first record to be input. This READ statement is called the **priming read**, because it gets the computer ready (primes the computer) to make the loop test. The second READ statement, called the **loop read**, is within the loop—usually the loop's last statement. It causes the next available input record to be read each time the loop steps are executed.

Both of these READ statements are essential for successful program operation. If we omitted the priming read in this example, the value of CODE would be undefined when the loop test was first attempted. If we omitted the loop read, no more records would be input, and the initial value of CODE would be tested over and over again. Because the trailer record would never be input, the loop steps would be repeated indefinitely. An infinite loop would result, eventually causing an error condition to be recognized by the system and signaled to the user or application program.

Look over this algorithm once again and concentrate on finding each control structure. We have a series of SIMPLE SEQUENCE statements and a DOWHILE loop, containing a nested IFTHENELSE structure, containing additional SIMPLE SEQUENCE statements. Convince yourself that this is, in fact, a proper program.

Automatic End-of-File Processing

It is not always necessary to physically place a trailer record at the end of the input. Most language-processor programs (interpreters or compilers) support a built-in function for recognizing when the end-of-file has been reached, even if a trailer record is not provided as input. The main design of the algorithm is not affected in such a case. The only change that needs to be made is the wording of the test to determine if there is any more input. Instead of checking for a specific trailer value, we can simply test for the generic name of the function, for example, "End-of-File" or, even more simply, EOF. Each programming language provides a specific way to implement this test; but at design time simply noting that a check for an EOF condition is needed is sufficient. If we design our algorithm in this way, we are utilizing what is called an **automatic end-of-file facility**.

Figures 5–4 and 5–5 illustrate the flowchart and pseudocode for one approach to this problem. As you can see, the steps in the algorithm are much like those shown in Figures 5–2 and 5–3. Both the priming read and the loop read are still present. We have, however, added an IFTHENELSE statement after the first READ statement to direct the computer to check whether or not there are any input records. Instead of checking for a specific value, the first IFTHENELSE tests whether or not end-of-file (EOF) has been reached. The test result will be true only if there is no input data. We sometimes call this the **empty file condition**. It may seem unnecessary to check for this condition, but the condition happens more often than you may expect. This part of the algorithm is analogous to the test made to determine if the header record read as input contains a positive number (see Figures 4–17 and 4–18). The second test—a check for a "not EOF" condition—occurs at the beginning of a DOWHILE structure. If "not EOF" is true, the loop steps in the true path of the structure are executed. Remember, having the true path within the loop is a DOWHILE loop requirement. The rest of the algorithm remains unchanged.

Sample Problem 5.2 (Defective Parts with Multiple Headings)

Problem:

Redo the solution to Sample Problem 5.1 to output headings on every page of the report, not just on the first page. In addition, number each page. Use automatic end-of-file processing in the solution.

Figure 5–4
Automatic End-of-File
(Flowchart)

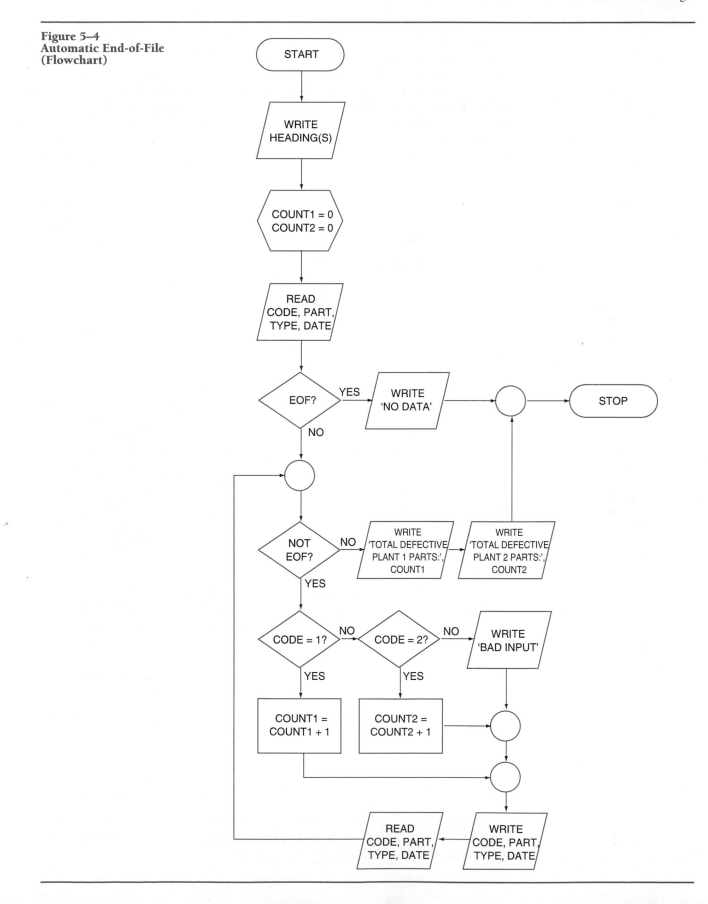

Figure 5–5
Automatic End-of-File
(Pseudocode)

```
Start
Write heading(s)
COUNT1 = 0
COUNT2 = 0
Read CODE, PART, TYPE, DATE
IF EOF THEN
    Write 'No data'
ELSE
    DOWHILE not EOF
        IF CODE = 1 THEN
                COUNT1 = COUNT1 + 1
        ELSE
                IF CODE = 2 THEN
                        COUNT2 = COUNT2 + 1
                ELSE
                        Write 'Bad Input'
                ENDIF
        ENDIF
        Write CODE, PART, TYPE, DATE
        Read CODE, PART, TYPE, DATE
    ENDDO
    Write 'Total defective plant 1 parts:', COUNT1
    Write 'Total defective plant 2 parts:', COUNT2
ENDIF
Stop
```

Solution:

To output headings at the top of every page, we will need to know when the bottom of every page (and hence the top of the next page) is reached. We will use a counter (LINECNT) to keep track of the number of detail lines that have been output on a page at any given time. LINECNT will be initialized to 0. Each time a detail line is output, we will increase the value of LINECNT by 1. When the value of LINECNT reaches a predetermined maximum (55 in this example), a new page will be started, and the heading lines will be the next lines written as output. The maximum value of LINECNT will determine the maximum number of detail lines printed on one page. We will use another counter (PAGECNT) to keep track of which page is currently being written as output. PAGECNT will be initialized to 1. Its value will be increased by 1 each time the heading lines are output. Its current value will be output whenever the heading lines are output. The program flowchart and pseudocode for this problem are shown in Figures 5–6 and 5–7.

The two counters PAGECNT and LINECNT are initialized at the beginning of the algorithm. PAGECNT is initialized to 1, since we want the first page of our report to be numbered 1. LINECNT is initialized to 55—not to 0 as you may have expected. If we had set LINECNT to 0, no headings would have been printed on the first page. We need to "fool the computer" into thinking that we are already at the bottom of a page before we even start. In this way, the first time the value of LINECNT is checked, it will cause a new page to be started and the desired headings to be output.

Figure 5–6
Defective Parts Problem with
Multiple Headings
(Flowchart)

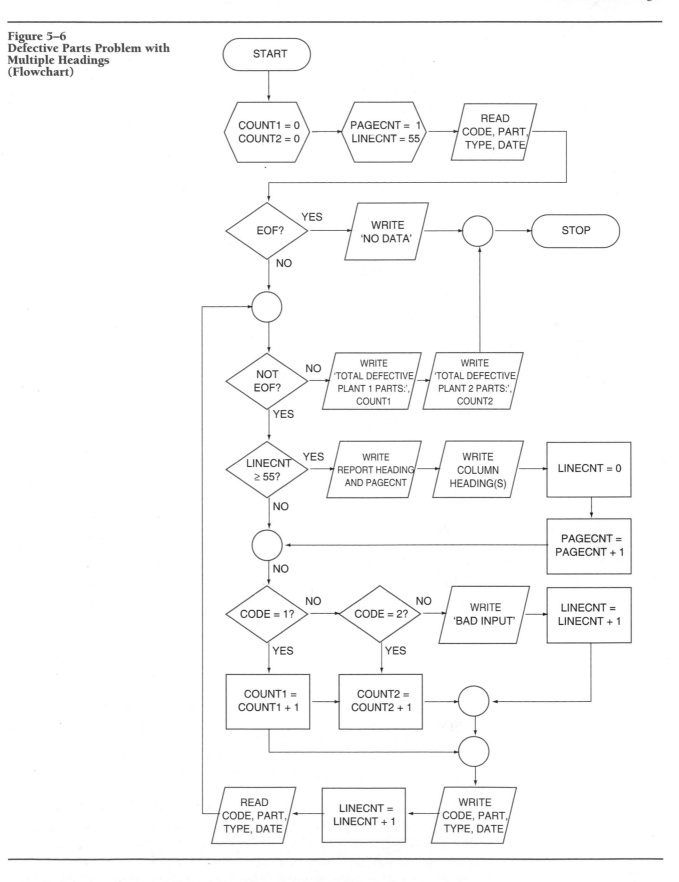

Figure 5–7
Defective Parts Problem with
Multiple Headings
(Pseudocode)

```
Start
COUNT1 = 0
COUNT2 = 0
PAGECNT = 1
LINECNT = 55
Read CODE, PART, TYPE, DATE
IF EOF THEN
    Write 'No data'
ELSE
    DOWHILE not EOF
        IF LINECNT ≥ 55 THEN
                Write report heading and PAGECNT on top of next page
                Write column heading(s)
                LINECNT = 0
                PAGECNT = PAGECNT + 1
        (ELSE)
        ENDIF
        IF CODE = 1 THEN
                COUNT1 = COUNT1 + 1
        ELSE
                IF CODE = 2 THEN
                        COUNT2 = COUNT2 + 1
                ELSE
                        Write 'Bad Input'
                        LINECNT = LINECNT + 1
                ENDIF
        ENDIF
        Write CODE, PART, TYPE, DATE
        LINECNT = LINECNT + 1
        Read CODE, PART, TYPE, DATE
    ENDDO
    Write 'Total defective plant 1 parts:', COUNT1
    Write 'Total defective plant 2 parts:', COUNT2
ENDIF
Stop
```

The first step within the DOWHILE loop checks the value of LINECNT to see if its value is equal to or exceeds 55. If so (as it will be the first time), the heading lines are output. In this example, we assume that both report and column headings are to be output. PAGECNT is written on the same line as the report heading. LINECNT is set back to 0 because we are starting a new page and no detail lines have yet been output on the page. PAGECNT is then increased by 1 so that the next page will be numbered correctly. The normal detail processing then occurs, a detail line is output, and LINECNT is increased by 1. If the plant code within the next input record is invalid, the "Bad Input" line is output, and LINECNT is again increased by 1. It is important to see that the detail processing must be done, whether or not the headings are output. This logic is provided for in the design by the use of the null ELSE clause located in the second IFTHENELSE statement in the algorithm. We might think of the logic as follows: Before a detail line can be output, we must check to see if any room exists on the current page. If so, we simply write out the detail line. If not, we start a new page, write the heading information, and then output the detail line. Note also that there are three IFTHENELSE statements within the loop. Can you see that the first two IFs are sequential and that the third IF is nested within the second IF?

The total lines will be output on the same page as the last detail line—even if the maximum number of lines (55 in this case) has been output. (Users of a report prefer to see totals in relation to the rest of the report, not alone on a separate page.)

Multiple Headings— Summary

Before we look at another algorithm, let us summarize what we have learned about multiple headings. We saw that two additional counters are needed when keeping track of how many lines on a given page have been output so far (LINECNT) and how many pages have been output so far (PAGECNT). These counters are initialized, incremented, and tested in much the same way in any problem that outputs multiple headings and page numbers. These functions are often used in application packages. For example, word processors use this type of logic to output page numbers, headers, and footers. Likewise, a database management program uses similar logic to generate a page overflow.

Figure 5–8 illustrates a flowchart showing the general processing requirements for multiple heading logic. Note that we have added another variable, MAXLINES, to the initialization step. MAXLINES acts as a **named constant**, a variable whose value does not change during program execution. This variable represents the maximum number of detail lines that will be output on one page. We set MAXLINES equal to 55 only once at the beginning of the algorithm. We can then refer to MAXLINES anywhere in the algorithm where we would otherwise refer to the constant 55. We set LINECNT equal to MAXLINES (not 55) and later we test for the value of MAXLINES (not 55).

The use of named constants makes an algorithm more flexible. For example, if the maximum number of lines per page needed to change to 40, we would need to change only one statement: MAXLINES = 55 to MAXLINES = 40. Without the use of the named constant MAXLINES, we would need to search through the algorithm and change every occurrence of the constant 55 to the constant 40. Such a process is tedious, time consuming, and error prone, particularly in more complex algorithms in which there are many references to a constant value.

Sample Problem 5.3 (Credits Problem)

Problem:

Design an algorithm to read individual records containing student names, addresses, and total numbers of accumulated credits as input. The names

Figure 5–8
 **Multiple Headings—
Summary**

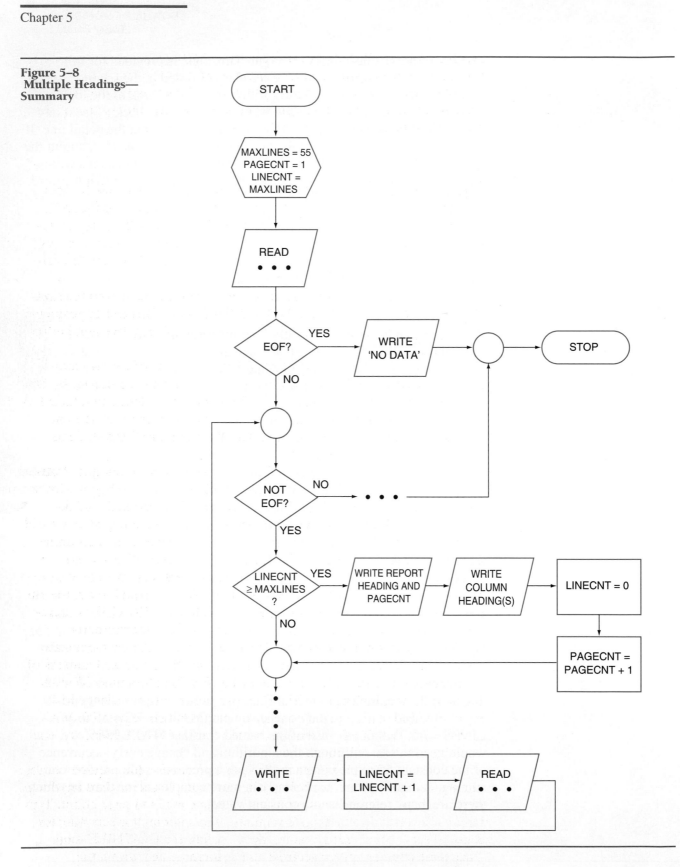

and addresses of all students who have earned 60 or more credits should be printed as output. For other student records, no action is required. Program execution should terminate when a trailer record containing a negative credits amount is input. Headings are to be written on every page, and the pages are to be numbered.

Solution:

Figures 5–9 and 5–10 illustrate the flowchart and pseudocode for this solution. The overall structure of this solution is the same as that of Sample Problem 5.2; however, there are a few minor differences in details. First, the two READ statements contain different variable names. The three variables—NAME, ADDRESS, and CREDITS—are the values that need to be input for each student.

A second difference is in the actual test that is made in both the first IFTHENELSE statement and the DOWHILE loop. This test corresponds to the requirement that a trailer record containing a negative number for the credits field be used.

Another difference in this solution is the omission of the two total lines, which are usually output after loop processing is complete. In this problem, however, we are not asked to compute or output any total information. For example, we do not need to count the number of student records or accumulate the total number of credits for all the students. (Doing so is left as an exercise.)

The major difference in this solution, as compared to Sample Problem 5.2, is the processing within the DOWHILE loop. To begin with, we check to see if the value of LINECNT equals or exceeds MAX-LINES. If so, the heading lines are output on a new page, just as we did in the previous algorithm. This IFTHENELSE construct is then immediately followed by another IFTHENELSE construct. The second IFTHENELSE checks to see if the current student's credits equal or exceed 60. If the value of CREDITS is greater than or equal to 60, the student name and address are output and the value of LINECNT is incremented by 1. You might wonder why the value of student credits (CREDITS) was not also output. If you re-examine the problem statement, you will notice that it specifies only that the name and address of each student be output. If the value of CREDITS is less than 60, nothing is to be output. The null ELSE clause in the design reflects this logic. It is interesting that this solution contains two sequential (not nested) IFTHENELSE statements within the DOWHILE loop, and each contains a null ELSE clause.

Notice that no counters are set to 0, as was done in the previous example. Remember, we are not being asked to count or accumulate anything other than the required computations for line count and page count. Typically, if a total of some type is required, a counter and/or accumulator must be set to some initial value, like 0, before the DOWHILE loop. That same counter and/or accumulator is incremented within the DOWHILE loop, and the final value of the counter and/or accumulator is output when the DOWHILE loop is exited.

Figure 5–9
Credits Problem (Flowchart)

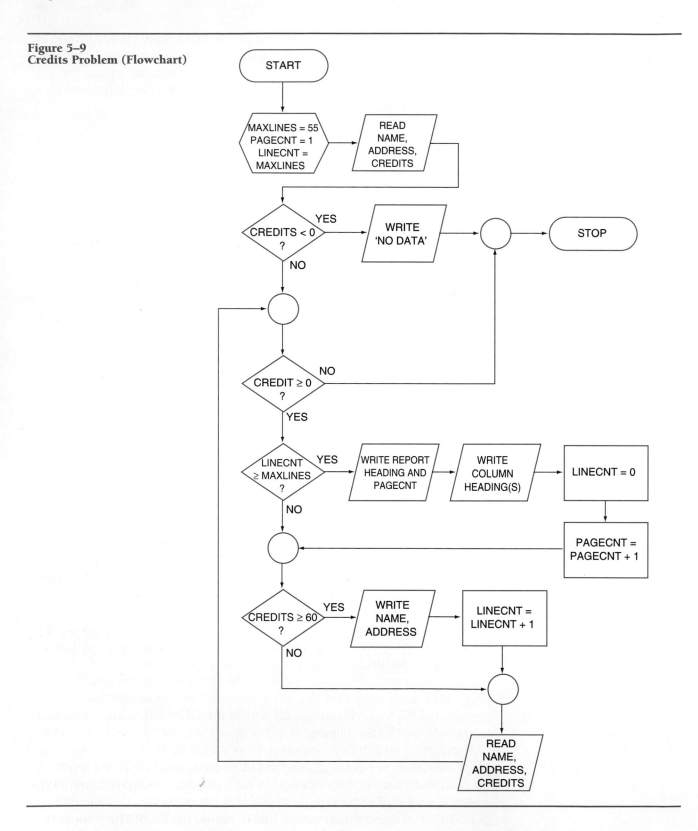

Figure 5–10
Credits Problem
(Pseudocode)

```
Start
MAXLINES = 55
PAGECNT = 1
LINECNT = MAXLINES
Read NAME, ADDRESS, CREDITS
IF CREDITS < 0 THEN
        Write 'No data'
ELSE
        DOWHILE CREDITS ≥ 0
            IF LINECNT ≥ MAXLINES THEN
                    Write report heading and PAGECNT on top of next page
                    Write column heading(s)
                    LINECNT = 0
                    PAGECNT = PAGECNT + 1
            (ELSE)
            ENDIF
            IF CREDITS ≥ 60 THEN
                    Write NAME, ADDRESS
                    LINECNT = LINECNT + 1
            (ELSE)
            ENDIF
            Read NAME, ADDRESS, CREDITS
        ENDDO
ENDIF
Stop
```

DOWHILE Loop Control—Summary

In Chapters 4 and 5 we introduced the DOWHILE control structure and four approaches to loop control. Chapter 4 focused on counter loops (simple and header record) and Chapter 5 focused on end-of-file control (trailer record and automatic). Figure 5–11 (flowcharts) and Figure 5–12 (pseudocode) illustrate the basic steps required to read and write several records. Each of the four approaches to loop control is shown; however, the initial test for an empty file is omitted for simplicity. Make sure you understand the differences among these four approaches.

Enrichment (Basic)

Figure 5–13 illustrates a listing of the program that solves the credits problem (see Figures 5–9 and 5–10).

The Basic code in Figure 5–13 is similar to the pseudocode in Figure 5–10. In this example, we use a *Read* statement to accept input values instead of an *Input* statement. A Read statement does not accept user input. Rather, *Data* statements containing the actual data to be input are listed at the end of the program. Each time the Read statement is executed, three data items are input from one of the Data statements. These three pieces of data are then stored in the appropriate variables listed in the Read statement. For example, the first time the Read statement is executed, "John" is

Figure 5–11
Types of DOWHILE Loop
Control (Flowcharts)

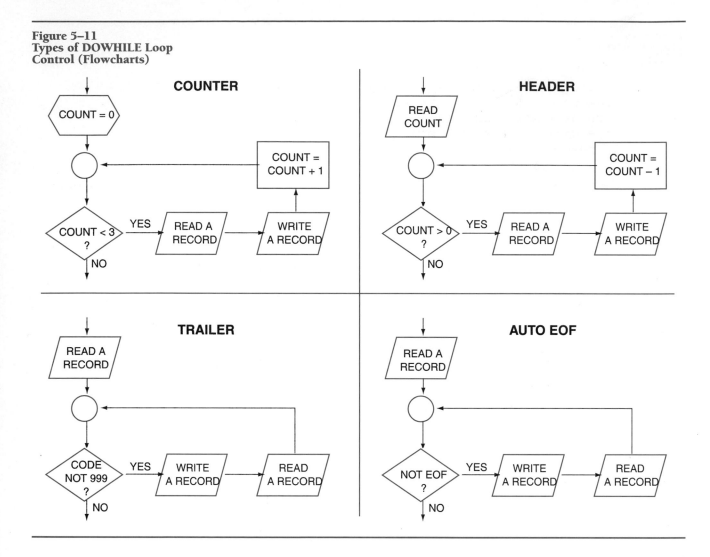

Figure 5–12
Types of DOWHILE Loop
Control (Pseudocode)

```
        COUNTER                       HEADER

COUNT = 0                    Read COUNT
DOWHILE COUNT < 3            DOWHILE COUNT > 0
   Read a record               Read a record
   Write a record              Write a record
   COUNT = COUNT + 1           COUNT = COUNT - 1
ENDDO                        ENDDO


        TRAILER                      AUTO EOF

Read a record                Read a record
DOWHILE CODE not 999         DOWHILE not EOF
   Write a record               Write a record
   Read a record                Read a record
ENDDO                        ENDDO
```

Figure 5–13
Credits Problem (Basic List)

```
MAXLINES = 55
PAGECNT = 1
LINECNT = MAXLINES
Read NAME$, ADDRESS$, CREDITS
IF CREDITS < 0 THEN
      PRINT "No data to process"
ELSE
      DO WHILE CREDITS >= 0
            IF LINECNT >= MAXLINES THEN
                  PRINT "CREDITS REPORT", "Page"; PAGECNT
                  PRINT
                  PRINT "NAME", "ADDRESS"
                  PRINT
                  LINECNT = 0
                  PAGECNT = PAGECNT + 1
            END IF
            IF CREDITS >= 60 THEN
                  PRINT NAME$, ADDRESS$
                  LINECNT = LINECNT + 1
            END IF
            READ NAME$, ADDRESS$, CREDITS
      LOOP
END IF
DATA "John", "111 Main St.", 60
DATA "Mary", "222 Oak St.", 70
DATA "Jane", "333 First Ave.", 50
DATA "Bill", "444 Cedar St.", 100
DATA "Terry", "555 Main St.", 65
DATA "Sue", "666 Oak Dr.", 59
DATA "Andy", "777 Star Ct.", 61
DATA "End", "No St.", -1
END
```

stored in the variable NAME$; "111 Main St." is stored in the variable AD-DRESS$; and 60 is stored in the variable CREDITS. The second time the Read statement is executed, "Mary" is stored in the variable NAME$; "222 Oak St." is stored in the variable ADDRESS$; and 70 is stored in the variable CREDITS. Note that the trailer record (the last Data statement) contains three values. Each time the Read statement is executed, three values must be input because the Read statement contains three variables. However, only the variable CREDITS is tested to determine when to exit the loop. The value of –1 is used to signal the end of the input. In this example, the Read and Data statements are used to receive input. We could also have used the Input statement to receive the data interactively from the user as we did in previous chapters.

The Basic statements for the most part, parallel the pseudocode with only minor differences in syntax. For example, in Basic the tests for "greater than or equal to" specify the logical operators > and = side by side as >=. Also, a "null else" clause in Basic is indicated by the absence of the keyword "Else."

**Figure 5–14
Credits Problem (Basic Run)**

```
CREDITS REPORT      Page 1

NAME         ADDRESS

John         111 Main St.
Mary         222 Oak St.
Bill         444 Cedar St.
Terry        555 Main St.
Andy         777 Star Ct.
```

Figure 5–14 illustrates the output that will be produced when the program is executed. Note that detail lines are output only for those records where the value of CREDITS is greater than or equal to 60. Note also that the trailer record is not output.

Enrichment (Visual Basic)

Figure 5–15 illustrates the graphical interface for the credits problem (see Figures 5–9 and 5–10). Although this example is based on the credits problem, some modifications to the original problem have been made. In this example, only the student name and number of credits are input. The address was left out to simplify the problem. A text box is used to input the student name and a new control, a *horizontal scroll bar*, is used to input the number of credits. The scroll bar is given a minimum value of 1 and a maximum value of 100; that is, only credit values between 1 and 100 can be input. In this way, no data validation is necessary. When the user clicks one of the small arrows to the left or right of the scroll bar, the value of credits is increased or decreased by one. This value is also displayed in a

**Figure 5–15
Credits Problem (Visual Basic—Screen 1)**

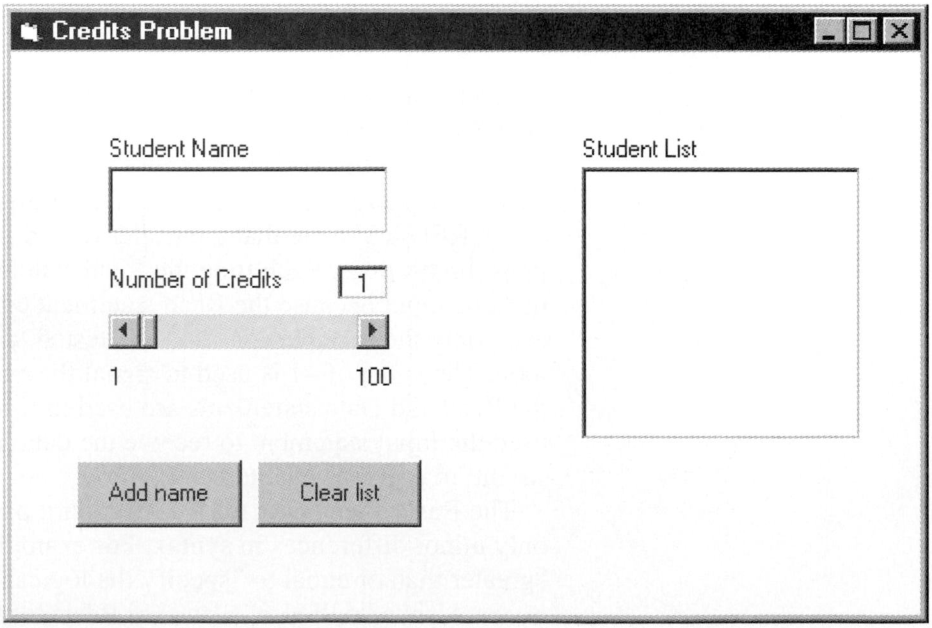

small label control above the scroll bar. Another new control, called a *list box*, is shown on the right of the screen. This control will be used to display the names of all students whose credits are greater than or equal to 60. Two command buttons are created, one to add a student name to the list and one to clear all student names from the list.

Figure 5–16 illustrates the screen after the user has entered the name and credits and has clicked the Add name button. Note that the student name John appears in the list box since the value of credits is equal to 60.

Figure 5–17 illustrates the screen after the user has entered another name and credits and has again clicked the Add name button. Note that the student name Mary appears in the list box since the value of credits is greater than 60. Note also that this second name is added to the list box; that is, Mary is listed in addition to John, not in place of John.

Figure 5–18 illustrates the screen after the user has entered another name and credits and has again clicked the Add name button. This time, the number of credits entered is 50. A message box is displayed informing the user that the student is ineligible and thus will not be added to the list. The user must then click the OK button in the message box before he or she can enter more names.

Figure 5–19 illustrates the screen after the user has entered several more names and credit amounts. Note that each name is added to the list box on a new line.

Figure 5–20 illustrates the screen after the user has clicked the Clear list button. As you can see, all the names have been removed from the list box.

Figure 5–16
Credits Problem (Visual Basic—Screen 2)

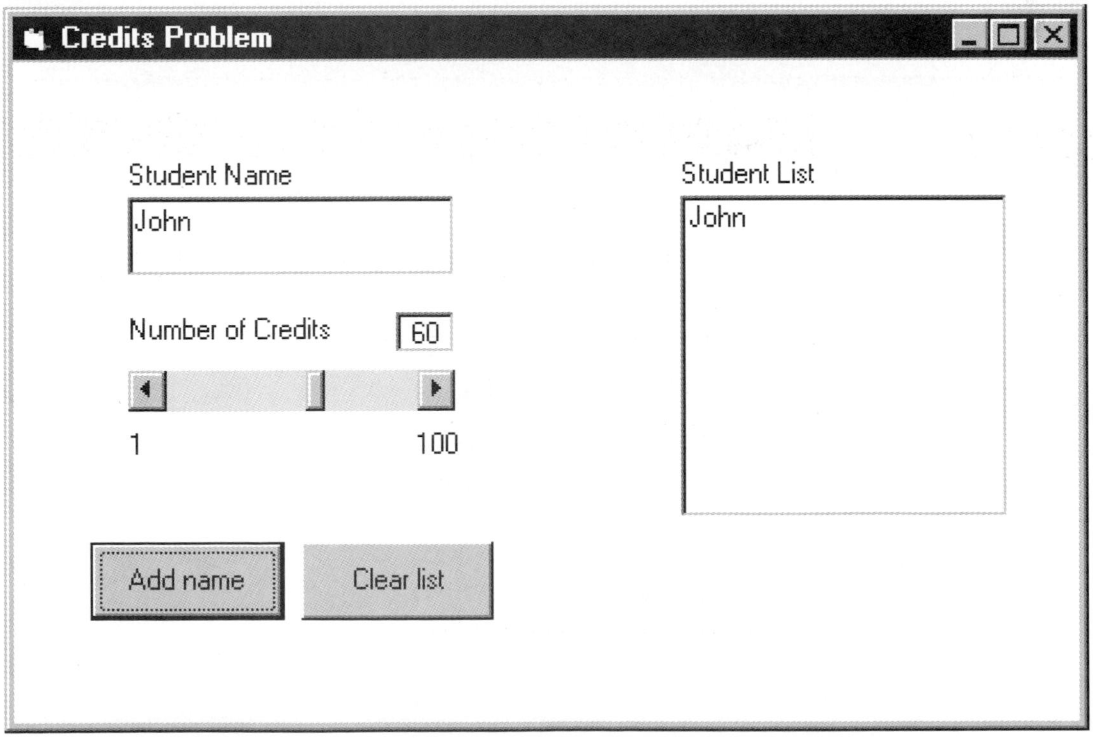

Figure 5–17
Credits Problem (Visual
Basic—Screen 3)

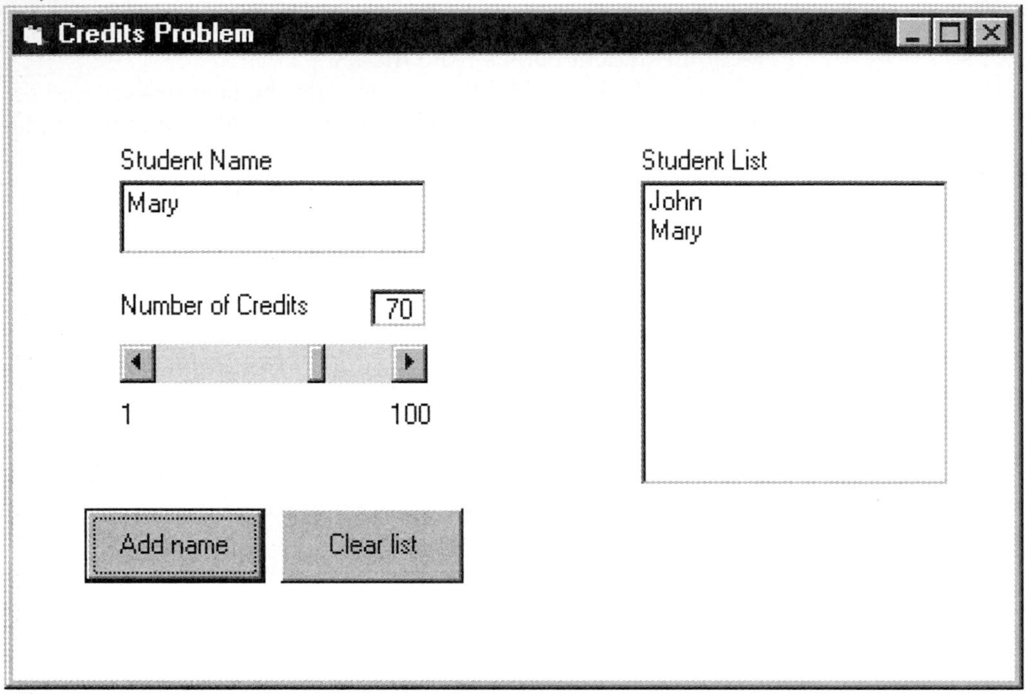

Figure 5–18
Credits Problem (Visual
Basic—Screen 4)

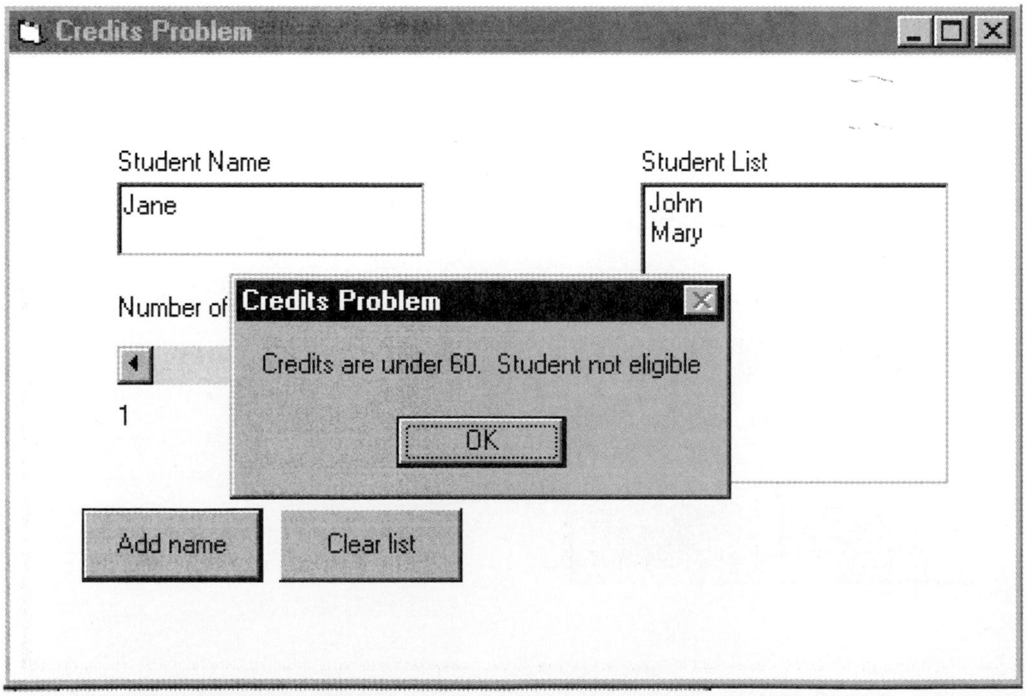

**Figure 5–19
Credits Problem (Visual
Basic—Screen 5)**

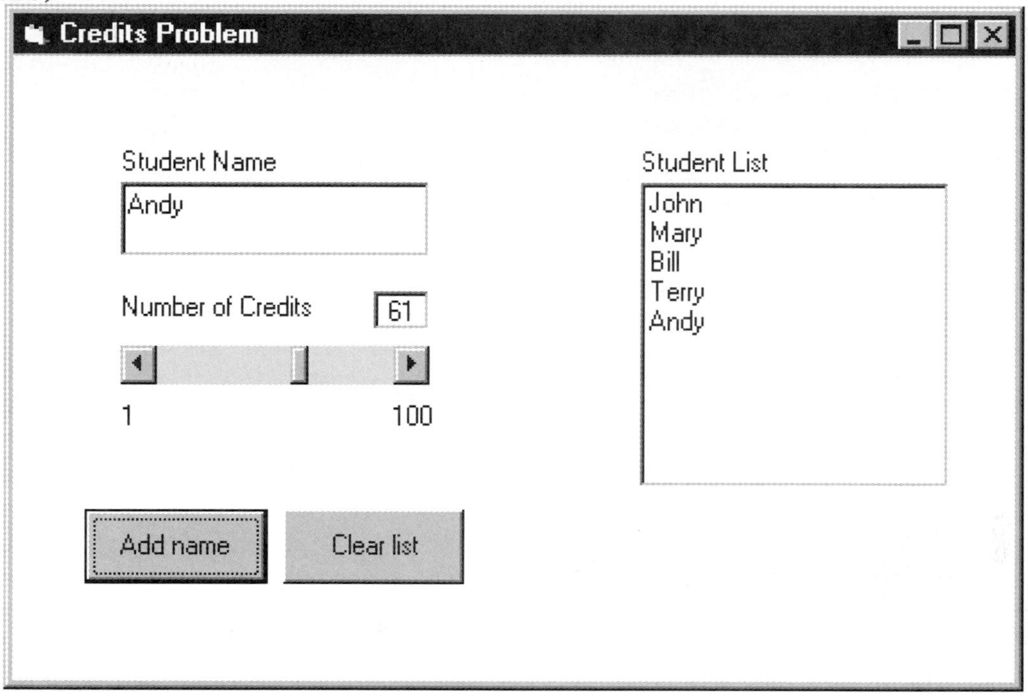

**Figure 5–20
Credits Problem (Visual
Basic—Screen 6)**

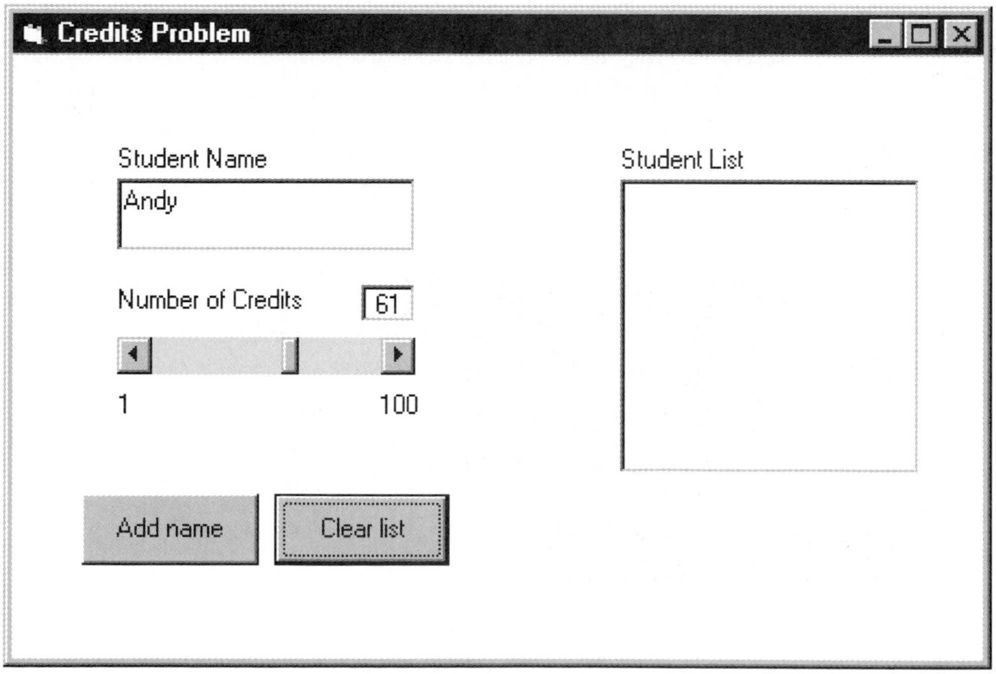

Figure 5–21
Credits Problem (Visual
Basic—cmd_ADD_Click)

```
Private Sub cmd_ADD_Click()

If hsb_CREDITS.Value >= 60 Then
        1st_NAMES.AddItem txt_NAME.Text
Else
        MsgBox "Credits are under 60. Student not eligible",,"Credits Problem"
End If

End Sub
```

Figure 5–22
Credits Problem (Visual
Basic—cmd_CLEAR_Click)

```
Private Sub cmd_CLEAR_Click()

1st_NAMES.Clear

End Sub
```

Figure 5–21 illustrates the program associated with the click event of the Add name button. A simple IFTHENELSE statement checks the value of credits, which is represented by the value property of the horizontal scroll bar. The standard name for this type of scroll bar begins with *hsb*. In Visual Basic a test for greater than or equal to is represented by the logical operators > and =, side by side as >=.

The true path of the IFTHENELSE statement causes a name to be added to the list box. The standard name for a list box begins with *lst*. The AddItem method adds the value of the text box to the end of the list box on a separate line. A *method* is a prewritten program that performs some special function. There are many methods in Visual Basic. Methods are invoked using the same dot notation that is used to specify properties. The false path of the IFTHENELSE statement causes the message box to be displayed.

Figure 5–22 illustrates the program associated with the click event of the Clear list button. The Clear method is used to delete all items (names in this case) from the list box.

Key Terms

trailer record logic	detail line	automatic end-of-file
trailer record	total line	facility
dummy data value	priming read	empty file condition
(sentinel value)	loop read	named constant
heading line		

Exercises

1. State in your own words the difference between header record logic and trailer record logic. How are the loops controlled in each type of logic?

2. Explain the differences among a heading line, a detail line, and a total line. Where should the WRITE statements associated with each type of line be positioned in reference to the main processing loop?

For the remainder of these exercises, include report and column headings as well as a page number on every page, with 55 detail lines per page, unless directed otherwise. Output an appropriate message if the input contains no records; include descriptive messages in the total lines; and construct both a flowchart and pseudocode for your solution.

3. Redo the solution to Sample Problem 3.1 to process data for any number of employees. Use the automatic end-of-file facility to control end-of-loop processing.

4. Redo Sample Problem 5.3 to include two total lines showing a count of the total number of students for which data was input and the total number of accumulated credits for all students.

5. Redo Exercise 16 in Chapter 2 but process multiple records. Assume that a length of 0 will be used to indicate the end of the input.

6. Redo Exercise 17 in Chapter 2 but process multiple records. Use automatic end-of-file logic to signal the end of the input.

7. Redo Exercises 5 and 6 to include the steps to compute and output the total number of input records processed.

8. Design an algorithm to read an arbitrary number of data records, each containing a name, age, and code. A code of 1 will indicate female; a code of 2 will indicate male; and a code of 0 will indicate that the end-of-file has been reached. For each record, write a detail line listing the person's name and age. In addition, compute and output the following values:

   - Number of males less than or equal to 21 years old
   - Number of females less than or equal to 21 years old
   - Average age of all persons over 21
   - Total number of people

9. Design an algorithm to input student records; each record contains a student name, registration code, and credits field. A code of 1 indicates that the student is a resident, and a code of 2 indicates that the student is a nonresident. Output a detail line for each student containing the student's name and a tuition amount computed as follows (12 or more credits means full-time status):

- FT resident $600.00 flat fee
- PT resident $50.00 per credit
- FT nonresident $1,320.00 flat fee
- PT nonresident $110.00 per credit

Five total lines are also to be output as follows:

- Total number of FT resident students
- Total number of PT resident students
- Total number of FT nonresident students
- Total number of PT nonresident students
- Total number of students

Use automatic end-of-file logic to signal the end of the input.

10. Design an algorithm to prepare a daily hotel charge report. Input consists of a series of records that contain a room number, the customer name, the cost of the room, and the cost of meals charged to the room. Output is a hotel charge report that will contain the room number, customer name, room charge, meal charges, and total charges. After all records have been processed, the total number of rooms rented, the total room charges, total meal charges, and a final total of all charges are to be printed. A room number of 000 will be used to signal the end of the input.

11. Design an algorithm to prepare a report of real estate sales and commissions. Input consists of a series of records that contain the address, city, selling price of houses that have been sold during the month, and the percentage used to compute the commission that the real estate company received. Output is to consist of a real estate sales and commissions report that will contain the address, city, selling price, and commission paid for each of the houses. After all records have been processed, the total number of houses sold, the total selling price of all houses, the average price of all houses sold, and the total commission are to be printed. Use automatic end-of-file logic to signal the end of the input.

12. Design an algorithm to prepare a job applicant report. Input consists of a series of records that contain the Social Security number, last name, first name, middle initial, verbal test score, science test score, math test score, and logic test score of each job applicant. Output is to consist of detail lines containing the contents of each input record as well as the average of the four test scores. In addition, averages for each of the four test score categories should be output at the end of the report. Use automatic end-of-file logic to signal the end of the input.

13. Design an algorithm to prepare a monthly report for a legal clinic. Input consists of a series of records that contain the name of the client, name of the attorney, and hours worked by the attorney on the case. Output is a monthly legal clinic report that lists the client's name, attorney, hours worked by the attorney on the case, and fee. The fee charged by the attorney is based on the hours worked. The first 20 hours are charged at the rate of $350.00 per hour. Hours in excess of 20 are charged at the rate of $200.00 per hour. After all records have been processed, the final totals are to be printed. Include the total clients, total hours billed, total hours billed at $350.00 per hour, total hours billed at $200.00 per hour, and total fees. End-of-file will be indicated when the hours worked input is 0.

14. Design an algorithm to compute and print the average earnings, lowest earnings, and highest earnings of a group of employees. Each input record will contain the name and earnings of one employee. No headings or page numbers are required. Use automatic end-of-file logic to signal the end of the input.

15. Design an algorithm for the following problem: You are a cashier in a candy store. Each of your customers buys exactly one item costing $1.00 or less. Each customer pays for the item with exactly $1.00. Your job is to give each customer the correct amount of change in some combination of pennies, nickels, dimes, and quarters. The combination must be the minimum number of coins. For example, if the item cost is $.38, the change would be 2 quarters, 1 dime, 0 nickels, and 2 pennies ($.62). The input is composed of customer records, each containing customer name and item cost. The output is composed of lines, each containing customer name, item cost, change, number of quarters, number of dimes, number of nickels, and number of pennies. A cost of $0.00 will be used to signal the end of the input.

16. Design an algorithm to prepare a property tax report. Input consists of a series of records that contain the property type field, which indicates the type of property owned (H–home; C–commercial), the name of the property owner, the home type field (N–nonresidence; R–residence), the commercial property type (L–commercial land; B–commercial building), the tract parcel number, and the assessed value. Output is to consist of the property owner, parcel number, assessed value, tax rate, and property tax for each property in the input. In addition, totals for the following fields are to be printed at the end of the report:

Home—primary residence

Home—nonresidence

Commercial building

Commercial land

Total property taxes

The property tax is determined in the following manner:

- If the property is a home, is used as the primary residence of the owner, and has an assessed value greater than $150,000.00, the tax rate is 2 percent of the assessed value.

- If the property is a home, is used as the primary residence of the owner, and has an assessed value equal to or less than $150,000.00, the tax rate is 1.4 percent of the assessed value.

- If the property is a home, but it is not the primary residence of the owner, and the assessed value is greater than $95,000.00, the tax rate is 2 percent of the assessed value.

- If the property is a home, but it is not the primary residence of the owner, and the assessed value is equal to or less than $95,000.00, the tax rate is 1.4 percent of the assessed value.

- If the property is a commercial building and the assessed value is greater than $200,000.00, the tax rate is 2.5 percent of the assessed value.

- If the property is a commercial building and the assessed value is equal to or less than $200,000.00, the tax rate is 2 percent of the assessed value.

- If the property is commercial land and the assessed value is greater than $60,000.00, the tax rate is 2.5 percent of the assessed value.

- If the property is commercial land and the assessed value is equal to or less than $60,000.00, the tax rate is 2 percent of the assessed value.

Use automatic end-of-file logic to signal the end of the input; assume all the input values are valid.

Modularization

Objectives	**Upon completion of this chapter you should be able to**

- Identify, and use in program design, the predefined-process and annotation program flowcharting symbols.
- Define the terms *module, modularization, and top-down design.*
- Understand, and use in program design, a structure chart.
- Design programs using top-down design, structured programming, and modularization techniques.

Introduction	

You may have noticed that our flowcharts and pseudocode are becoming more complicated. The flowcharts can barely fit on one page, and the pseudocode shows several levels of indentation. This happens when several structures are used together in the same algorithm. For example, in our solution to the payroll problem in Chapter 4, an IFTHENELSE is nested inside a DOWHILE loop (see Figures 4–8 and 4–9). The algorithms in Chapter 5 become even more complex when three, four, or even more structures are needed in a problem solution.

This apparent complexity can be lessened by the use of a technique called **modularization**. Consider an alternative solution to the payroll problem, shown in Figures 6–1, 6–2, and 6–3.

The flowchart in Figure 6–1 is much simpler and easier to follow than the one in Figure 4–8. The counter is still there—initialized, tested, and incremented as before—but the steps to compute the pay for one employee are replaced by a single box identified in two ways: It is given a descriptive name (Process 1 employee) and a label (B000). The box itself is called a predefined-process symbol. The **predefined-process symbol** is used to identify a series of steps shown on another flowchart. These steps are given a label for reference purposes. Together they make up a **module**. This modularization technique allows us to concentrate on the overall processing the program needs to do, instead of being overwhelmed by the details early on.

Because each module in our program is identified by a label, we should also give a label to the main processing module, that is, the one we start with. We will call this module the **overall control module** (also known as the **driver module**), and we give it an identifying label of A000.

Figure 6–1
Payroll Problem—10
Employees (Overall Control)

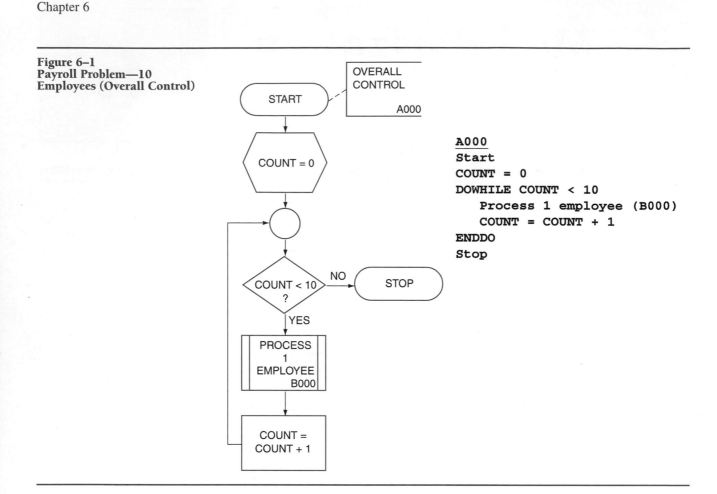

We indicate this next to the START symbol in another flowcharting symbol called the annotation symbol. The **annotation symbol** can be used to further explain any step in the flowchart. It is a documentation symbol, not an additional processing step. The choice of module descriptive names and labels is either left up to the programmer or dictated by company standards. As with variable names, module names should be chosen to indicate what the modules actually do.

A module that is referenced in a predefined-process symbol needs to be described in another flowchart or in pseudocode representation. Module B000 is shown in Figures 6–2 and 6–3. Notice that the flowchart does not begin with START because this module is not used at the start of program execution. Instead, we identify the module by placing its label in the first symbol. Thus, the label indicates a linkage between the two flowcharts. Similarly, when the steps within the module are complete, a STOP is not appropriate because we have not yet completed the algorithm. We need to return control (go back) to module A000, which will continue processing.

To summarize, when a program is divided into several parts, or modules, several flowcharts or pseudocode representations will be needed. The overall control module can be designed first, leaving details until later. This approach is often called **top-down design**. Only the overall control module contains a START and a STOP. We refer to the overall control module as a **calling module** because it references other modules. We refer to the other modules as **called modules** because they are referenced (called) by the main

Figure 6–2
Payroll Problem—Process
Employee Record (Flowchart)

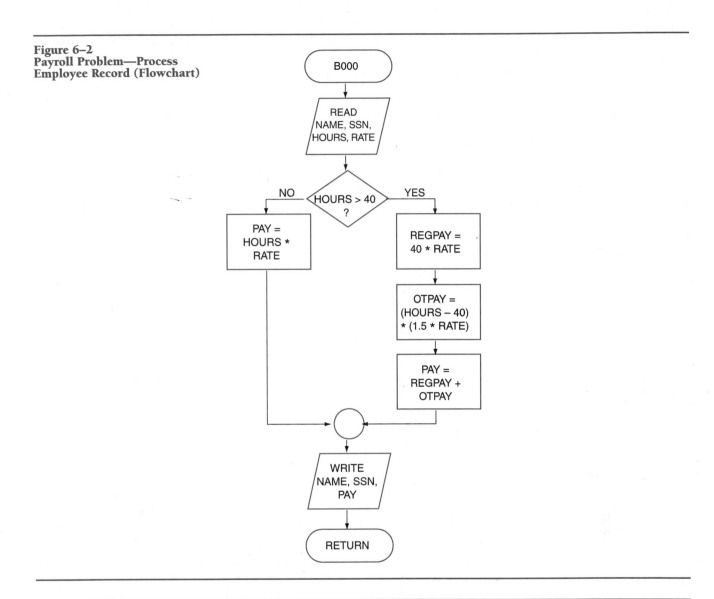

Figure 6–3
Payroll Problem—Process
Employee Record
(Pseudocode)

```
B000
Enter
Read NAME, SSN, HOURS, RATE
IF HOURS > 40 THEN
    REGPAY = 40 * RATE
    OTPAY = (HOURS - 40) * (1.5 * RATE)
    PAY = REGPAY + OTPAY
ELSE
    PAY = HOURS * RATE
ENDIF
Write NAME, SSN, PAY
Return
```

module. The called modules are identified by labels (reference numbers) at the beginning of their flowcharts, and each one will return control (RE-TURN) back to the calling module (the main module, in this case) as its last step. This logic can also be indicated in the pseudocode by the appropriate statements. It is important to understand that, when module B000 returns control to module A000, the step in A000 following the predefined-process symbol will be executed next. A000 does not restart from the beginning. It is possible for a module to be both a calling module and a called module. We will see examples of this later in the book.

Structure Charts

It might be valuable at this point to step back and look at the big picture. We already know that an information-processing system may be composed of a single program or several programs. This general information is shown in a system flowchart. A system flowchart typically shows the flow of work within a system; that is, it shows what inputs are needed for what processes, and what processes produce what outputs. Figure 6–4 shows the system flowchart for our payroll problem. It includes only one program.

We also have seen in this chapter that a program can be composed of one or more modules. Each module is a segment of logically related code, a part of a complete program that gets executed to solve a problem. As much as possible, each module should be independent of all other modules. It should constitute a logical unit of work, performing one or a small number of functions of the overall problem-solving task.

Figure 6–5 shows a new graphic representation, namely a **structure chart**, or **hierarchy chart**, which shows the potential flow of control within one program. In other words, a structure chart shows the relationships of all modules within a program. It is similar to an organizational chart in a company. The top box shows the overall control module or driver module. All lower-level boxes represent modules that may be given control during program execution.

The example in Figure 6–5 is very simple because the program we have just looked at has only two modules. Now let's look at a more complex generic structure chart shown in Figure 6–6.

In this example there are six modules, each indicated by one box on the structure chart. For simplicity, only module numbers are used. Module

Figure 6–4
Payroll Problem (System Flowchart)

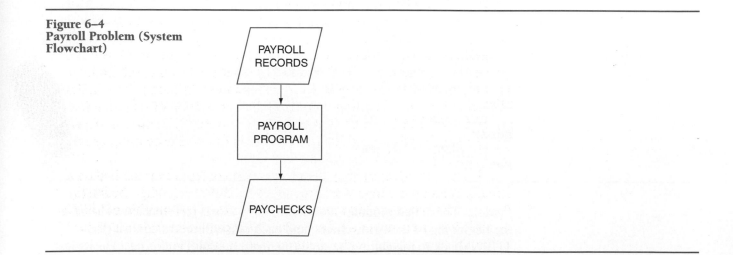

**Figure 6–5
Payroll Problem (Structure
Chart)**

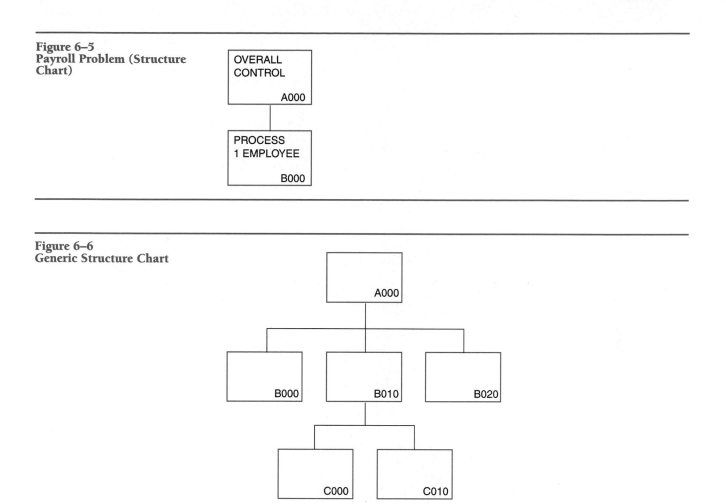

**Figure 6–6
Generic Structure Chart**

A000, the overall control module, may call on three lower-level modules (B000, B010, and B020) at various points in its processing. This structure is likewise shown in the program flowchart in Figure 6–7.

The structure chart in Figure 6–6 does not show any module references below B000 and B020. We can conclude that these two modules do not call any lower-level modules during execution (see Figure 6–8). However, module B010 may refer to two lower-level modules (level 3 since B010 is itself level 2) during execution (see Figure 6–9). Neither of the level 3 modules (C000 and C010) calls any lower-level modules (see Figure 6–10). It is important to understand that each called module returns control to its calling module at the completion of its execution. For example, C000 and C010 return control to B010; B000, B010, and B020 return control to A000. When A000 is finished executing, processing for that program is complete.

It is also important to note that no execution order is ever implied on a structure chart other than which module may call which other module(s) during execution. A called module may be invoked any number of times or, in some instances, not at all. On a structure chart, left-to-right order does not imply first, second, and so forth in execution sequence.

Figure 6–7
Generic Program Flowchart
(Overall Control)

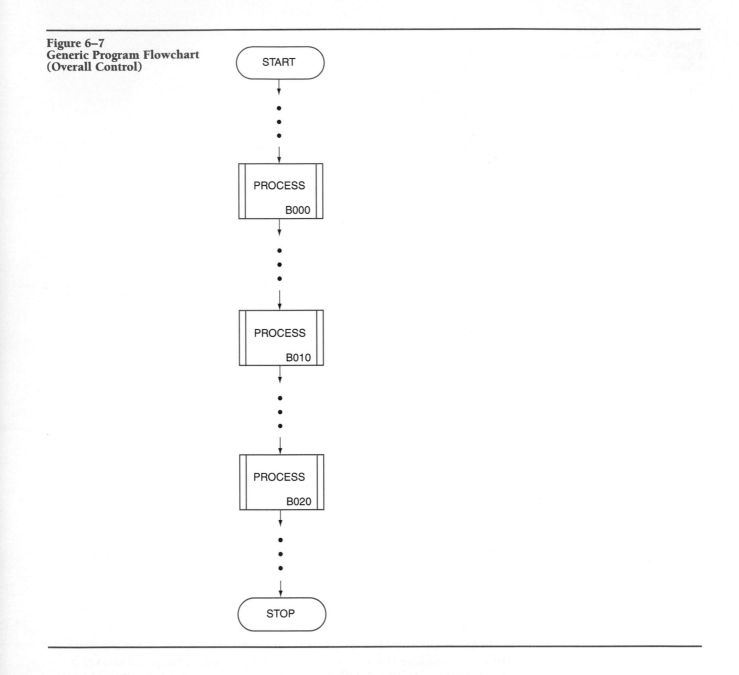

Figure 6–8
Generic Program Flowchart
(Modules B000 and B020)

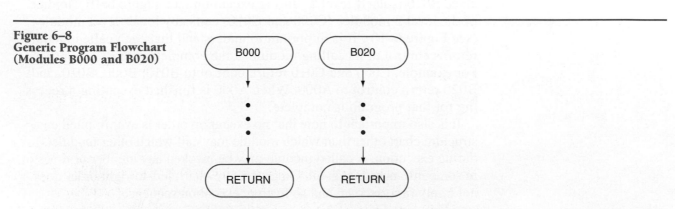

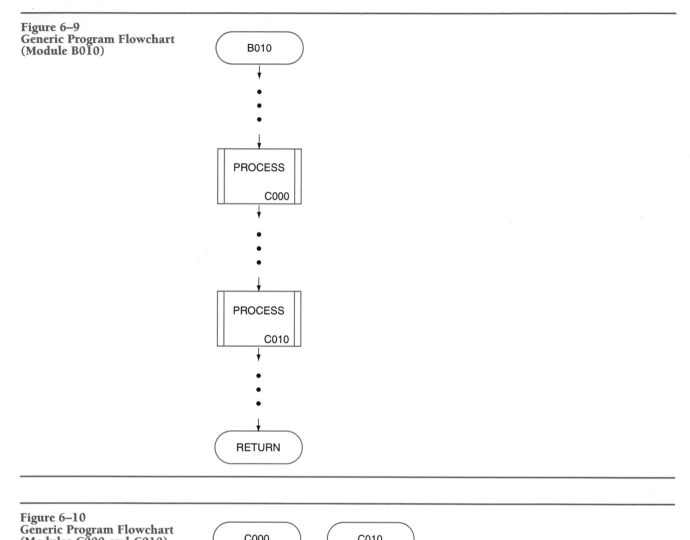

Figure 6–9
Generic Program Flowchart
(Module B010)

Figure 6–10
Generic Program Flowchart
(Modules C000 and C010)

Sample Problem 6.1 (Averaging Problem Using Modules)

Problem:

Redo Sample Problem 4.4 using a modular approach as outlined in the structure chart in Figure 6–11.

Solution:

The overall control module is shown in Figure 6–12 (flowchart) and Figure 6–13 (pseudocode). This module (A000) reads the header record and checks to make sure that N is a positive number. A000 also controls the

Figure 6–11
Averaging Problem (Structure Chart)

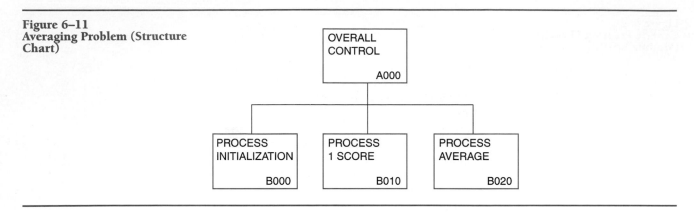

Figure 6–12
Averaging Problem—Overall Control (Flowchart)

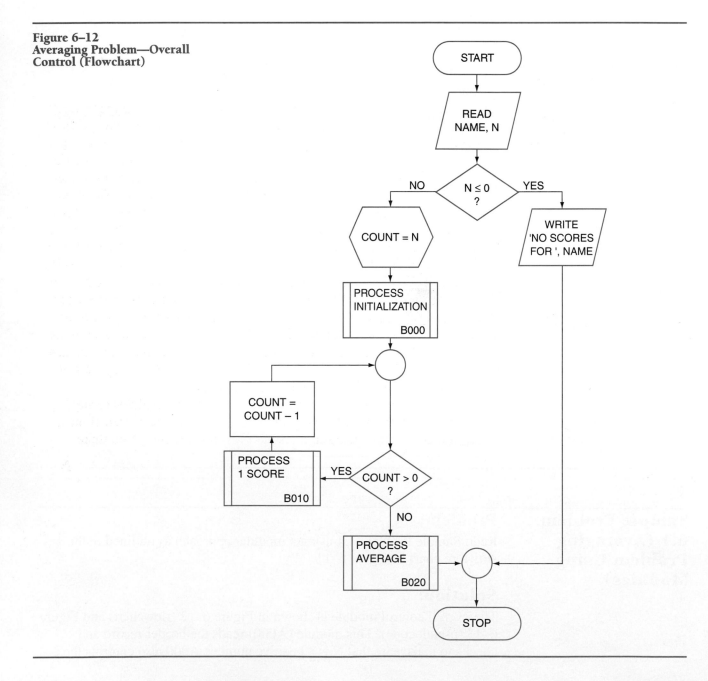

Figure 6–13
Averaging Problem—Overall
Control (Pseudocode)

```
A000
Start
Read NAME, N
IF N ≤ 0 THEN
   Write 'No scores for ', NAME
ELSE
   COUNT = N
   Process initialization (B000)
   DOWHILE COUNT > 0
      Process 1 score (B010)
      COUNT = COUNT - 1
   ENDDO
   Process average (B020)
ENDIF
Stop
```

DOWHILE loop and may call three lower-level modules during the course of its processing. Module B000 is called prior to entering the loop to initialize the variable named SUM to 0. This module, which is fairly trivial, is shown in Figure 6–14. We could have initialized COUNT in this module as well, but the general approach in this book is to show all the steps related to loop control in the overall control module. Another module (B010) is referenced inside the loop. As the name of the module implies, it performs the processing associated with one score. Figure 6–15 shows the details of B010. As you can see, this module is also fairly trivial. An individual score is first input (SCORE), and then it is accumulated using the variable SUM. At this point B010 returns control to A000, where the counter controlling the number of times the DOWHILE loop is executed is decremented and the test is again made to see if any scores remain. Again, we could have decremented COUNT in module B010, but we left this step in overall control because it is part of the loop control process. When the count reaches 0, the loop is exited and a third module (B020) is called to compute and output the average. This module is shown in Figure 6–16.

This problem solution could have been broken into modules in any of several ways. In this example, we have, as before, one overall control module; but this time the problem solution is broken down into three

Figure 6–14
Averaging Problem—Process
Initialization

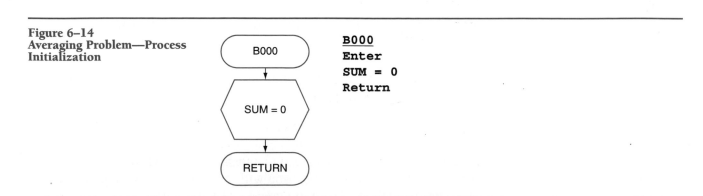

```
B000
Enter
SUM = 0
Return
```

Figure 6–15
Averaging Problem—Process
1 Score

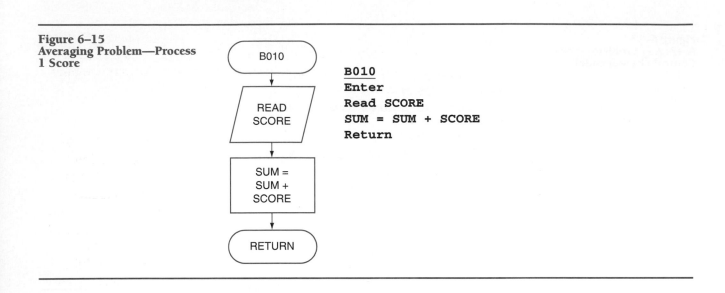

```
B010
Enter
Read SCORE
SUM = SUM + SCORE
Return
```

Figure 6–16
Averaging Problem—Process
Average

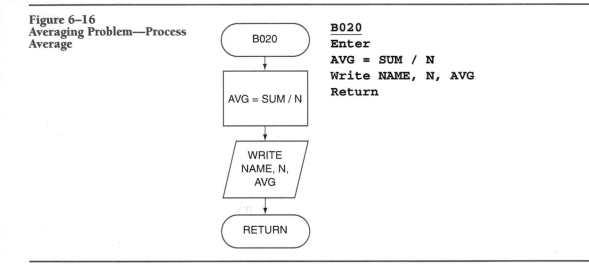

```
B020
Enter
AVG = SUM / N
Write NAME, N, AVG
Return
```

lower-level modules. Let us look at each module and how it interacts within the program as a whole. The overall control module (A000) has control when processing begins. During processing, A000 may give control (once or more than once) to any of the other three modules. When any of these modules completes its processing, it returns control to A000. Because all three modules are at the same level, they cannot communicate directly with each other. Each can be given control only from A000, and each can return control only to A000. B000, B010, and B020 are all relatively trivial modules— you might wonder why we even bother to design them as separate modules. Typically, a complete problem solution involves steps like printing headings, writing totals, and doing initialization. Each of these tasks can become complex, and, more importantly, each can be performed in a relatively generic fashion, independent of other processing in the algorithm. At some later point, we may even need to break the processing of an individual input record into several parts, thus creating a third level on the structure chart. As you study each of these modules, you will see that the steps within the modules are precisely those shown in Figures 4–17 and 4–18.

Sample Problem 6.2 (Defective Parts with Multiple Headings Using Modules)

Problem:

Redo Sample Problem 5.2 to construct a modular design. Assume that automatic end-of-file will be used to indicate when there are no more regular data records to process.

Solution:

This problem solution can be broken into modules in any of several ways. The structure chart in Figure 6–17 illustrates one possible construction.

In this example we have, as before, one overall control module; but this time the problem solution is broken down into four lower-level modules. The modular breakdown is very similar to the one used in Sample Problem 6.1; however, in the present case we have included an additional module to process the headings. This module is numbered C000 and is controlled from B010 instead of directly from A000. You might notice that modules B010 and B020 have been given general names, whereas in Sample Problem 6.1 the names were more specific. The Process 1 score module in Sample Problem 6.1 took care of detail processing; the Process average module in Sample Problem 6.1 took care of total processing.

Now let's look at the flowchart (see Figure 6–18) and the pseudocode (see Figure 6–19) for module A000.

As you can see, there are three predefined-process symbols in the flowchart, each referring to one of the modules. The initialization module (B000) is called first. It execution and return of control to A000 is followed immediately by the step to input the first record. After the first record is read, an IFTHENELSE statement directs the computer to check if the end-of-file has been reached. If so, the trailer record has been read first, either by accident or because there are no data records to process. A special message is output and program execution stops. In the present example, if end-of-file has not been reached, then we can assume the record is a normal data record and needs to be processed. The DOWHILE loop is shown as before, but this time the details are left to module B010. Any current input record that is not the trailer record will be processed. Then the next record will be

Figure 6–17
Defective Parts Problem (Structure Chart—Three Levels

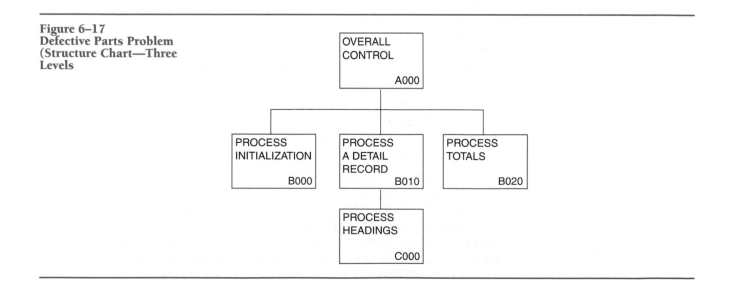

Figure 6–18
Defective Parts Problem—
Multiple Headings—Overall
Control (Flowchart)

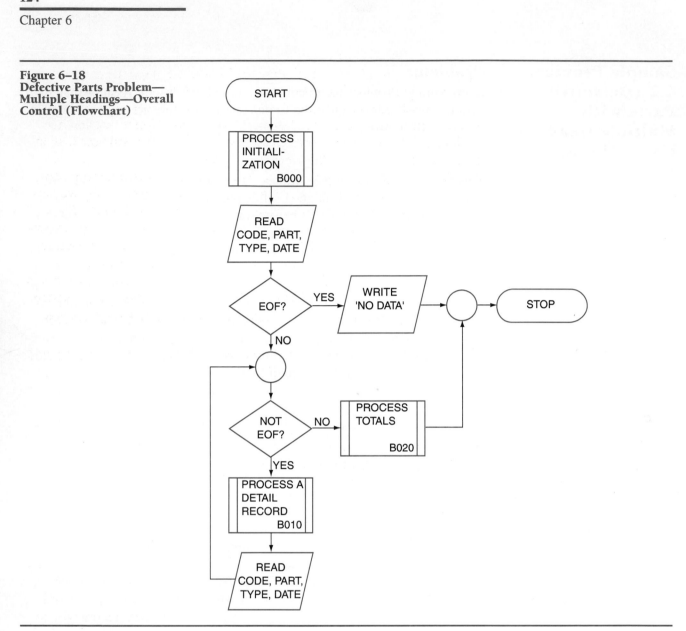

Figure 6–19
Defective Parts Problem—
Multiple Headings—Overall
Control (Pseudocode)

```
A000
Start
Process initialization (B000)
Read CODE, PART, TYPE, DATE
IF EOF THEN
   Write 'No data'
ELSE
   DOWHILE not EOF
      Process a detail record (B010)
      Read CODE, PART, TYPE, DATE
   ENDDO
   Process totals (B020)
ENDIF
Stop
```

input. In this type of logic, each time the loop is executed, the record already input is processed and then the next record is read. Can you find the priming and loop reads in A000? After all records have been processed, the loop will be exited and the total lines will be output. Notice that another module (B020) is referred to when the loop is exited. We will now look at the details of each of the three modules called by A000.

B000 (see Figure 6–20) and B020 (see Figure 6–21) are relatively trivial; however, since each of these tasks can become more complex, we have defined each as a separate module. B010 (see Figures 6–22 and 6–23) is more significant and probably a more obvious module. As we begin execution of module B010, LINECNT is checked. If its value equals or exceeds 55, a headings module is called to output the appropriate headings. The headings module (C000) is shown in Figure 6–24. An annotation symbol is positioned next to the first WRITE statement in Figure 6–24 to emphasize that the report headings will be written on the top of a new page. As you

Figure 6–20
Defective Parts Problem—
Multiple Headings—Process
Initialization

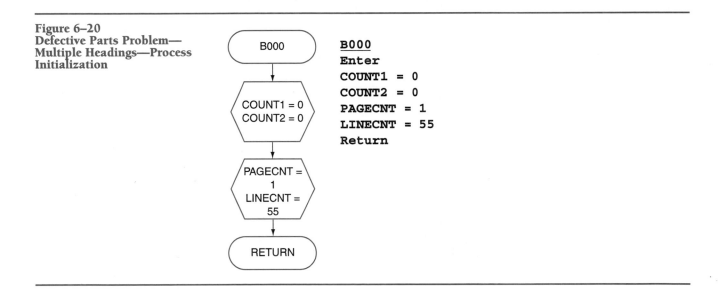

```
B000
Enter
COUNT1 = 0
COUNT2 = 0
PAGECNT = 1
LINECNT = 55
Return
```

Figure 6–21
Defective Parts Problem—
Multiple Headings—Process
Totals

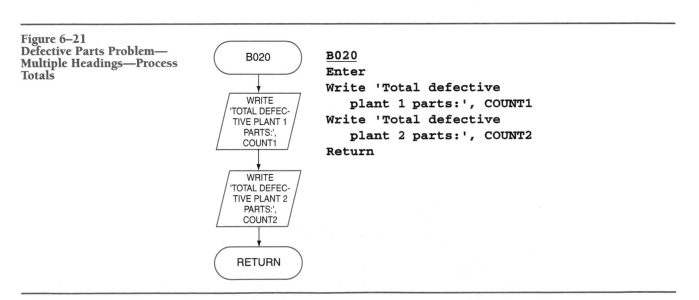

```
B020
Enter
Write 'Total defective
    plant 1 parts:', COUNT1
Write 'Total defective
    plant 2 parts:', COUNT2
Return
```

Figure 6–22
Defective Parts Problem—
Multiple Headings—Process
Detail Record (Flowchart)

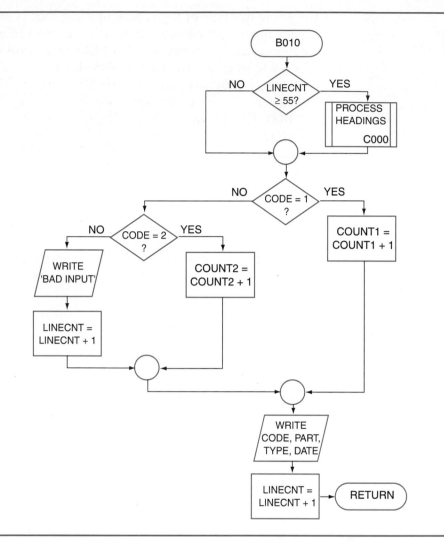

Figure 6–23
Defective Parts Problem—
Multiple Headings—Process
Detail Record (Pseudocode)

```
B010
Enter
IF LINECNT ≥ 55 THEN
    Process headings (C000)
(ELSE)
ENDIF
IF CODE = 1 THEN
    COUNT1 = COUNT1 + 1
ELSE
    IF CODE = 2 THEN
        COUNT2 = COUNT2 + 1
    ELSE
        Write 'Bad Input'
        LINECNT = LINECNT + 1
    ENDIF
ENDIF
Write CODE, PART, TYPE, DATE
LINECNT = LINECNT + 1
Return
```

Figure 6–24
Defective Parts Problem—
Multiple Headings—Process
Headings

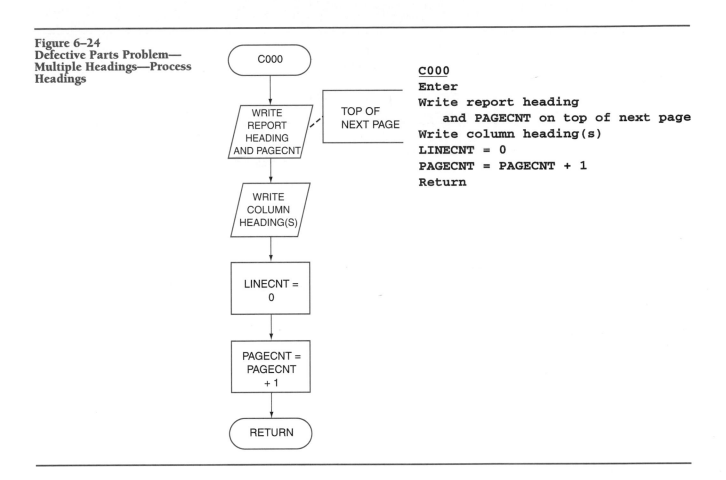

study each of these modules, you will see that the steps within the modules are like those shown in Figures 5–6 and 5–7.

Sample Problem 6.3 (Credits Problem Using Modules)

Problem:

Redo Sample Problem 5.3 to construct a modular design. Assume a trailer record will mark the end of the input file.

Solution:

The structure chart for a solution to this problem is shown in Figure 6–25.

The four modules have exactly the same names as four of the modules used in our previous design (see Figure 6–17)—only the totals module is missing. Figures 6–26 and 6–27 illustrate the flowchart and pseudocode for the overall control module (A000). The overall structure of this module is the same as that of A000 in our previous solution; however, there are a few minor differences in details. First, the two READ statements contain different variable names. The three variables—NAME, ADDRESS, and CREDITS—are the values that need to be input for each student.

A second difference is in the actual test that is made in both the first IFTHENELSE statement and the DOWHILE loop. This test corresponds to the requirement that a trailer record containing a negative number for the credits field be used to indicate the end-of-file.

Figure 6–25
Credits Problem (Structure
Chart)

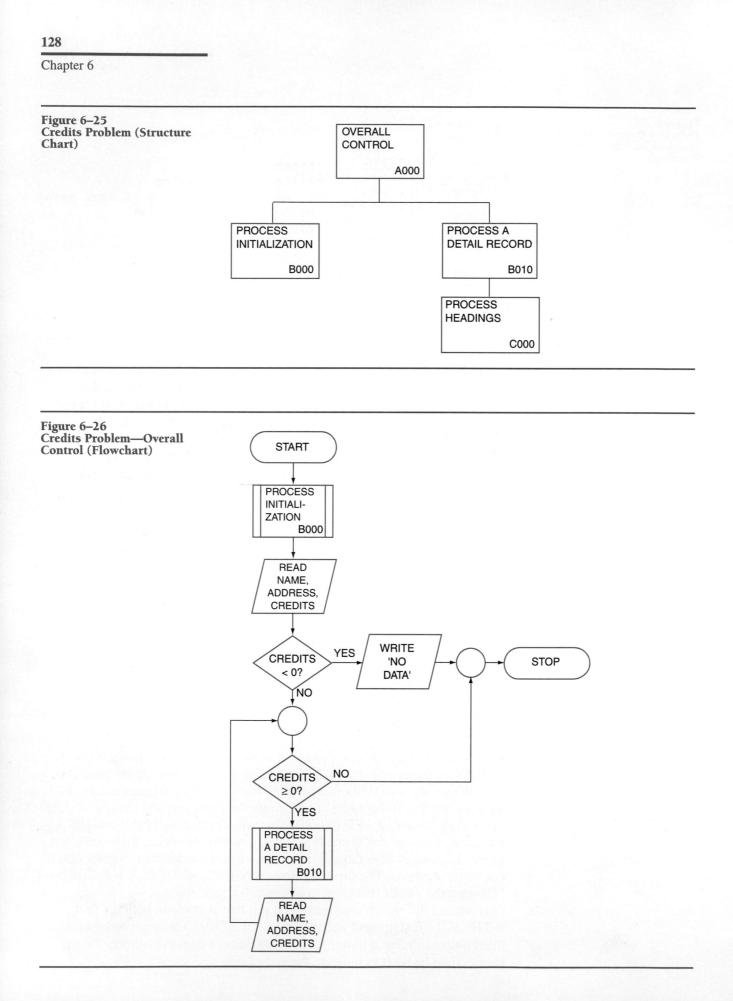

Figure 6–26
Credits Problem—Overall
Control (Flowchart)

**Figure 6–27
Credits Problem—Overall
Control (Pseudocode)**

```
A000
Start
Process initialization (B000)
Read NAME, ADDRESS, CREDITS
IF CREDITS < 0 THEN
   Write 'No data'
ELSE
   DOWHILE CREDITS ≥ 0
      Process a detail record (B010)
      Read NAME, ADDRESS, CREDITS
   ENDDO
ENDIF
Stop
```

The only other difference in this solution is the omission of a totals module, which is usually invoked after loop processing is complete. In this problem, however, we are not asked to compute or output any total information. For example, we do not need to count the number of student records or accumulate the total number of credits for all the students.

The detail-processing module (B010) is shown in Figure 6–28 (flowchart) and Figure 6–29 (pseudocode). To begin with, we check

**Figure 6–28
Credits Problem—Process
Detail Record (Flowchart)**

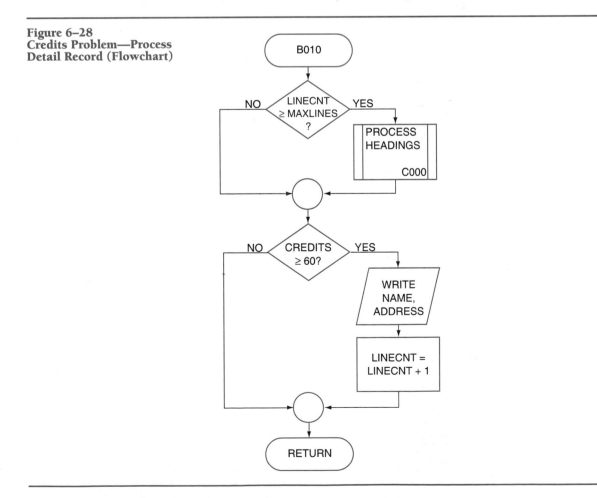

Figure 6–29
Credits Problem—Process
Detail Record (Pseudocode)

```
B010
Enter
IF LINECNT ≥ MAXLINES THEN
    Process headings (C000)
(ELSE)
ENDIF
IF CREDITS ≥ 60 THEN
    Write NAME, ADDRESS
    LINECNT = LINECNT + 1
(ELSE)
ENDIF
Return
```

to see if the value of LINECNT equals or exceeds MAXLINES. If so, the headings module is called to output the headings on a new page. This IFTHENELSE construct is then immediately followed by another IFTHENELSE construct. The second IFTHENELSE checks to see if the current student's credits equal or exceed 60. Remember, the data for the current student was read as input by A000. If the value of CREDITS is greater than or equal to 60, the student name and address are output and the value of LINECNT is incremented by 1. Note that the steps in module B010 are exactly the same as those shown in Figures 5–9 and 5–10.

Finally, the initialization module (B000) and the headings module (C000) are shown in Figures 6–30 and 6–31. Again, the steps in these two modules are exactly the same as those shown in Figures 5–9 and 5–10.

Remember that in this algorithm we are not asked to count or accumulate anything other than the required computations for line count and page count. Typically, if a total of some type is required, a counter and/or accumulator must be set to some initial value, like 0, in the initialization module. That same counter and/or accumulator is incremented in the detail-processing module, and a total module is designed to output the final value of the counter and/or accumulator.

Figure 6–30
Credits Problem—Process
Initialization

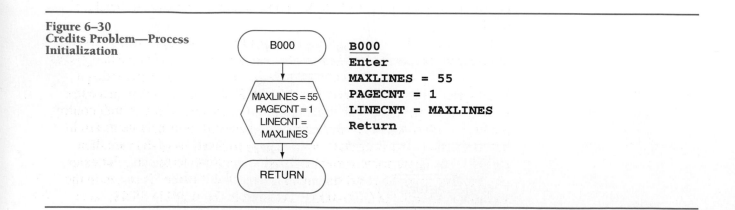

```
B000
Enter
MAXLINES = 55
PAGECNT = 1
LINECNT = MAXLINES
Return
```

**Figure 6–31
Credits Problem—Process
Headings**

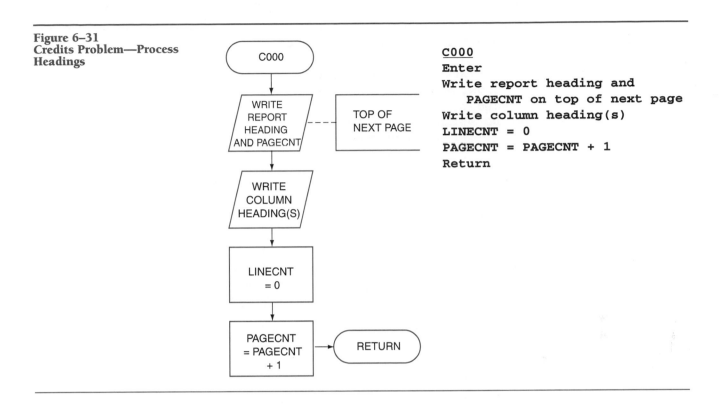

```
C000
Enter
Write report heading and
    PAGECNT on top of next page
Write column heading(s)
LINECNT = 0
PAGECNT = PAGECNT + 1
Return
```

Enrichment (Basic)

Figure 6–32 illustrates a listing of the program that solves the payroll problem (see Figures 6–1 through 6–3). In Basic, each module (except the overall control module) is considered a subprogram and is identified by a *Sub* statement followed by the name of the module. The *End Sub* statement denotes the end of each module or subprogram. In addition, a *Rem* statement identifying the module number is listed at the beginning of each module. A Rem statement is simply a remark or comment and is ignored by the computer. We use Rem statements to clarify program statements, making them easier to follow.

In Basic, all subprograms that are accessed in a program must be stated or "declared" at the beginning of the program. Thus, the first statement is a *Declare Sub* statement for the called module. The logic within this overall control module is similar to the pseudocode in Figure 6–1. Note that in Basic we use the keyword *Call*, followed by the module name, to invoke a module.

In this example, we use a *Read* statement to accept input values instead of an *Input* statement. The Read statement is placed in module B000 just as in the program design shown in Figures 6–2 and 6–3. Remember that a Read statement does not accept user input. Rather, *Data* statements containing the actual data to be input are listed at the end of the overall control module. Each time the Read statement is executed, four data items are input from one of the Data statements. These four pieces of data are then stored in the appropriate variables listed in the Read statement. For example, the first time the Read statement is executed, "Joyce" is stored in the variable NAME$; "000–00–0000" is stored in the variable SSN$; 40 is

Figure 6–32
Payroll Problem (Basic List)

```
DECLARE SUB ProcessEmployee()

REM A000 - Overall Control Module
COUNT = 0
CLS
DO WHILE COUNT < 10
        CALL ProcessEmployee
        COUNT = COUNT + 1
LOOP
DATA "Joyce", "000-00-0000", 40, 10.00
DATA "John", "111-11-1111", 60, 7.00
DATA "Mary", "222-22-2222", 70, 8.00
DATA "Jane", "333-33-3333", 50, 12.00
DATA "Bill", "444-44-4444", 10, 5.00
DATA "Terry", "555-55-5555", 20, 9.00
DATA "Sue", "666-66-6666", 30, 8.00
DATA "Andy", "777-77-7777", 50, 10.00
DATA "Steve", "888-88-8888", 40, 6.00
DATA "Tedd", "999-99-9999", 40, 15.00
END


REM B000
SUB ProcessEmployee
READ NAME$, SSN$, HOURS, RATE
IF HOURS > 40 THEN
        REGPAY = 40 * RATE
        OTPAY = (HOURS - 40) * (1.5 * RATE)
        PAY = REGPAY + OTPAY
ELSE
        PAY = HOURS * RATE
END IF
PRINT NAME$, SSN$, PAY
END SUB
```

stored in the variable HOURS; and 10.00 is stored in the variable RATE. The second time the Read statement is executed, "John" is stored in the variable NAME$; "111–11–1111" is stored in the variable SSN$; 60 is stored in the variable HOURS; and 7.00 is stored in the variable RATE.

Every time the Read statement is executed, four values must be input because the Read statement contains four variables. Because the loop is a counter-controlled loop and is set up to execute 10 times, 10 DATA statements must be included, each containing four values. If fewer than 10 Data statements are included, an error will occur. If more than 10 Data statements are included, the extra statements will be ignored. Note that even though the Read statement is placed in module B000, the Data statements must be placed in the overall control module, A000. We also could have used the Input statement to receive the data interactively from the user as we did in previous chapters.

The statements for each module or subprogram are listed after the overall control module. In this example there is only one lower-level module,

Figure 6–33
Payroll Problem (Basic Run)

Joyce	000-00-0000	400
John	111-11-1111	490
Mary	222-22-2222	680
Jane	333-33-3333	660
Bill	444-44-4444	50
Terry	555-55-5555	180
Sue	666-66-6666	240
Andy	777-77-7777	550
Steve	888-88-8888	240
Tedd	999-99-9999	600

B000. The statements in B000 are very similar to the pseudocode in Figure 6–3. Figure 6–33 illustrates the output that will be produced when the program is executed.

Enrichment (Visual Basic)

Figure 6–34 illustrates the graphical user interface for the payroll problem (see Figures 6–1 through 6–3). In this example, four text boxes are created to accept user input. A label is created to hold the computed pay. Two command buttons are created, one to compute the pay and one to end execution of the program.

Figure 6–35 illustrates the screen after the user has entered the input values and Figure 6–36 illustrates the screen after the user has clicked the Compute Pay button. The value for the pay is displayed in the label.

Figure 6–34
Payroll Problem (Visual Basic—Screen 1)

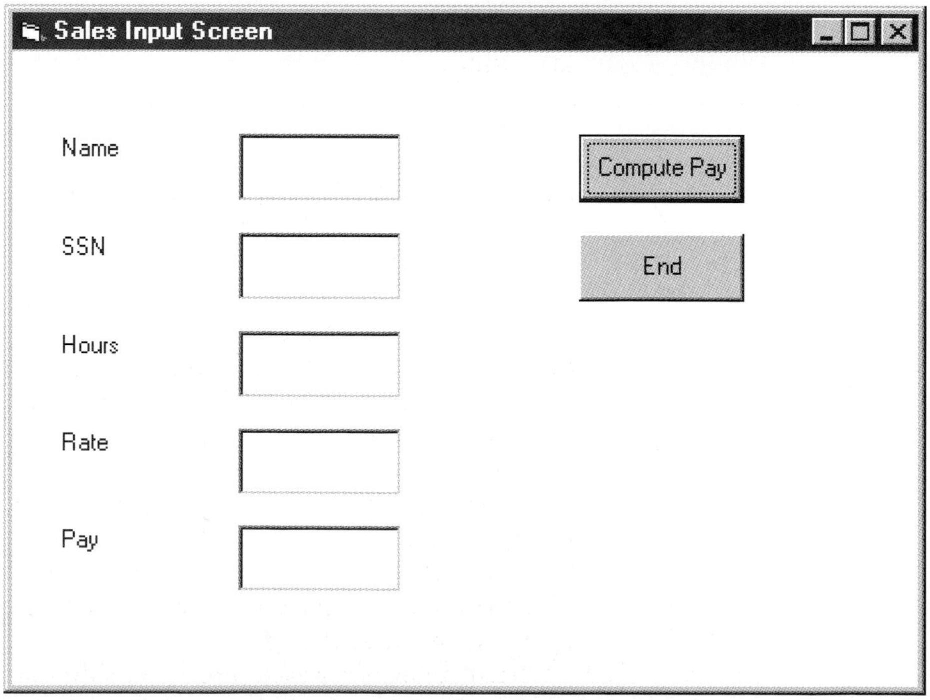

Figure 6–35
Payroll Problem (Visual Basic—Screen 2)

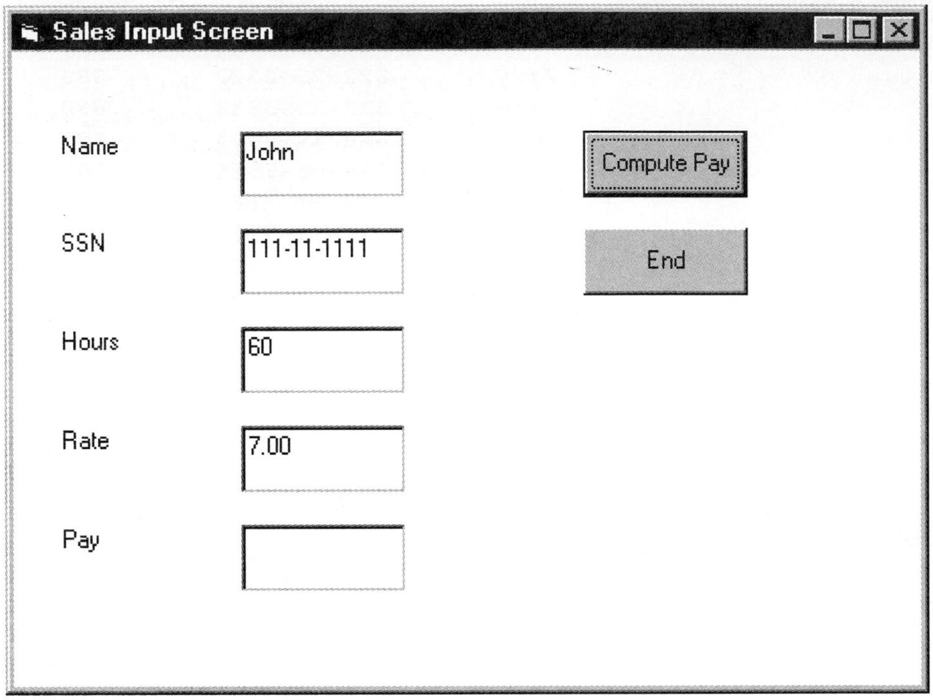

Figure 6–36
Payroll Problem (Visual Basic—Screen 3)

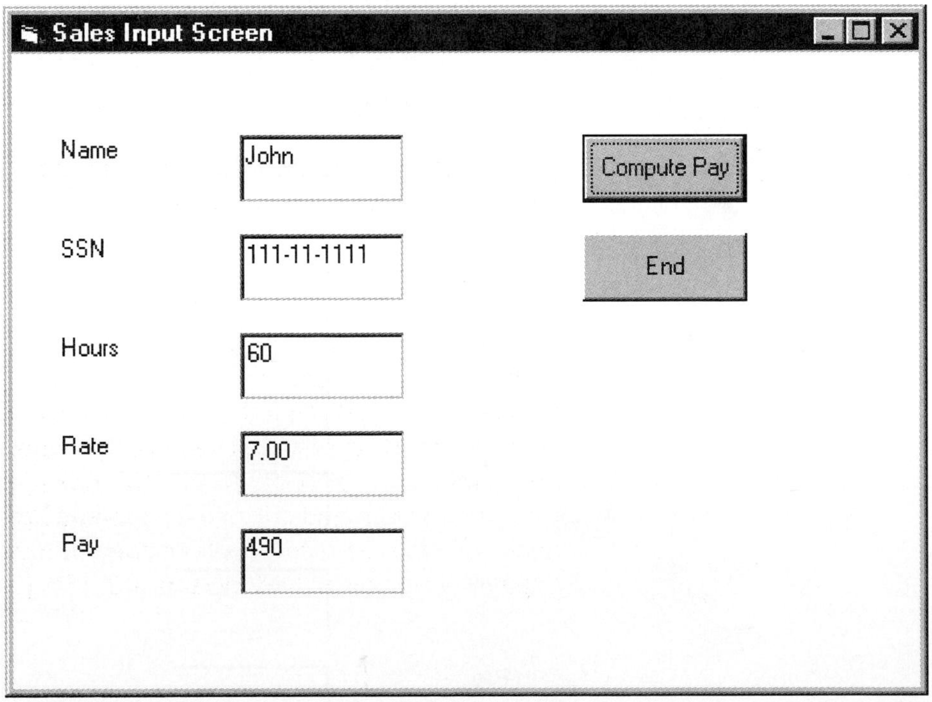

Figure 6–37
Payroll Problem (Visual Basic—cmd_COMPUTEPAY_Click)

```
Private Sub cmd_COMPUTEPAY_Click()

If txt_HOURS.Text > 40 Then
        RegPay = 40 * txt_RATE.Text
        OtPay = (txt_HOURS.Text - 40) * (1.5 * txt_RATE.Text)
        lbl_PAY.Caption = RegPay + OtPay
Else
        lbl_PAY.Caption = txt_HOURS.Text * txt_RATE.Text
End If

End Sub
```

Figure 6–38
Payroll Problem (Visual Basic—cmd_END_Click)

```
Private Sub cmd_END_Click()

End

End Sub
```

Figure 6–37 illustrates the code associated with the click event of the Compute Pay button. Note that this segment of code very closely parallels the pseudocode in Figure 6–3. The four text boxes are used to accept the input values from the user so the Read statement is not necessary. At the same time, these four text boxes display the input values, and the label control at the bottom of the screen displays the computed pay. Thus, the Write statement is also not necessary. The IF statement and the statements within the IF statement are the statements in Figure 6–3 except that the variable names are replaced by the control names.

An End button was also included in this example. Figure 6–38 illustrates the code associated with the click event of the End button. Remember, when the user clicks the End button, the Visual Basic End statement is executed and the program execution terminates.

Note that the program code contains no loop structure. The original design contained a DOWHILE loop in the overall control module that executed 10 times. In visual programs, the actions of the user can sometimes replace the loop, as is the case in this example. Every time the user enters four input values and clicks the Compute Pay button, a new value of Pay is displayed. The user decides how many times Pay will be computed—10 times, 3 times, or perhaps 50 times. No loop needs to be explicitly coded.

Key Terms

modularization	overall control module	calling module
predefined-process symbol	driver module	called module
module	annotation symbol	structure (hierarchy) chart
	top-down design	

Exercises

For each of the following exercises, design an algorithm using a modular approach. Include a structure chart as well as a program flowchart and corresponding pseudocode for each module in your solution. Include report and column headings as well as a page number on every page, with 55 detail lines per page, unless directed otherwise. Output an appropriate message if the input contains no records and include descriptive messages in the total lines.

1. Redo the solution shown in Figure 6–1, Figure 6–2, and Figure 6–3 incorporating the specifications listed above in the general exercise directions. In addition, output the total amount paid at the regular rate for all employees, the total amount paid at the overtime rate for all employees, and the total pay for all employees. Process any number of employees, not just 10. Design the algorithm using both a header record logic approach and an automatic end-of-file approach.)

2. The ABC company needs a weekly payroll report for its salespeople. Input to the program is the salesperson's name, number, and weekly sales. Output is the salesperson's name, number, and pay. Each salesperson receives a base pay of $300.00 as well as a 10 percent commission on his or her total sales up to and including $500.00. Any sales over $500.00 merit a 15 percent commission for the employee. (For example, if sales = $600.00, then pay = $300.00 + $50.00 [or 10 percent * 500] + $15.00 [or 15 percent * 100] = $365.00.) Design the algorithm using both a header record logic approach and an automatic end-of-file approach.

3. You are in a pumpkin patch looking for the great pumpkin. Each input record contains the weight of one pumpkin. You are to find the largest weight and the average weight of all the pumpkins. You are to output each weight, as well as the largest weight and the average weight. Design the algorithm using both a header record logic approach and a trailer record logic approach. (Use a dummy record with a negative pumpkin weight to denote the end-of-file.)

4. Redo the solution to Sample Problem 6.1 to process data for any number of students. Assume that the total number of students will be the first input, followed by the name record for the first student. This name record will then be followed by the individual scores for that student. After the scores for one student have been read, a name record for the next student will follow, and so on. Be sure to check that both the total number of students and the number of assignments completed for each student are positive numbers. Report headings, column headings, and page numbers are not required.

5. Redo Exercise 8 in Chapter 5 using a modular approach.

6. Redo Exercise 9 in Chapter 5 using a modular approach.

7. Redo Exercise 10 in Chapter 5 using a modular approach.

8. Redo Exercise 11 in Chapter 5 using a modular approach.

9. Redo Exercise 12 in Chapter 5 using a modular approach.

10. Redo Exercise 13 in Chapter 5 using a modular approach.

11. Redo Exercise 14 in Chapter 5 using a modular approach.

12. Redo Exercise 15 in Chapter 5 using a modular approach.

13. Redo Exercise 16 in Chapter 5 using a modular approach.

CASE Control Structure

Objectives

Upon completion of this chapter you should be able to

- Distinguish between a master file and a transaction file.
- Identify, and use in program design, the CASE control structure.
- Distinguish between numeric and alphabetic data.

Introduction

Often, a computer program must be designed and coded to handle a wide variety of inputs. We must provide flexibility in a solution algorithm, incorporating within the program an ability to process not only a variable number of inputs, but also whatever type of input is provided.

In business applications such as accounts receivable or employee payroll, large numbers of records are kept for reference purposes as relatively permanent data. Such data is not highly subject to change. Usually, it is needed for numerous business operations of the firm. Together, the data records constitute a **master file**. Current activities, or transactions, to be processed against the master file are called a **transaction file** or a **detail file**.

Assume, for example, that a firm's customer master file contains customer records. Each customer record contains several fields. The fields are customer number, name, address, telephone number, and credit rating. Customer transaction records to be processed against this file include fields containing address changes, corrections to telephone numbers, and the like.

Inventory Control Example

As another example, consider an inventory-control master file that contains stock status records of the numbers of various kinds of parts available for manufacturing planning. Each record contains several data items, located in specific fields of the record. These data items indicate quantity in stock, quantity on order, and so on.

Transactions to be processed against this master file originate daily. They are assigned transaction numbers and grouped together to form a transaction file. A transaction file is a temporary file containing data that is used to update a master file. In some applications, transactions are collected and processed as a group, say at the end of each day. In other applications, transactions are entered directly, as they occur, through any number of online input devices. In our example, a one-digit code is placed in a field of each transaction record to indicate the type of activity, as follows:

Code	Activity
1	Receipts (parts that arrive in response to previous orders)
2	Orders (requests for additional parts to be included in stock)
3	Withdrawals (also called issues; depletions from stock)
4	Adjustments (changes to stock levels for reasons other than those above; for example, transfers of parts to other manufacturing locations)

Our task is to design, code, and test a program to process the transaction records against the inventory-control master file. Part of the program flowchart and corresponding pseudocode that we might construct in developing a solution algorithm are shown in Figures 7–1 and 7–2.

**Figure 7–1
Master File Update Using
Nested IFs (Partial
Flowchart)**

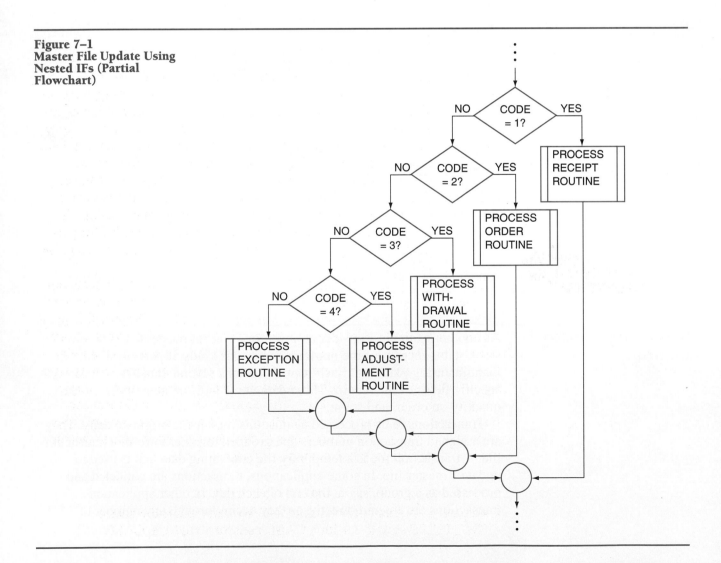

**Figure 7–2
Master File Update Using
Nested Ifs (Partial
Pseudocode)**

```
        •
        •
        •
IF CODE = 1 THEN
    Process receipt routine
ELSE
    IF CODE = 2 THEN
        Process order routine
    ELSE
        IF CODE = 3 THEN
            Process withdrawal routine
        ELSE
            IF CODE = 4 THEN
                Process adjustment routine
            ELSE
                Process exception routine
            ENDIF
        ENDIF
    ENDIF
ENDIF
        •
        •
        •
```

Solution 1: Nested IFTHENELSE Control Structure

Do you recognize which control structure is being used? You should see a nested IFTHENELSE pattern. If one of the tests for the activity type (specified as CODE) yields a true outcome, the routine or module for that activity type is executed. No additional tests are made (they're obviously unnecessary). Instead, the entire nested IFTHENELSE control structure is exited. Program execution continues with the processing step that follows the nested IFTHENELSE. The number of tests performed depends on the activity type. Only if all four tests are made and yield false outcomes is the exception routine (error routine) executed. (Recall that not executing all tests each time is a key difference between the nested IFTHENELSE pattern and a sequence of separate IFTHENELSEs, which we call the sequential IFTHENELSE pattern.)

Although the nested IFTHENELSE pattern seems to meet our decision-making needs, it is often difficult to work with. In this example we have to make four tests, which amounts to a nesting level of four. What will happen if we inadvertently omit a program statement corresponding to one of the closing ENDIFs? Or if we want to remove just one of the tests of activity type at a later time? Suppose we need to add a test for another activity type within the nested IFTHENELSE structure. The program coding to implement this structure must be done very carefully, or errors will most certainly occur. And what if there are 10, or even 100, possible activity types to be tested? Obviously, many pages of flowcharting, or an unmanageable number of pseudocode indentations, will be required.

Solution 2: CASE Control Structure

Fortunately, another option is available. We can replace the nested IFTHENELSE structure with a **CASE control structure**. CASE

Figure 7–3
Master File Update Using
CASE (Partial Flowchart)

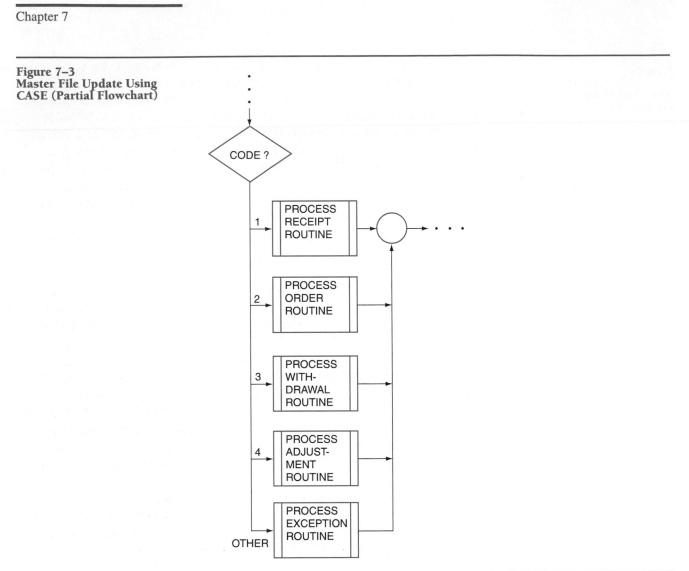

generalizes the basic IFTHENELSE pattern, extending it from a two-val-ued operation to a multiple-valued one. With one CASE control structure, we can represent all of the tests shown in Figure 7–1 (see Figure 7–3). Once we understand how this structure is derived, and see that it consists only of basic patterns, we can use it where we might otherwise resort to a nested IFTHENELSE.

One note of caution is warranted: Although the CASE control structure serves as an alternative for the nested IFTHENELSE control structure, it cannot be used in place of a sequential IFTHENELSE pattern. Both se-quential and nested IFTHENELSE patterns were initially discussed in Chapter 3. Make sure you remember the difference between them and con-sider the CASE control structure only when the logic needed in an algo-rithm is the nested IFTHENELSE pattern.

Another point worth mentioning is that you should not confuse the use of the term *CASE* here with our earlier use of the letters *CASE* as an acronym for the phrase *computer-assisted software engineering*. Here CASE is simply the name of a particular kind of control structure.

The program logic in Figure 7–3 is the same as that shown in Figure 7–1, but it appears in a much more understandable form. By the simple use of parallel flowlines, the same possible outcomes are documented. Generally, when this documentation technique is used, the first test to be made should be shown by the topmost flowline on the flowchart (or the leftmost one, if the parallel flowlines are vertical rather than horizontal). For processing efficiency, the test most likely to reveal a true outcome should be made first, the next most likely one second, and so on; this helps to minimize the number of tests actually carried out on any *one pass* through this portion of the program. In most cases, the word *other* appears by the last (bottom or rightmost) flowline to indicate what processing should occur if none of the preceding conditions is true. This path usually involves some type of error processing.

We express the CASE control structure in pseudocode form using the keywords CASENTRY, CASE, and ENDCASE, as shown in Figure 7–4. Notice that three levels of indentation are used, no matter how many tests are made. The text following the keyword CASENTRY identifies the variable data on which tests are to be made. Each CASE keyword is then slightly indented from the position of CASENTRY. The text following each CASE keyword indicates the details of a particular test. For example, CASE 1 in Figure 7–4 represents a test to see if the variable CODE is equal to 1. The statement(s) following each CASE line are then indented a few spaces from the position of the CASE line. These statements specify what processing is to be done in a particular case. For example, the receipt routine will be executed when CODE is equal to 1.

The keyword ENDCASE is specified at the physical end of the CASE structure to denote that all required tests (cases) have been stated. It is lined up with the keyword CASENTRY. This positioning is consistent with the positioning of the keywords ENDIF and ENDDO in other structures we've

Figure 7–4
Master File Update Using
CASE (Partial Pseudocode)

```
        .
        .
        .
CASENTRY CODE
   CASE 1
      Process receipt routine
   CASE 2
      Process order routine
   CASE 3
      Process withdrawal routine
   CASE 4
      Process adjustment routine
   CASE other
      Process exception routine
ENDCASE
        .
        .
        .
```

already learned. Note that, in the flowchart, the exit point of the structure is shown by the connector symbol, as was done with the IFTHENELSE structure. Even though the CASE structure implies several tests, only one exit connector (or one ENDCASE) is indicated, because we are representing only one CASE control structure. The top-to-bottom arrangement of the conditions or tests in the CASE structure dictates the actual order that will be used by the computer to make the tests.

We should make one final note before leaving this example. We have illustrated the CASE control structure with an example of part of the logic required to process a master file. In Chapter 14 we will expand this discussion and show a complete modular design for a master file update procedure.

Sample Problem 7.1 (Op Code Problem)

Problem:

Design an algorithm to accept three values as input. The first input value will be one of four operation codes, either A (addition), S (subtraction), M (multiplication), or D (division). The other two input values will be numbers. The computer is to perform a computation on the two numbers as determined by the operation code. The result of the computation, as well as the original inputs, is to be written as output. An error message should be output if the operation code is invalid.

Solution:

A plan for the solution to this problem is given in flowchart and pseudocode forms in Figures 7–5 and 7–6, respectively.

For simplicity, this solution is written to process only one set of input; that is, we have not included loop processing logic. A modular design including the handling of several inputs is provided as an exercise.

In this example, values for three variables are input. One is an operation code (OP), and two are numbers (N1 and N2). In previous examples, most of the variables were numeric quantities such as N1 and N2 in this example. The variable OP is used to represent one of four operation codes. The operation codes are not numbers; they are the letters A, S, M, and D. A variable can hold only one type of data; that is, we cannot use a single variable name to represent both numeric quantities and alphabetic data. Alphabetic data cannot be used in computations. Alphabetic data values are called **character strings**. We usually designate the value of a character string by enclosing it in single or double quotations marks, thereby distinguishing it from a variable name: "A" (character string), then, is very different from A (variable name). Do not confuse the term *character string* (discussed in this chapter) with the term *character-string constant* (discussed in Chapter 3). A character-string constant is something like NAME = "Mary" that is set once and never changes during program execution. A value that is input during processing, like the value for OP in this example, is a character string, but it is not a character-string constant.

There are several types of variables besides numeric and character variables. We will not address those other variable types in this text. However, it is important for you to understand that most programming languages

Figure 7–5
Op Code Problem
(Flowchart)

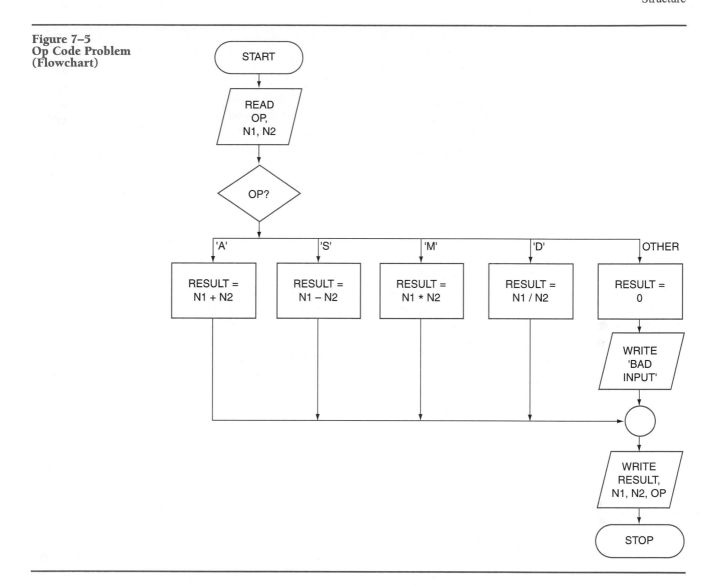

Figure 7–6
Op Code Problem
(Pseudocode)

```
Start
Read OP, N1, N2
CASENTRY OP
   CASE 'A'
      RESULT = N1 + N2
   CASE 'S'
      RESULT = N1 - N2
   CASE 'M'
      RESULT = N1 * N2
   CASE 'D'
      RESULT = N1 / N2
   CASE other
      RESULT = 0
      Write 'Bad Input'
ENDCASE
Write RESULT, N1, N2, OP
Stop
```

require that each variable used in a program be defined to the computer as being of a specific type. In practice, the programmer names each variable in the program and "declares" it to be of a specific type.

Now look again at Figure 7–5. After the data is input, a CASE control structure is used to determine which computation needs to be done. There are five paths shown, one for each operation code and one "other" path to handle the case in which the operation code input is invalid. For example, someone may key in an E instead of an S because the E and S keys are close together on the keyboard. Notice the order in which the tests will be made. If the operation codes are evenly distributed in the input file, the order in which the tests are made will not matter. In this example, the natural order is used. When dealing with simple arithmetic, most people think in order of add, subtract, multiply, and divide. We do, however, need to place the "other" path last. All the tests must fail before we conclude that the code is invalid.

The variable RESULT is used in all five cases to hold the answer. It may seem strange that RESULT is given a value of 0 when the operation code is invalid. If you look at the second WRITE statement in this algorithm (the one after ENDCASE), you will see why. The value of RESULT is output, regardless of which path is taken. If we do not give RESULT some "dummy" value in the "other" path, then RESULT will be undefined to the computer when an invalid record is processed. If an attempt is made to output a variable with an undefined value, an error will occur. Notice also that if the operation code is invalid, two lines of output will be written: first, the Bad Input line and then the detail line, which contains 0 in the RESULT field.

Sample Problem 7.2 (Sales Problem without Modules)

Problem:

Redo Sample Problem 3.4 using the CASE control structure.

Solution:

A solution to this problem is shown in flowchart and pseudocode forms in Figures 7–7 and 7–8, respectively.

In this solution we replace the nested IFTHENELSE control structure with the CASE control structure to determine the value of CLASS. There are four possible valid values for CLASS (1,2,3,4), and these values are tested in order within the CASE structure. The "other" path represents the processing required if CLASS is invalid. This example illustrates how several structures can be contained within each path of the CASE control structure. For example, the CASE 1 path contains a nested IFTHENELSE structure, and the CASE 2 path contains a simple IFTHENELSE structure. It is even possible to have a DOWHILE control structure within one or more of the CASE paths. As you can see, the problem solution can become very complex and hard to follow unless the steps within each CASE path are grouped together in a separate module, as was shown in Figures 7–3 and 7–4. This modular breakdown is certainly desirable, though not always required. Now let's look at the next problem to see how we might modularize the same solution.

Figure 7–7
Sales Problem without
Modules (Flowchart)

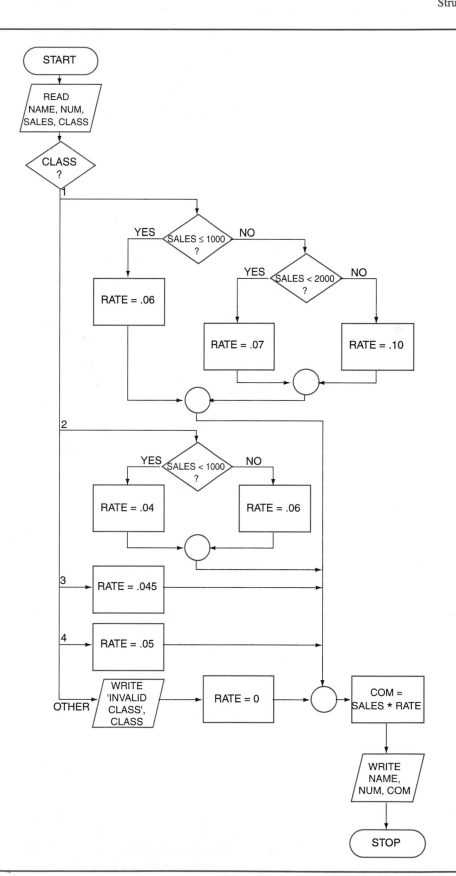

Figure 7–8
Sales Problem without
Modules (Pseudocode)

```
Start
Read NAME, NUM, SALES, CLASS
CASENTRY CLASS
    CASE 1
        IF SALES ≤ 1000 THEN
            RATE = .06
        ELSE
            IF SALES < 2000 THEN
                RATE = .07
            ELSE
                RATE = .10
            ENDIF
        ENDIF
    CASE 2
        IF SALES < 1000 THEN
            RATE = .04
        ELSE
            RATE = .06
        ENDIF
    CASE 3
        RATE = .045
    CASE 4
        RATE = .05
    CASE other
        Write 'Invalid Class', CLASS
        RATE = 0
ENDCASE
COM = SALES * RATE
Write NAME, NUM, COM
Stop
```

Sample Problem 7.3 (Sales Problem Using Modules)

Problem:

Redo Sample Problem 7.2 and place all the steps within each CASE path in separate modules.

Solution:

A structure chart for this problem is shown in Figure 7–9. There are five level 2 modules, each representing the steps to process one value of CLASS.

The overall control module (A000) is shown in flowchart and pseudocode forms in Figures 7–10 and 7–11, respectively.

You can see that the problem solution appears simpler with a modular design; it allows us to focus more clearly on the CASE control structure and not on the complex details within each path. These details are shown in modules B000 (see Figure 7–12), B010 (see Figure 7–13), B020 (see Figure 7–14), B030 (see Figure 7–15) and B040 (see Figure 7–16). Each module represents the processing for a specified value of CLASS.

Figure 7–9
Sales Problem Using Modules
(Structure Chart)

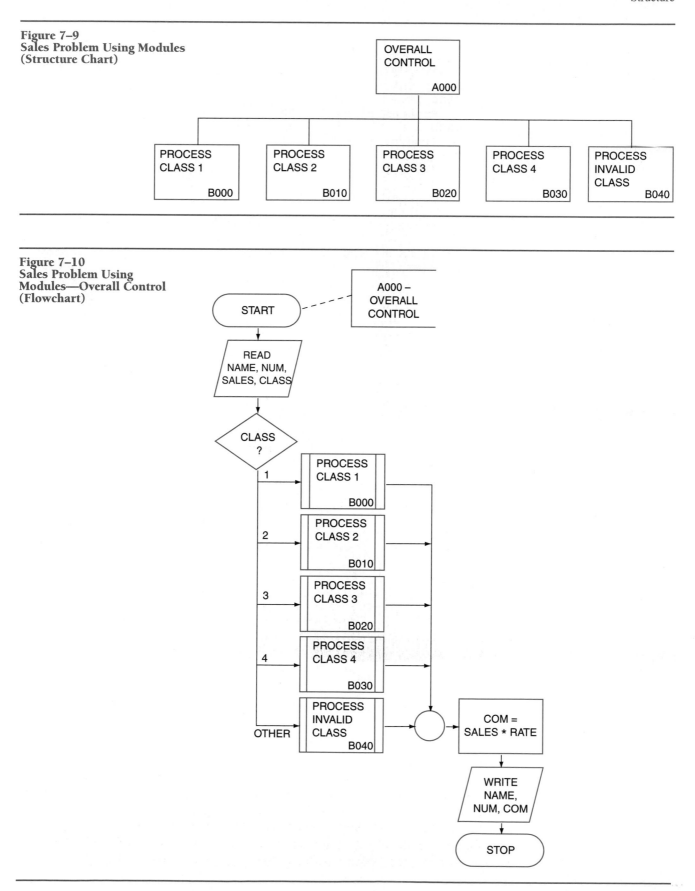

Figure 7–10
Sales Problem Using
Modules—Overall Control
(Flowchart)

Figure 7–11
Sales Problem Using
Modules—Overall Control
(Pseudocode)

```
A000
Start
Read NAME, NUM, SALES, CLASS
CASENTRY CLASS
    CASE 1
        Process class 1 (B000)
    CASE 2
        Process class 2 (B010)
    CASE 3
        Process class 3 (B020)
    CASE 4
        Process class 4 (B030)
    CASE other
        Process invalid class (B040)
ENDCASE
COM = SALES * RATE
Write NAME, NUM, COM
Stop
```

Figure 7–12
Sales Problem Using
Modules—Process Class 1

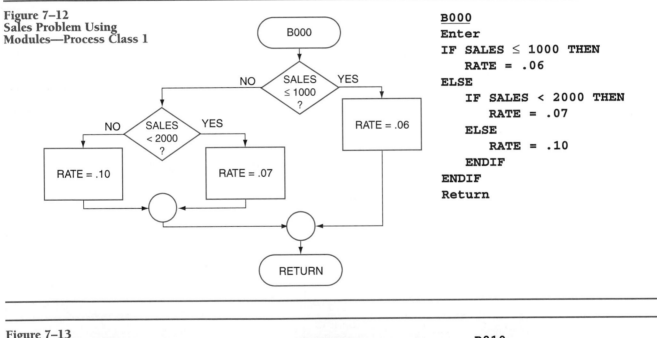

```
B000
Enter
IF SALES ≤ 1000 THEN
    RATE = .06
ELSE
    IF SALES < 2000 THEN
        RATE = .07
    ELSE
        RATE = .10
    ENDIF
ENDIF
Return
```

Figure 7–13
Sales Problem Using
Modules—Process Class 2

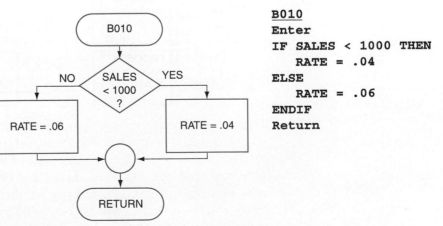

```
B010
Enter
IF SALES < 1000 THEN
    RATE = .04
ELSE
    RATE = .06
ENDIF
Return
```

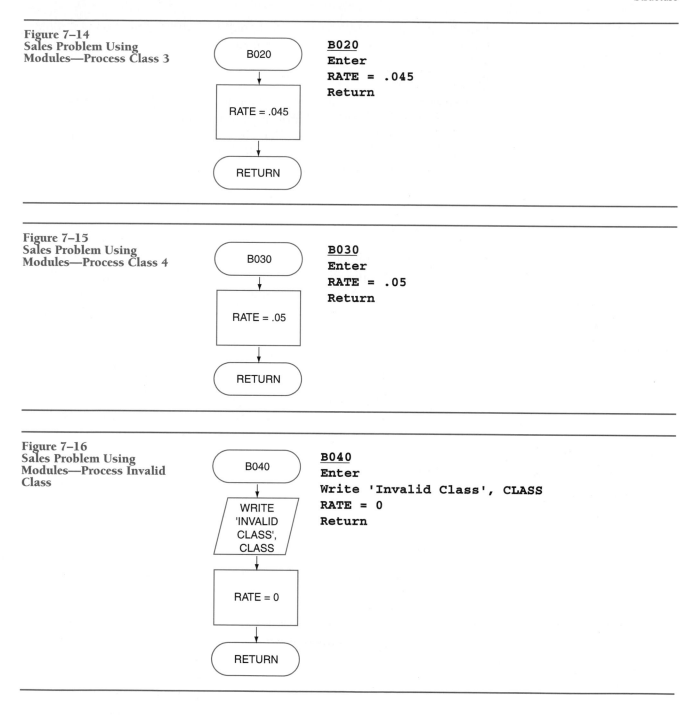

Figure 7–14
Sales Problem Using
Modules—Process Class 3

```
B020
Enter
RATE = .045
Return
```

Figure 7–15
Sales Problem Using
Modules—Process Class 4

```
B030
Enter
RATE = .05
Return
```

Figure 7–16
Sales Problem Using
Modules—Process Invalid
Class

```
B040
Enter
Write 'Invalid Class', CLASS
RATE = 0
Return
```

Sample Problem 7.4 (Sales Problem with Totals)

Problem:

Redo Sample Problem 7.3 to process several records. Detail lines are to be output on a daily sales commission report. In addition, report and column headings are to be output on every page of the report. The pages are to be numbered, and up to 55 detail lines are to be included on a page. Five total lines are to be output on the last page of the report. These lines are to specify the total number of records containing a class of 1, 2, 3, and 4, as well as the total number of records containing an invalid class. All totals should have identifying labels. The end-of-file will be indicated by a special trailer record containing an employee number of 0000.

Solution:

A structure chart representing all the modules used in this solution is shown in Figure 7–17. The relationships of modules A000, B000, B010, B020, and C000 should look familiar, since this same design was used in Chapter 6. However, the detail-processing module is now broken into several additional third-level modules. These modules represent the steps within each CASE path. As you will see, the CASE control structure is now part of the detail-processing module.

The overall control module is shown in Figures 7–18 and 7–19, and is, again, consistent with the general requirements of trailer record logic. There is no indication, however, of a CASE structure in this module. Remember, A000 specifies only the main processing loop. The details of the solution (using the CASE in this example) are placed in another module.

The initialization module (see Figure 7–20) contains no unusual processing, so it should look somewhat familiar to you. Notice that, in B000, we initialize five variables to 0. These variables are used to accumulate the total number of records within each class.

Now look at the detail-processing module (see Figures 7–21 and 7–22). First, a check is made to see if the headings need to be output and a new page started. Then CLASS is checked using a CASE control structure. Four paths (1,2,3,4) process all the valid values of CLASS, and the fifth path handles invalid class records. The processing steps within each CASE path are handled in separate modules as before. After the appropriate third-level module is executed, commission is computed by multiplying the sales (input in A000) by the commission rate (computed in one of the previous modules). The actual detail line is then output in all five cases and LINECNT is updated.

Figure 7–17
Sales Problem with Totals
(Structure Chart)

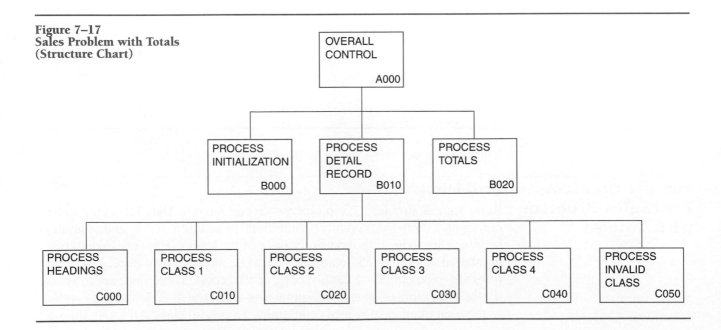

Figure 7–18
Sales Problem with Totals—
Overall Control (Flowchart)

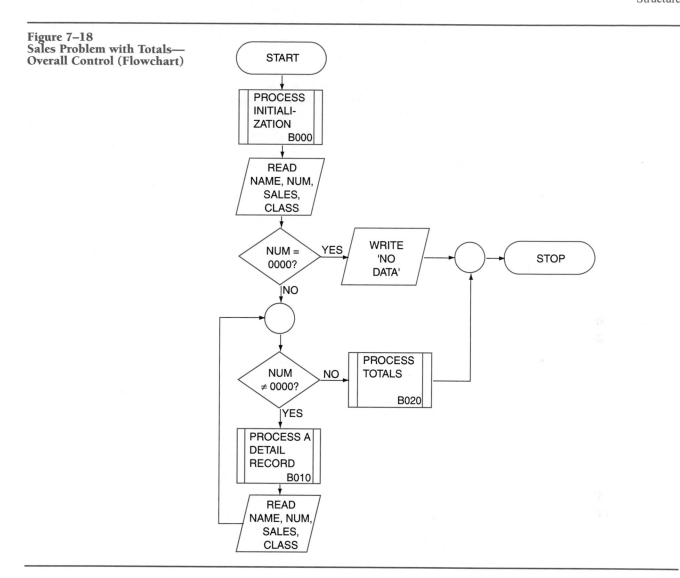

Figure 7–19
Sales Problem with Totals—
Overall Control (Pseudocode)

```
A000
Start
Process initialization (B000)
Read NAME, NUM, SALES, CLASS
If NUM = 0000 THEN
    Write 'No data'
ELSE
    DOWHILE NUM ≠ 0000
        Process a detail record (B010)
        Read NAME, NUM, SALES, CLASS
    ENDDO
    Process totals (B020)
ENDIF
Stop
```

Figure 7–20
Sales Problem with Totals—
Process Initialization

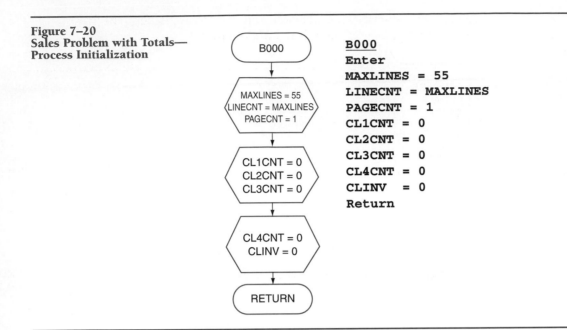

```
B000
Enter
MAXLINES = 55
LINECNT = MAXLINES
PAGECNT = 1
CL1CNT  = 0
CL2CNT  = 0
CL3CNT  = 0
CL4CNT  = 0
CLINV   = 0
Return
```

Figure 7–21
Sales Problem with Totals—
Process a Detail Record
(Flowchart)

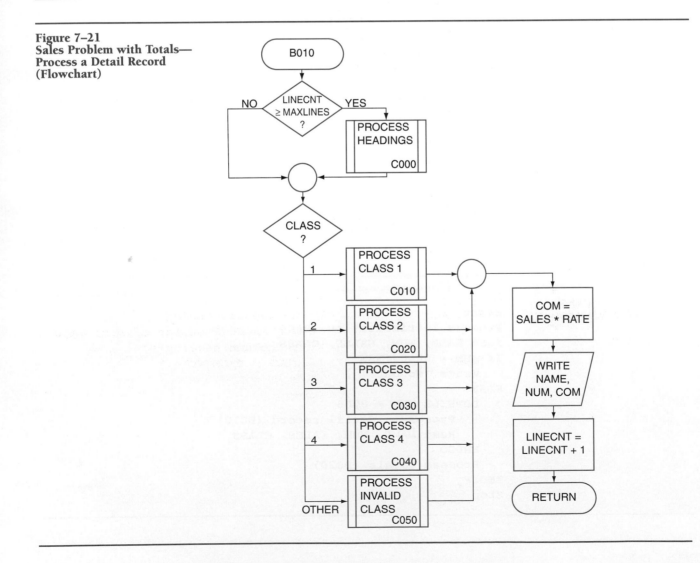

The headings module (C000), shown in Figure 7–23, contains no unusual processing, so you should find it likewise familiar.

Figure 7–22
Sales Problem with Totals—
Process Detail Record
(Pseudocode)

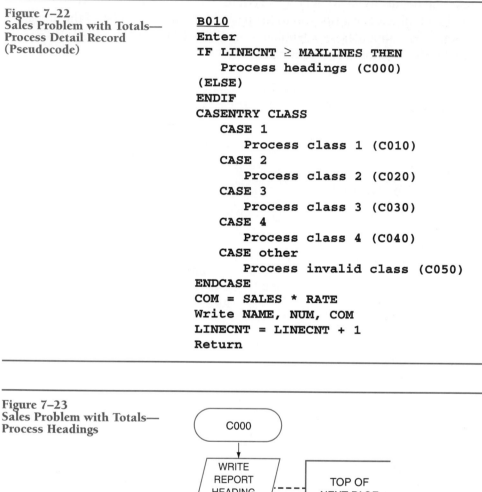

```
B010
Enter
IF LINECNT ≥ MAXLINES THEN
   Process headings (C000)
(ELSE)
ENDIF
CASENTRY CLASS
   CASE 1
      Process class 1 (C010)
   CASE 2
      Process class 2 (C020)
   CASE 3
      Process class 3 (C030)
   CASE 4
      Process class 4 (C040)
   CASE other
      Process invalid class (C050)
ENDCASE
COM = SALES * RATE
Write NAME, NUM, COM
LINECNT = LINECNT + 1
Return
```

Figure 7–23
Sales Problem with Totals—
Process Headings

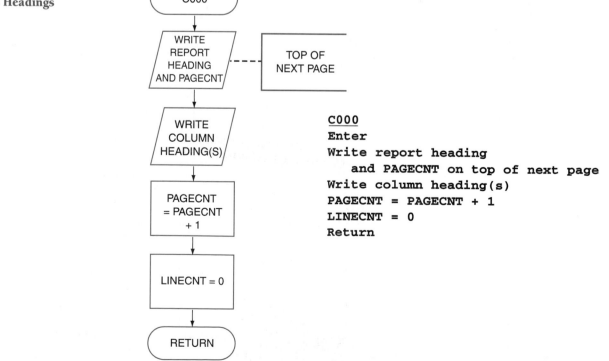

```
C000
Enter
Write report heading
   and PAGECNT on top of next page
Write column heading(s)
PAGECNT = PAGECNT + 1
LINECNT = 0
Return
```

Modules C010 (see Figure 7–24), C020 (see Figure 7–25), C030 (see Figure 7–26), C040 (see Figure 7–27), and C050 (see Figure 7–28) contain the steps required to compute the commission rate for each class. Each module begins by incrementing a counter (set to 0 in B000) to keep track of the number of records within the particular class. The rest of the steps within these modules are identical to modules B000, B010, B020, B030, and B040 in Sample Problem 7.3.

Figure 7–24
Sales Problem with Totals—
Process Class 1

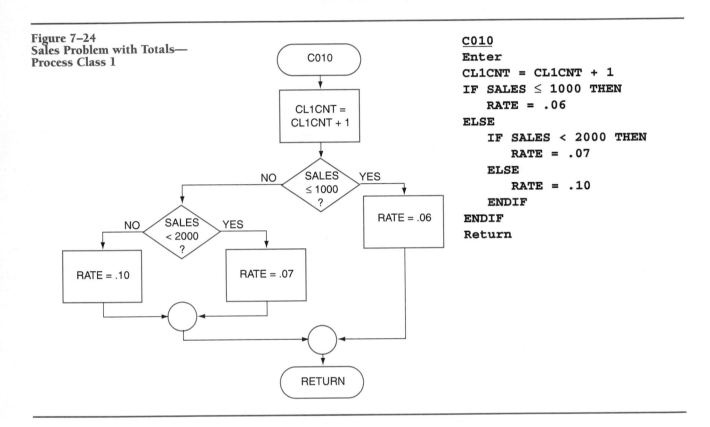

```
C010
Enter
CL1CNT = CL1CNT + 1
IF SALES ≤ 1000 THEN
    RATE = .06
ELSE
    IF SALES < 2000 THEN
        RATE = .07
    ELSE
        RATE = .10
    ENDIF
ENDIF
Return
```

Figure 7–25
Sales Problem with Totals—
Process Class 2

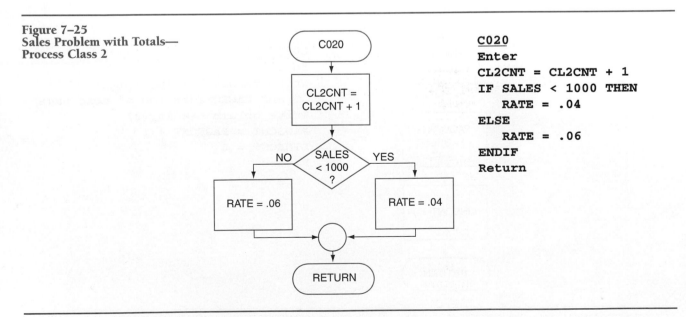

```
C020
Enter
CL2CNT = CL2CNT + 1
IF SALES < 1000 THEN
    RATE = .04
ELSE
    RATE = .06
ENDIF
Return
```

Figure 7–26
Sales Problem with Totals—
Process Class 3

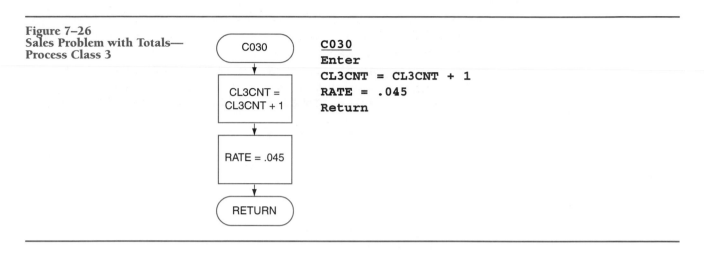

```
C030
Enter
CL3CNT = CL3CNT + 1
RATE = .045
Return
```

Figure 7–27
Sales Problem with Totals—
Process Class 4

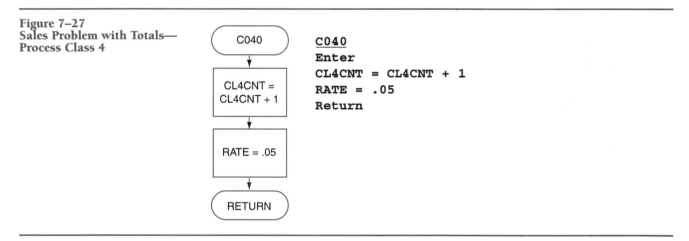

```
C040
Enter
CL4CNT = CL4CNT + 1
RATE = .05
Return
```

Figure 7–28
Sales Problem with Totals—
Process Invalid Class

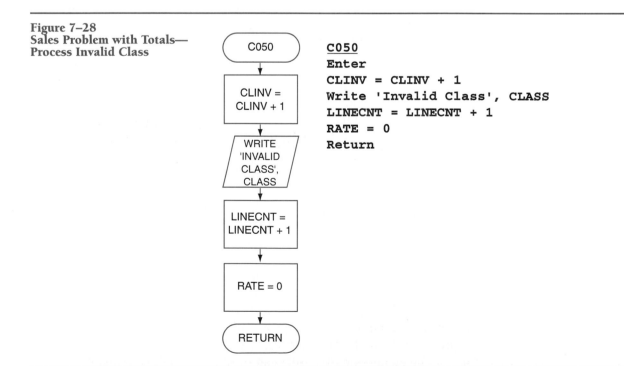

```
C050
Enter
CLINV = CLINV + 1
Write 'Invalid Class', CLASS
LINECNT = LINECNT + 1
RATE = 0
Return
```

Figure 7–29
Sales Problem with Totals—
Process Totals

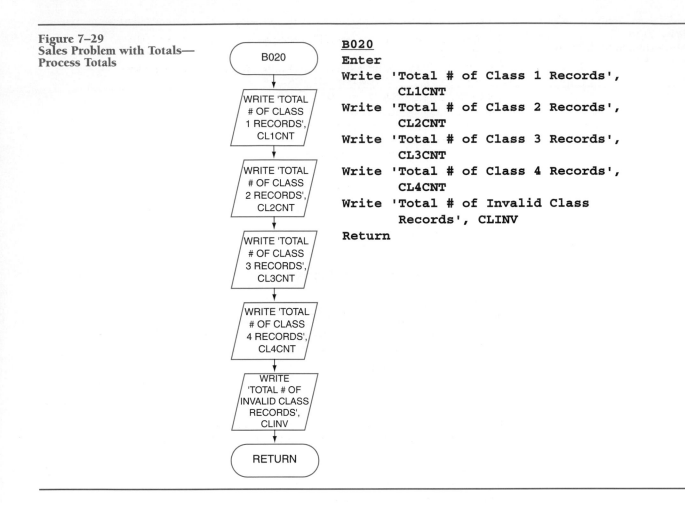

```
B020
Enter
Write 'Total # of Class 1 Records',
      CL1CNT
Write 'Total # of Class 2 Records',
      CL2CNT
Write 'Total # of Class 3 Records',
      CL3CNT
Write 'Total # of Class 4 Records',
      CL4CNT
Write 'Total # of Invalid Class
      Records', CLINV
Return
```

The totals module (B020), shown in Figure 7–29, simply writes out the values of the five counters with appropriate labels.

Enrichment (Basic)

Figure 7–30 illustrates a listing of the program that solves the op code problem (see Figures 7–5 and 7–6). In this example, the Input statement is used to request the op code and the two numbers from the user. Two separate Input statements are used for clarity. Note that the variable name for the op code is OP$ because this variable will be holding alphabetic data (A, S, M, or D). As you can see, the Basic program is similar to the pseudocode in Figure 7–6. One difference, however, is the syntax used in Basic for the CASE control structure. The keywords SELECT CASE and END SELECT are used in place of the pseudocode keywords CASENTRY and ENDCASE. In addition, the keyword ELSE replaces the word *other* in the last CASE path. We do, however, employ the same indentation standards within the SELECT CASE structure as we do in the pseudocode. The four output values have been labeled and have been written on separate lines for readability.

Figure 7–31 illustrates the output that will be produced when the program is executed and the user enters a valid op code, A in this case.

Figure 7–32 illustrates the output that will be produced when the program is executed and the user enters an invalid op code, B in this case.

Figure 7–30
Op Code Problem (Basic List)

```
PRINT "Enter an operation code - A, S, M or D";
INPUT OP$
PRINT "Enter two numbers, separated by a comma";
INPUT N1,N2
SELECT CASE OP$
     CASE "A"
          RESULT = N1 + N2
     CASE "S"
          RESULT = N1 - N2
     CASE "M"
          RESULT = N1 * N2
     CASE "D"
          RESULT = N1 / N2
     CASE ELSE
          RESULT = 0
          PRINT
          PRINT "Bad input"
END SELECT
PRINT
PRINT "The answer is "; RESULT
PRINT "The two numbers are "; N1; "and "; N2
PRINT "The operation code is "; OP$
END
```

Figure 7–31
Op Code Problem (Basic Run—Valid Data)

```
Enter an operation code - A, S, M, or D? A
Enter two numbers, separated by a comma? 5,7

The answer is  12
The two numbers are  5 and  7
The operation code is A
```

Figure 7–32
Op Code Problem (Basic Run—Invalid Data)

```
Enter an operation code - A, S, M, or D? B
Enter two numbers, separated by a comma? 3,8

Bad input

The answer is  0
The two numbers are  3 and  8
The operation code is B
```

Note that an extra line of output is printed—the error message—and that a dummy value of 0 is printed for the result.

Enrichment (Visual Basic)

Figure 7–33 illustrates the graphical interface for the op code problem (see Figures 7–5 and 7–6). In this example, three text boxes are created to accept user input. A command button is created to compute the result, and a label is created to hold the computed result.

Figure 7–34 illustrates the screen after the user has entered the input values in each of the text boxes.

Figure 7–33
Op Code Problem (Visual Basic—Screen 1)

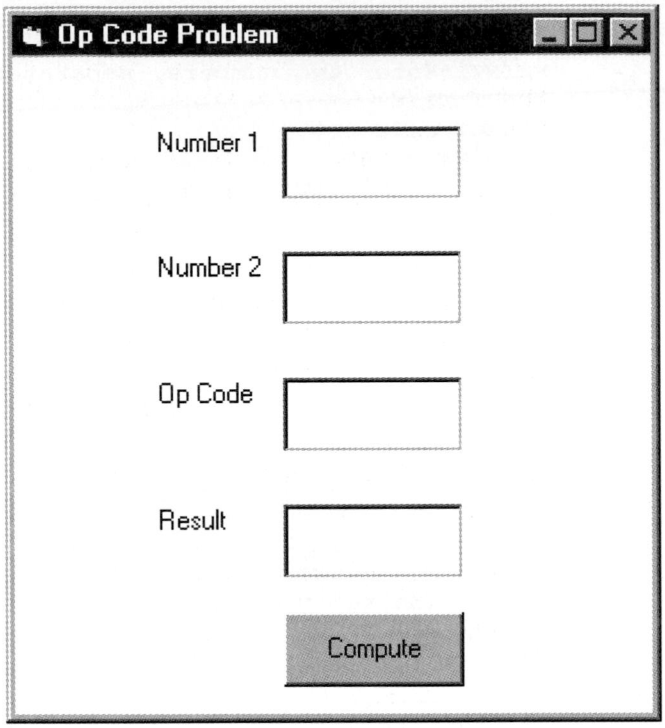

Figure 7–34
Op Code Problem (Visual Basic—Screen 2)

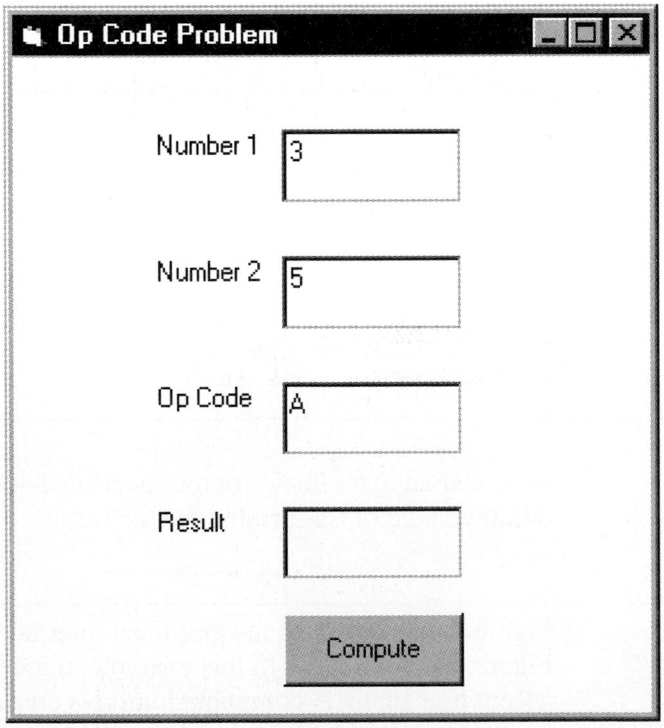

Figure 7–35 illustrates the screen after the user has clicked the Compute button. The value for the result is displayed in the label.

Figure 7–36 illustrates the screen after the user has entered an invalid op code. Note that the label control still contains an "8," the result of the last computation.

Figure 7–37 illustrates the screen after the user has clicked the Compute button. A message box is displayed indicating that the op code was invalid.

Figure 7–38 illustrates the screen after the user has read the error message and has clicked the OK button within the message box. Note that the label control holding the old result has been cleared. Note also that the insertion point is now located in the third text box. This will make it easier for the user to delete the invalid op code and key in a new one. A special method is used to reposition the insertion point. This method will be discussed shortly.

Figure 7–39 illustrates the program that is associated with the click event of the Compute button. As you can see, the Visual Basic code is similar to the pseudocode in Figure 7–6. One difference, however, is the syntax used in Visual Basic for the CASE control structure. The keywords SELECT CASE and END SELECT are used in place of the pseudocode keywords CASENTRY and ENDCASE. In addition, the keyword ELSE replaces the word *other* in the last CASE path. We do, however, employ the same indentation standards within the SELECT CASE structure as we do in the pseudocode.

The appropriate computation is done in each path of the SELECT CASE structure. Since the two numbers to be used in the computation are located in text boxes, the text property of each text box is used to reference the

Figure 7–35
Op Code Problem (Visual Basic—Screen 3)

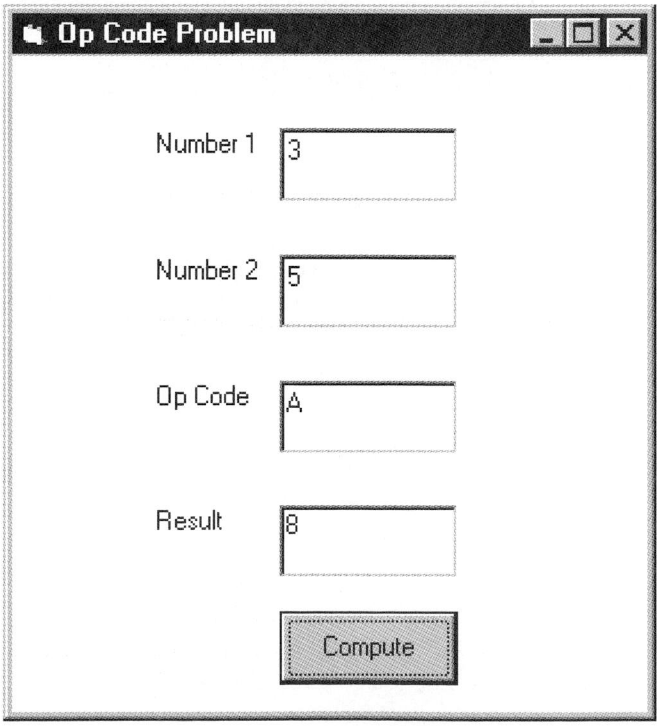

Figure 7–36
Op Code Problem (Visual Basic—Screen 4)

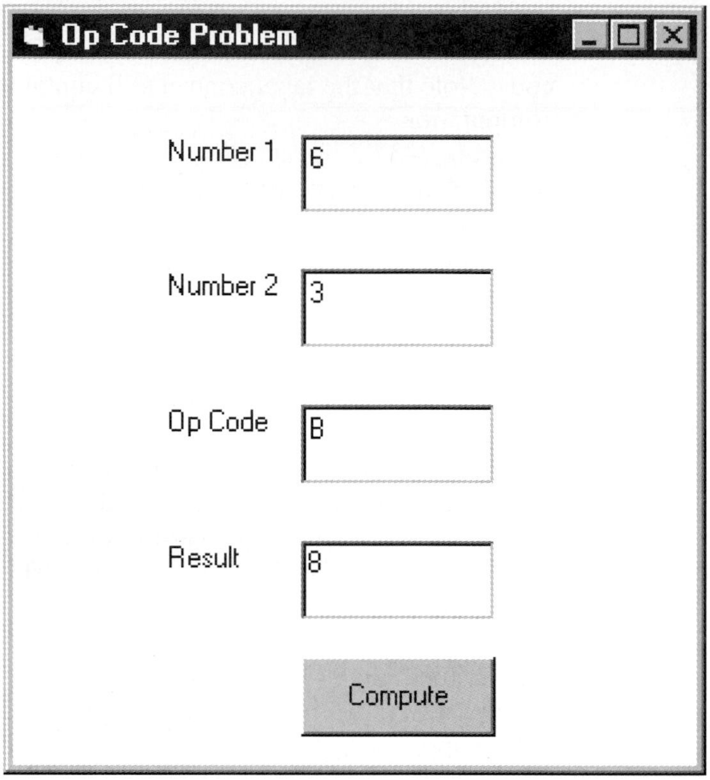

Figure 7–37
Op Code Problem (Visual Basic—Screen 5)

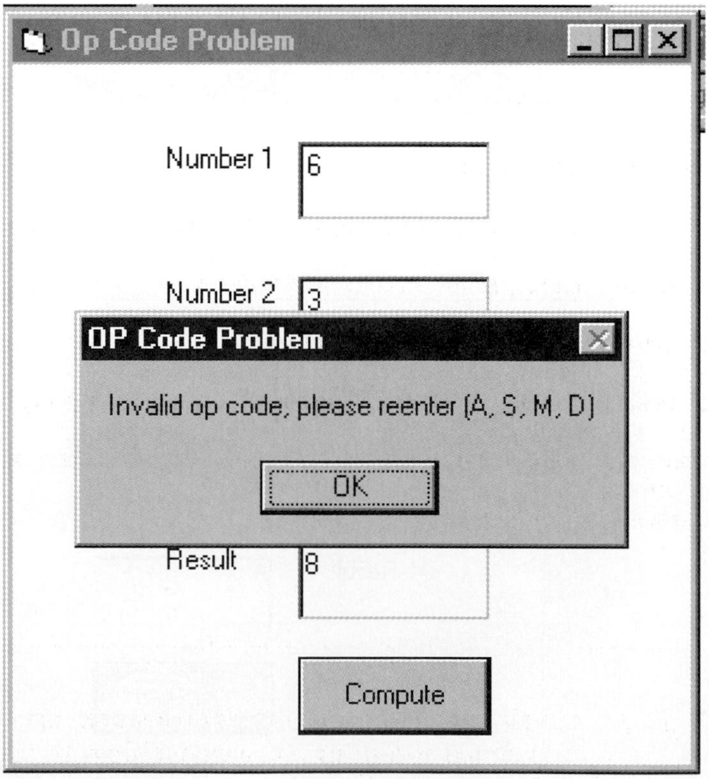

Figure 7–38
Op Code Problem (Visual Basic—Screen 6)

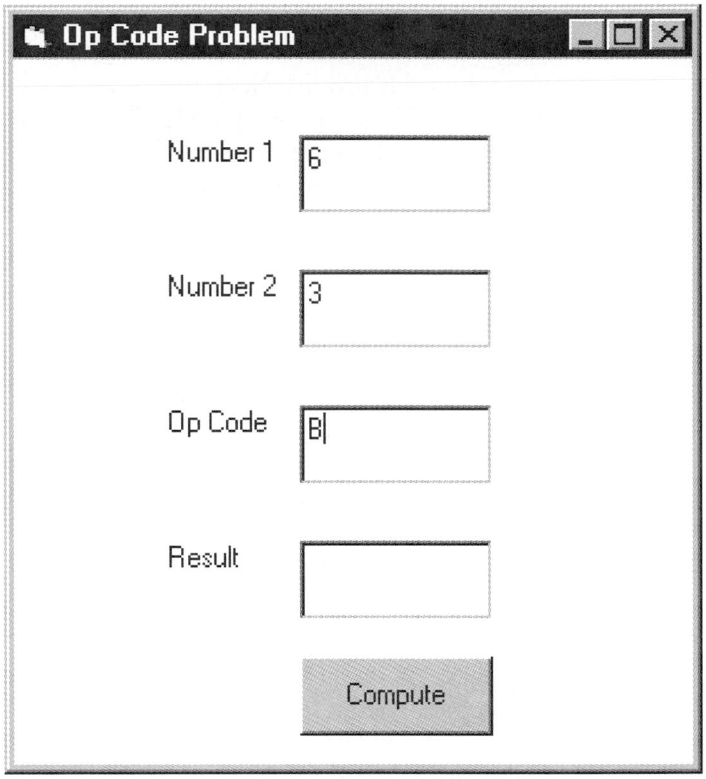

Figure 7–39
Op Code Problem (Visual Basic—cmd_COMPUTE_Click)

```
Private Sub cmd_COMPUTE_Click()

Select Case txt_OPCODE.Text
    Case "A"
        lbl_RESULT.Caption = Val(txt_NUMBER1.Text) + Val(txt_NUMBER2.Text)
    Case "S"
        lbl_RESULT.Caption = Val(txt_NUMBER1.Text) - Val(txt_NUMBER2.Text)
    Case "M"
        lbl_RESULT.Caption = Val(txt_NUMBER1.Text) * Val(txt_NUMBER2.Text)
    Case "D"
        lbl_RESULT.Caption = Val(txt_NUMBER1.Text) / Val(txt_NUMBER2.Text)
    Case Else
        MsgBox "Invalid op code, please reenter (A, S, M, D)",,"OP Code Problem"
        txt_OPCODE.SetFocus
        lbl_RESULT.Caption = ""
End Select

End Sub
```

value entered in the text box. However, data that is entered into a text box is stored as a string and cannot always be used in computations. A special function, *Val*, is needed to convert the data from string form to numeric form. The Val function is used twice in each of the four assignment statements to facilitate this conversion.

In the CASE ELSE path, a message box is displayed if the op code that was entered is invalid. Next, the *SetFocus* method is executed for the text box control holding the op code. This method causes the *focus* (the insertion point in this case) to be placed in the text box—txt_OPCODE. The SetFocus method can be executed for any control that can receive the focus. For example, the statement cmd_COMPUTE.SetFocus would place the focus on the Compute button. In this case, the border of the command button would be darkened, since command buttons do not have insertion points. Finally, the label holding the old result is cleared by assigning the *null string* (" ") to the caption property of the label control.

Now let's look at another solution to this same problem. Figure 7–40 illustrates a different graphical interface for the op code problem. In this solution, two text boxes are used to input the numbers as before. The third text box (used to input the op code) and the command button have been replaced by four option buttons. Each option button represents one of the four arithmetic computations that will be performed on the two numbers. These option buttons will be used to accept the user input for op code. The user will enter the numbers into the two text boxes and then click the appropriate option button. At this point the computation will be done and the result displayed in the label. Note that appropriate program statements will now have to be placed in the click event of each option button.

Figure 7–41 illustrates the screen after the user has entered the two numbers. No option button has been selected and no result computed.

Figure 7–42 illustrates the screen after the user has clicked the first option button (Add). As soon as the Add option was selected, the click event of the first option button was executed. This event caused the two numbers to be added and the result to be placed in the label. If the user now selects the

Figure 7–40
Op Code Problem (Visual Basic—Screen 7)

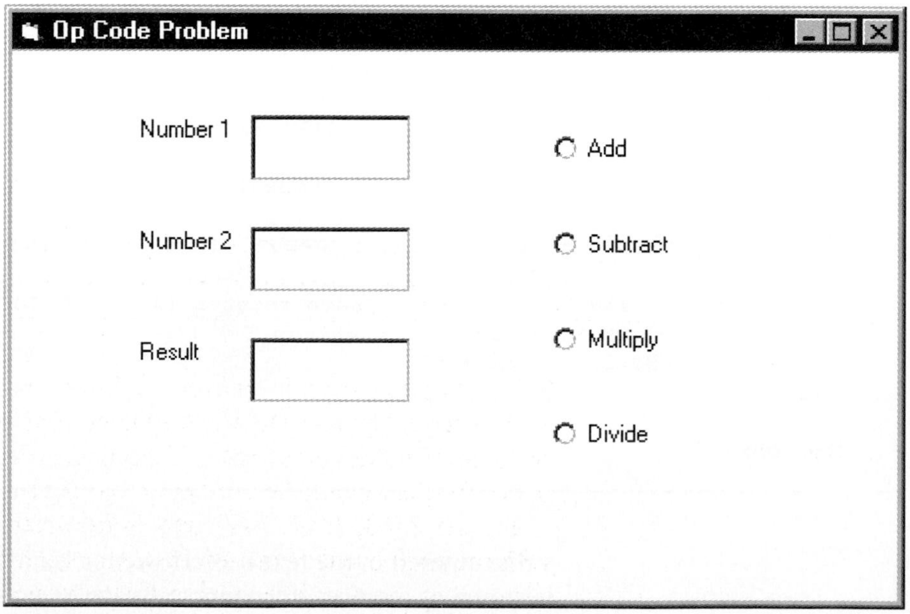

Figure 7–41
Op Code Problem (Visual Basic—Screen 8)

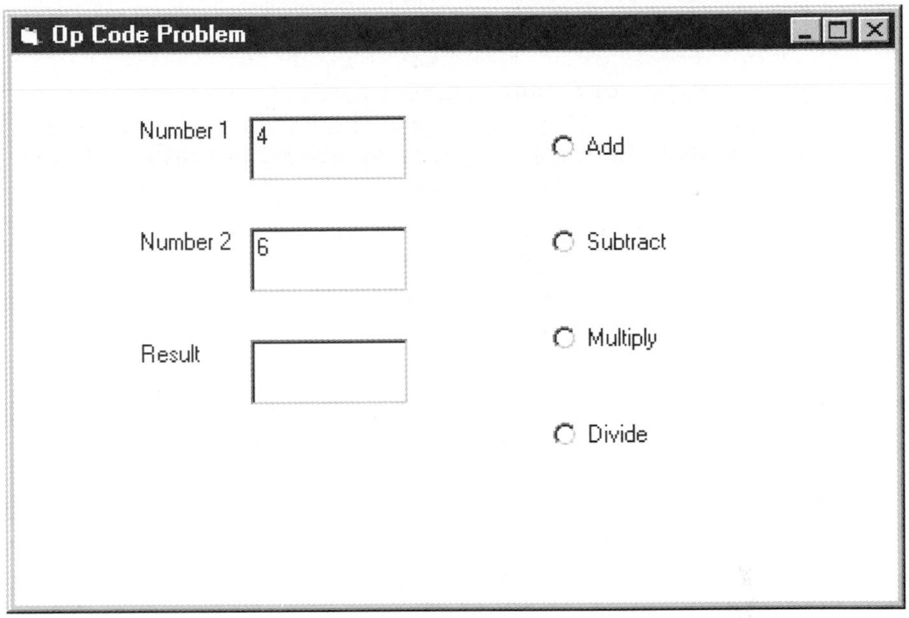

Figure 7–42
Op Code Problem (Visual Basic—Screen 9)

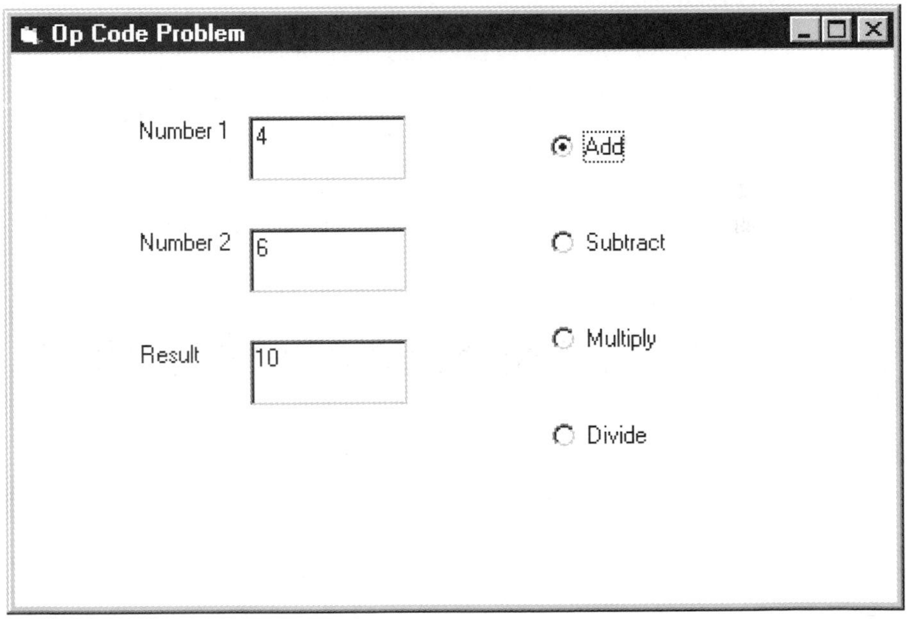

Subtract option, the label will immediately display a –2, since the click event of the second option button will execute. The code in this event subtracts the two numbers. Similar events will occur for Multiply and Divide. Remember, option buttons should be used only to represent mutually exclusive choices.

Figures 7–43, 7–44, 7–45, and 7–46 illustrate the programs associated with the click event of the option buttons. This example illustrates how we can change the interface to simplify the program. Notice that we did not need to use the SELECT CASE control structure in this solution. Do you see why?

Figure 7–43
Op Code Problem (Visual Basic—opt_ADD_Click)

```
Private Sub opt_ADD_Click()

lbl_RESULT.Caption = Val(txt_NUMBER1.Text) + Val(txt_NUMBER2.Text)

End Sub
```

Figure 7–44
Op Code Problem (Visual Basic—opt_SUBTRACT_Click)

```
Private Sub opt_SUBTRACT_Click()

lbl_RESULT.Caption = Val(txt_NUMBER1.Text) - Val(txt_NUMBER2.Text)

End Sub
```

Figure 7–45
Op Code Problem (Visual Basic—opt_MULTIPLY_Click)

```
Private Sub opt_MULTIPLY_Click()

lbl_RESULT.Caption = Val(txt_NUMBER1.Text) * Val(txt_NUMBER2.Text)

End Sub
```

Figure 7–46
Op Code Problem (Visual Basic—opt_DIVIDE_Click)

```
Private Sub opt_DIVIDE_Click()

lbl_RESULT.Caption = Val(txt_NUMBER1.Text) / Val(txt_NUMBER2.Text)

End Sub
```

Key Terms

master file
transaction (detail) file

CASE control structure
character string

Exercises

1. (a) Distinguish between master files and transaction files.
   (b) Describe in detail some examples of each.

2. Describe three types of flowcharting situations in which the use of connector symbols is necessary or advisable.

3. **(a)** Using ANSI-approved flowcharting symbols, sketch the logic of a CASE control structure.
   (b) State the logic of the CASE control structure in pseudocode form.
   (c) What pseudocode keywords did you use in your response to Exercise 3(b)?
   (d) What indentations did you use in your response to Exercise 3(b)? Why?

4. Does the order of the tests specified in a CASE control structure matter? If so, why?

5. **(a)** Using a modular design, redo Sample Problem 7.1 to process any number of input records. Print a report heading and column headings on every page, as well as a page number, with 55 detail lines per page. Assume the first record input will specify how many records are to follow. Construct a structure chart. Create a flowchart and pseudocode for each module in the design.
   (b) Use automatic end-of-file processing instead of a header record to solve Exercise 5(a).

6. Simulate the execution of the solution to Exercise 5(b) assuming the following input values:

OP	N1	N2
A	3	5
M	2	7
A	4	3
S	3	1
D	1 0	5
D	1 2	3
B	1	4
S	4	4
A	8	2

What will the computer provide as output?

7. Redo Sample Problem 7.4 to compute and output the total sales within each class, as well as the total records within each class. Also compute and output the total number of records processed.

8. Construct a structure chart, as well as a flowchart and pseudocode, for the following problem. Assume each input record contains a taxpayer's name, the value of a personal property belonging to the taxpayer, and a code defining the type of personal property owned. Each type of property is taxed at a unique rate. The codes, property types, and tax rates follow:

Code	Property Type	Tax Rate
1	Bike	2 percent of value
2	Car	4 percent of value
3	Truck	5 percent of value

Your program is to compute the tax for each property and to output a line specifying the taxpayer's name, value of property, and tax. The program should output counts of the numbers of bikes, cars, and trucks for which taxes are computed, with appropriate labels. Include a report heading and column headings, as well as a page number on every page, with 55 detail lines per page. Also include an initial IF and a modular design. Write an error message if the input contains an invalid code value. Assume a code of 0 indicates the end of the input file.

9. Simulate the execution of the solution to Exercise 8 assuming the following input values:

Name	Value	Code
J. GREENE	350.00	1
A. SMITH	9750.00	3
P. WOOLEY	10500.00	3
M. MANLEY	500.00	1
B. COURTNEY	15000.00	2
C. JONES	12225.00	2
L. BLACK	100.00	1
F. KINGSMAN	300.00	1
V. HENLEY	8250.00	4
T. MORROW	6000.00	3
END-OF-FILE	0.00	0

What will the computer provide as output?

10. Construct a structure chart, as well as a flowchart and pseudocode, for the following problem. The NVCC National Bank needs a program to compute the monthly balances in customers' checking accounts. Each customer input record contains customer name, account number, previous account balance, transaction amount, and a code specifying the type of transaction. The code can be interpreted as follows:

CODE	**TYPE OF TRANSACTION**
D	deposit
W	withdrawal

Any other value of CODE is invalid. Use the CASE control structure to check CODE, and output an error message if CODE is invalid. Compute the new monthly balance for each customer by adding the transaction amount to the previous account balance if the transaction is a deposit, or subtracting the transaction amount from the previous account balance if the transaction is a withdrawal. Output the customer name, account number, previous monthly balance, and new computed monthly balance. Include a report heading and column headings, as well as a page number, on every page, with 50 detail lines per page. Also include an initial IF and a modular design. A code of S will be used to signal the end of the input.

11. Construct a structure chart, as well as a flowchart and pseudocode, for the following problem, which is a modification of Exercise 10. The NVCC National Bank needs a program to compute the monthly balances in customers' checking accounts. Each customer input record contains customer name, account number, previous account balance, and number of transactions. The customer record is then followed by the transaction records for that customer. Each transaction record contains the amount of the actual transaction and a code specifying the type of transaction. The code can be interpreted as follows:

CODE	TYPE OF TRANSACTION
1	deposit
2	withdrawal

Any other value of CODE is invalid. Use the CASE control structure to check CODE, and output an error message if CODE is invalid. Compute the new monthly balance for each customer by adding the transaction amount to the previous account balance if the transaction is a deposit (CODE = 1), or subtracting the transaction amount from the previous account balance if the transaction is a withdrawal (CODE = 2). Complete this processing for each transaction. Finally, output the customer name, account number, previous monthly balance, and new computed monthly balance. If the number of transactions is less than or equal to 0, the new balance will be equivalent to the previous balance. Also, assume that the proper number of transaction records always follows the customer name record. Include a report heading and column headings, as well as a page number, on every page, with 50 detail lines per page. Also include an initial IF and a modular design. Use automatic end-of-file to signal the end of the input.

12. Construct a program flowchart and corresponding pseudocode to prepare automobile liability insurance estimates for customers. The input consists of a series of records that contain the name of the customer, the age of the customer, and a risk code. A code of 1 indicates a high-risk driver with recent moving violations. A code of 2 indicates a low-risk driver with no recent moving violations. If a record does not contain a code of 1 or a code of 2, the code is invalid, and an error message is to be printed on the report. The output is to consist of insurance estimates. The report is to contain the customer's name, age, risk code, a message identifying the customer as a high insurance risk or a low insurance risk, and the cost of the insurance coverage. If the customer is less than 25 years of age and has a risk code of 1 (high risk), the cost of insurance is $1000.00. If the customer is 25 years of age or more and has a risk code of 1 (high risk), the cost of the insurance is $800.00. If the customer is 25 years of age or more and has a risk code of 2 (low risk), the cost of insurance is $500.00. If the customer is less than 25 years of age and has a risk code of 2 (low risk), the cost of insurance is $650.00. If the risk code is invalid, the message "RISK CODE IS INVALID" should appear on the report. After all records have been processed, the total number of customers, the total number of low-risk

drivers, the total number of high-risk drivers, and the total number of invalid risk types should be printed. A sample format of the report follows:

```
                    INSURANCE REPORT
NAME                    AGE    RISK        RISK        INSURANCE
                               CODE        TYPE          COST
T.  L.  ELTON           42      1        HIGH RISK        800.00
C.  C.  FOX             18      2        LOW RISK         650.00
L.  R.  GUMM            26      2        LOW RISK         500.00
M.  M.  MATT            23      1        HIGH RISK       1000.00
C.  R.  NUMIS           21      RISK CODE IS INVALID
TOTAL CUSTOMERS         05
TOTAL LOW RISK          02
TOTAL HIGH RISK         02
TOTAL INVALID RISK TYPE    01
```

Include a report heading and column headings, as well as a page number, on every page, with 45 detail lines per page. Also include an initial IF and a modular design. Use automatic EOF to signal the end of the input and use the CASE control structure to check the risk code.

DOUNTIL Control Structure

8

Objectives

Upon completion of this chapter you should be able to

- Identify, and use in program design, the DOUNTIL control structure.
- Distinguish between the logic of a DOWHILE pattern and the logic of a DOUNTIL pattern.

Introduction

By now you have acquired some familiarity with the three basic patterns of structured programming: SIMPLE SEQUENCE, IFTHENELSE, and DOWHILE. You can express these patterns in flowchart and pseudcode forms.

In Chapter 7 we saw how a series of tests may be set up as a nested IFTHENELSE control structure, and we learned a shorthand notation to substitute in its place: the CASE control structure. The CASE structure more conveniently describes the decision-making logic used in a nested IFTHENELSE. Because the CASE structure consists entirely of basic structured-programming patterns, we can use it with confidence that we are maintaining structure in our solution algorithms. The computer-program representation of the algorithm, then, will be a well-structured program.

Another common combination of the basic patterns of structured programming is a SIMPLE SEQUENCE followed by a DOWHILE. This combination is shown in its general form in Figure 8–1. First, statement e is executed. Then we enter a DOWHILE loop. We test for the condition "not q" (indicated by q with a line over it, or $\overline{q}$). We execute statement e while condition $\overline{q}$ is true. Note that when condition q is false, we have not $\overline{q}$—which is the same as q. We exit the loop at that time. If you find this negative logic confusing, don't be discouraged—many system designers and programmers also find it confusing. Remember, negative logic is frequently found when a DOWHILE loop is used, because the loop steps can be executed only when the tested condition is true.

Because of the difficulties that may arise when using negative logic, many programmers prefer to use the **DOUNTIL control structure**, which is represented in Figure 8–2. This program logic can be summarized as follows: Do statement e until condition q is known to be true.

Figure 8–1
SIMPLE SEQUENCE and
DOWHILE Loop—Generic

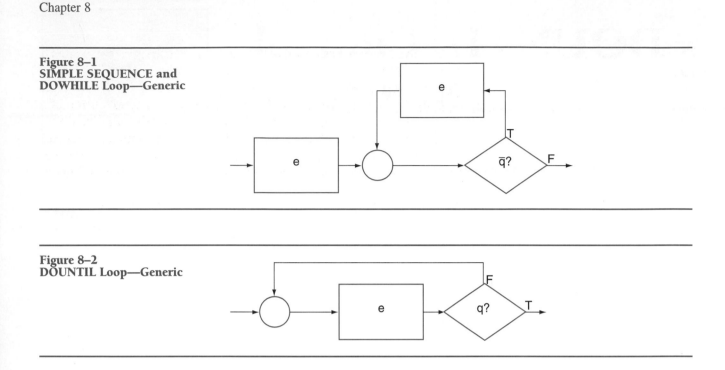

Figure 8–2
DOUNTIL Loop—Generic

To demonstrate the differences between DOWHILE logic and DOUNTIL logic, let's consider how they might apply to a real-life situation. Suppose, for example, that you are employed as a sandwich maker in a neighborhood delicatessen. Your employer may say: "Make one beef sandwich. Continue to make (do) beef sandwiches while you are not too tired to do so." This is the pattern of logic shown in Figure 8–1. Alternatively, your employer may say: "After you have made one beef sandwich, continue to make (do) beef sandwiches until you are too tired to do so." This is the pattern of logic shown in Figure 8–2.

What is the significant difference between the DOWHILE pattern that we have learned to use and this combination of SIMPLE SEQUENCE and DOWHILE known as DOUNTIL? We saw earlier that the DOWHILE pattern is a **leading-decision program loop**: The test for the loop-terminating condition is made immediately upon entering the loop. In contrast, the DOUNTIL pattern is a **trailing-decision program loop**: The test for the loop-terminating condition is not made until the other processing steps in the loop have been executed. This means that no matter what the outcome of the test within the loop, the processing steps that precede the test will always be performed at least once before the test is made. So before using DOUNTIL we must be sure that we want to perform the functions within the loop at least once. How many more times we perform the functions within a trailing-decision loop depends on the outcomes of any successive tests. The execution of the processing steps in the loop and subsequent condition testing continue until the tested condition is known to be true. Then the loop is exited.

Note that a DOWHILE control structure is exited when the tested condition is false, but a DOUNTIL is exited when the tested condition is true. It is essential to set up all loop constructs in a solution algorithm using either a DOWHILE or DOUNTIL control structure. Otherwise, the computer-program representation of the algorithm will not be a well-structured program.

DOUNTIL Counter Loops

In Chapter 4 we discussed numerous DOWHILE loops that were executed once, re-executed, or exited on the basis of the current value of a loop counter. (See the tests for COUNT < 6 in Figure 4–4, for COUNT < 10 in Figure 4–8, or any of the other examples in that chapter.) The execution of a DOUNTIL loop can be controlled in a similar manner. We initialize the loop counter to its starting value before entering the loop. With each execution of the loop, we increase (or decrease) the value of the counter; then we test its value to determine whether or not the processing steps within the loop should be re-executed.

The flowchart in Figure 8–3 shows how to use a DOUNTIL loop to read six data values, add them together in an accumulator, and write their sum. The corresponding pseudocode is shown in Figure 8–4.

Figure 8–3
Adding Six Numbers Using
DOUNTIL (Flowchart)

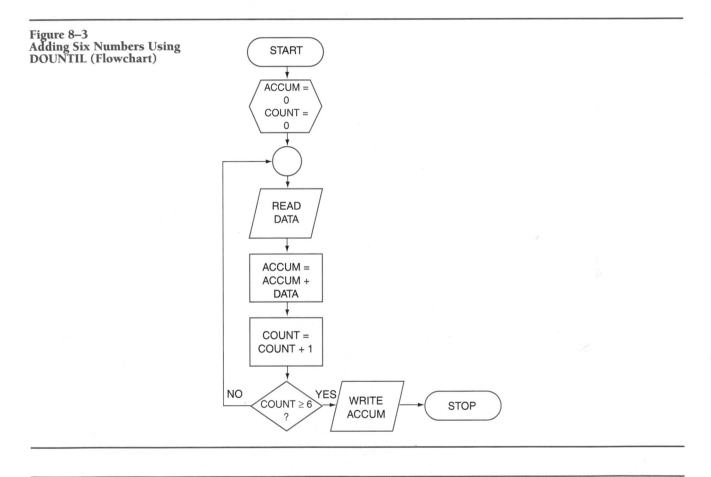

Figure 8–4
Adding Six Numbers Using
DOUNTIL (Pseudocode)

```
Start
ACCUM = 0
COUNT = 0
DOUNTIL COUNT ≥ 6
    Read DATA
    ACCUM = ACCUM + DATA
    COUNT = COUNT + 1
ENDDO
Write ACCUM
Stop
```

We achieved exactly the same result using a DOWHILE program loop in Figure 4–4. Compare the tests for loop termination in these algorithms. As mentioned earlier, the DOWHILE program loop is exited when the tested condition is false; the DOUNTIL program loop is exited when the tested condition is true. It is for this reason that the test in Figure 8–3 (COUNT ≥ 6) is the negation of the test in Figure 4–4 (COUNT < 6). Look at the test in Figure 8–3 again. Do you think it could be made even simpler?

Now look at the pseudocode in Figure 8–4. It may appear that the condition (COUNT ≥ 6) is being tested before the loop steps are executed. However, this is not the case. In pseudocode, the tested condition is placed next to the keyword DOUNTIL, which precedes the loop steps. It's important to understand that even though the test is written before the loop steps, it is not actually made until the loop steps have been executed one time.

Sample Problem 8.1 (Property— Counter- Controlled)

Problem:

With computer help, a table is to be generated that shows the storage costs for personal property at a large warehouse. The table covers property values from $1000 through $20,000, in increments of $100. Storage costs are computed at 5 percent of property values. Each table entry is to consist of a property value and a corresponding storage cost, as follows:

$1000	$50
$1100	$55
$1200	$60
.	.
.	.
.	.

Solution:

A program flowchart of an algorithm to solve this problem is given in Figure 8–5. It contains a DOUNTIL pattern and numerous examples of SIMPLE SEQUENCE. The same algorithm is shown in pseudocode form in Figure 8–6.

Here, the loop-control variable is not a specially introduced data item. Instead, it is one of the data items discussed in the problem statement: the personal property value (PROP). We initialize PROP to its starting value, $1000, before we enter the DOUNTIL loop. Upon entering the loop, we use this value immediately in computing a storage cost and print a table entry accordingly. Next, we increase the value of PROP by $100. Finally, we test the value of PROP to see whether it exceeds $20,000. The loop is re-executed if the tested condition is false; it is exited if the tested condition is true. Note that one table entry has been computed and printed before the value of the loop-control variable is tested.

How do we know to set PROP to $1000 initially, increase it by $100, and test for an upper limit of $20,000? We derive these steps from the requirements of the problem statement. Because the problem statement

**Figure 8–5
Property—Counter-
Controlled (Flowchart)**

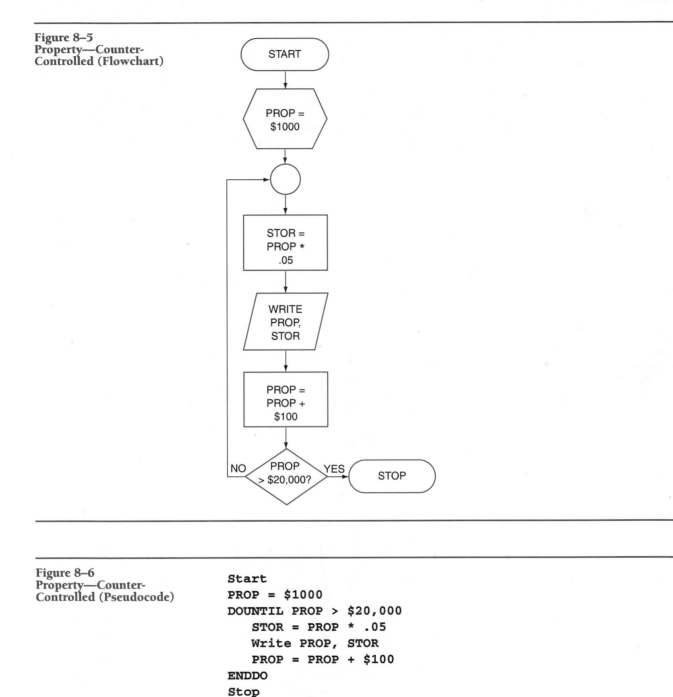

**Figure 8–6
Property—Counter-
Controlled (Pseudocode)**

```
Start
PROP = $1000
DOUNTIL PROP > $20,000
   STOR = PROP * .05
   Write PROP, STOR
   PROP = PROP + $100
ENDDO
Stop
```

specified exactly which property values to process, we did not have to read any property values as input. In this example the loop steps actually generated the property values that needed to be processed. What would have happened if we had tested for PROP = $20,000, or PROP ≥ $20,000, instead of PROP > $20,000? The loop would not have been re-executed when PROP had a value of $20,000. No table entry would have been created to show the storage cost for personal properties valued at $20,000, so the requirements of the problem statement would not have been satisfied.

Sample Problem 8.2 (Property— Header Record Logic)

Problem:

Redo the solution to Sample Problem 8.1 to accept any number of property values as input. Assume that the first input record will contain a number indicating how many property values are to be processed. Succeeding records (if any) will contain property values. Include report and column headings, as well as page numbers on every page of the output, with 55 detail lines per page. Construct a flowchart and pseudocode for this revised algorithm.

Solution:

This solution has been designed using four modules, as indicated in the structure chart in Figure 8–7. The flowchart and pseudocode for A000 are shown in Figures 8–8 and 8–9.

In this algorithm we are using header record logic to determine when the end-of-file has been reached. The main processing loop is a DOUNTIL loop. First, B000 is executed to initialize MAXLINES, the line count, and the page count. Then the header record is input, and the variable represented by N is tested to make sure it is positive—that is, to determine if additional records follow. If N proves to be greater than 0, the main processing loop is entered. Because we are using a DOUNTIL loop, the connector is followed by the actual processing steps within the loop. After one record containing a property value is processed in B010, the loop-control variable (N) is decremented by 1 and the DOUNTIL loop test is made. Remember, this test must be the last step in a DOUNTIL control structure. In terms of the logic used, a DOUNTIL test is more straightforward than typical DOWHILE tests. The DOUNTIL loop requirement specifies that the loop will be executed when the tested condition is false and exited when the tested condition is true. In most cases, negative logic is not necessary. We would normally test for $N > 0$ if the DOWHILE loop was used. Because we are using the DOUNTIL loop, we test for the opposite condition, $N \leq 0$. Because N must be a positive number at this point, it will get increasingly smaller and will reach 0 before it can ever become negative. Thus, we can simplify the loop test

Figure 8–7
Property—Header Record Logic (Structure Chart)

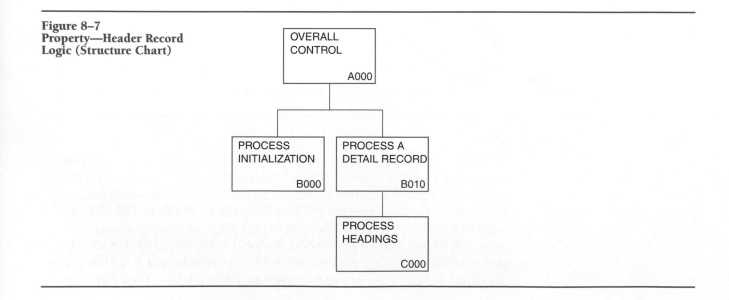

Figure 8–8
Property—Header Record
Logic—Overall Control
(Flowchart)

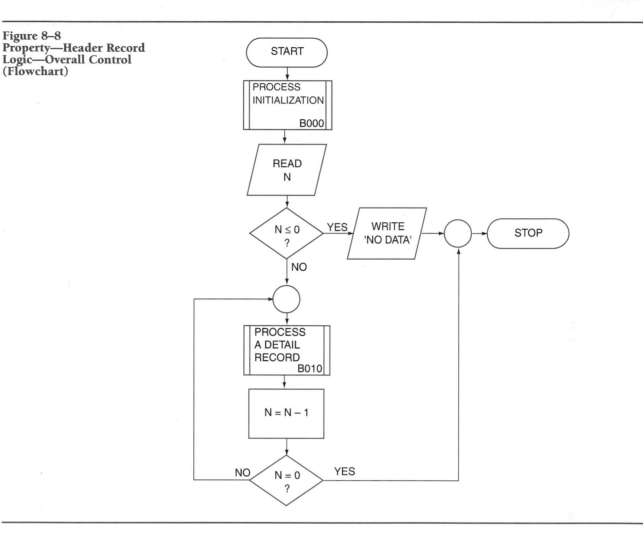

Figure 8–9
Property—Header Record
Logic—Overall Control
(Pseudocode)

```
A000
Start
Process initialization (B000)
Read N
IF N ≤ 0 THEN
   Write 'No data'
ELSE
   DOUNTIL N = 0
      Process a detail record (B010)
      N = N - 1
   ENDDO
ENDIF
Stop
```

further and change the $N \leq 0$ test to an $N = 0$ test. Consequently, the wording within the DOUNTIL test is less confusing.

The initialization module (B000) is shown in Figure 8–10. The detail-processing module (B010) is shown in Figure 8–11. In B010, LINECNT is first checked to determine if a new page needs to be started and headings output. An input record containing one property value is then read; the corresponding storage cost is computed; a detail line is written; and LINECNT is incremented. Notice that several steps in the original algorithm are omitted in this one. PROP is not initialized to $1000; PROP is not increased by $100; and PROP is not tested against $20,000. Do you see why these steps are no longer necessary?

Figure 8–10
Property—Header Record
Logic—Process Initialization

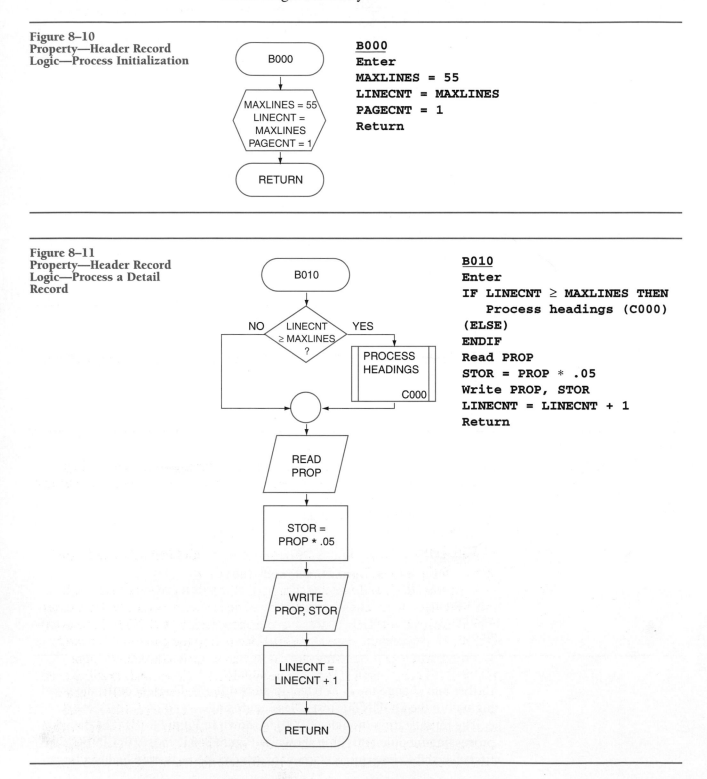

```
B000
Enter
MAXLINES = 55
LINECNT = MAXLINES
PAGECNT = 1
Return
```

Figure 8–11
Property—Header Record
Logic—Process a Detail
Record

```
B010
Enter
IF LINECNT ≥ MAXLINES THEN
    Process headings (C000)
(ELSE)
ENDIF
Read PROP
STOR = PROP * .05
Write PROP, STOR
LINECNT = LINECNT + 1
Return
```

**Figure 8–12
Property—Header Record
Logic—Process Headings**

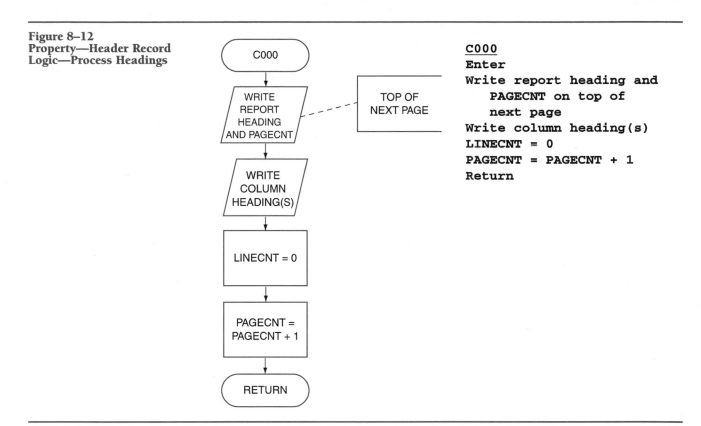

```
C000
Enter
Write report heading and
    PAGECNT on top of
    next page
Write column heading(s)
LINECNT = 0
PAGECNT = PAGECNT + 1
Return
```

The headings module (C000) is shown in Figure 8–12. This logic should look familiar to you.

Sample Problem 8.3 (Property— Trailer Record Logic)

Problem:

Redo the solution to Sample Problem 8.2 to illustrate trailer record logic instead of header record logic. A property value of 0 will serve as the trailer record, denoting end-of-file.

Solution:

A program flowchart and pseudocode for the overall control module are shown in Figure 8–13 and Figure 8–14.

The first READ statement (priming read) reads a property value as input. The value is checked immediately to determine whether or not the end-of-file has been reached. If not, the main processing DOUNTIL loop is executed. Notice that the loop steps are done before the condition is tested. The order of these processing steps is the same for both types of loops (DOWHILE and DOUNTIL). The second READ (loop read) is still placed after the predefined-process symbol that references module B010. Again, the important differences between these two loops is the position of the test. In a DOWHILE loop, the test is always placed first, that is, right after the connector. In a DOUNTIL loop, the test is always placed last. The detail-processing module for this algorithm is shown in Figure 8–15. Can

Figure 8–13
Property—Trailer Record
Logic—Overall Control
(Flowchart)

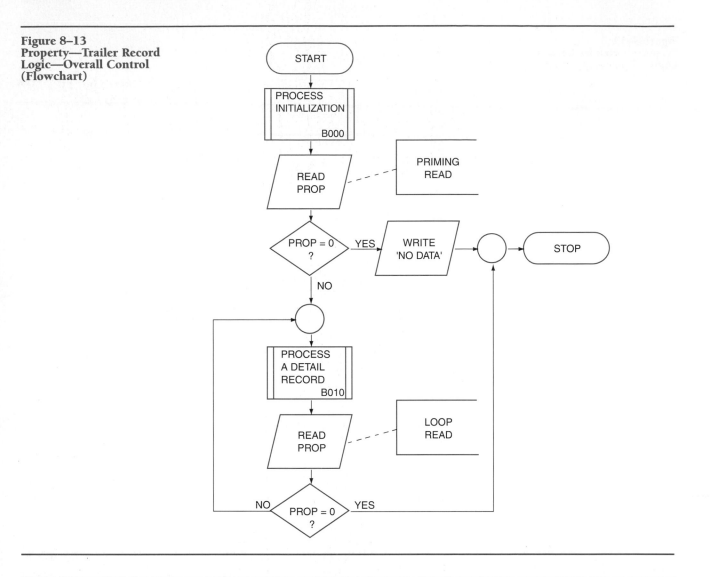

Figure 8–14
Property—Trailer Record
Logic—Overall Control
(Pseudocode)

```
A000
Start
Process initialization (B000)
Read PROP
IF PROP = 0 THEN
    Write 'No data'
ELSE
    DOUNTIL PROP = 0
        Process a detail record (B010)
        Read PROP
    ENDDO
ENDIF
Stop
```

you see why a READ statement is not needed in this module? Why was a READ statement placed in the detail-processing module in Figure 8–11? Modules B000 and C000 contain the same processing steps as they did in Sample Problem 8.2.

**Figure 8–15
Property—Trailer Record
Logic—Process a Detail
Record**

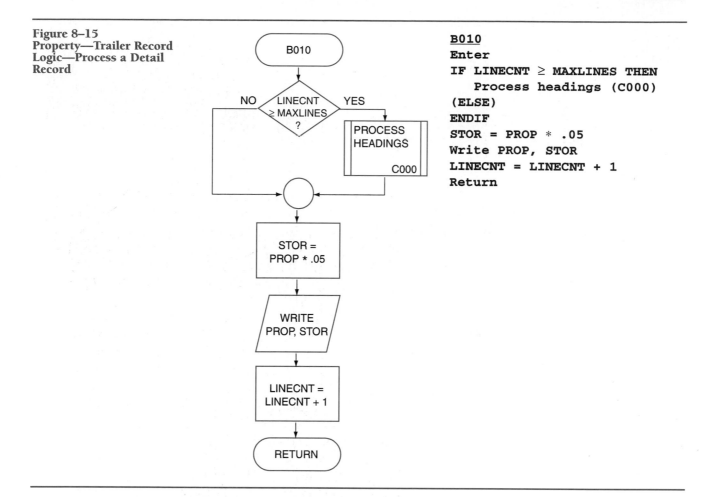

```
B010
Enter
IF LINECNT ≥ MAXLINES THEN
     Process headings (C000)
(ELSE)
ENDIF
STOR = PROP * .05
Write PROP, STOR
LINECNT = LINECNT + 1
Return
```

DOWHILE Versus DOUNTIL

Before we leave this topic, it is important to discuss some very subtle differences between DOWHILE and DOUNTIL. Look at Figure 8–16. Two flowchart and pseudocode representations are shown. Each shows one of the loop types (DOWHILE or DOUNTIL) and how it is used in conjunction with the typical logic needed when automatic end-of-file processing is used. Both algorithms appear to work; however, only one is correct in all situations. Consider what would happen if end-of-file was reached the first time a READ was attempted. If the DOWHILE loop is used, the NOT EOF condition will be false the first time the test is made and the loop steps will never be executed. If the DOUNTIL loop is used, the module will be executed before the EOF test is even made. This attempt to process a normal input record will probably cause an error. If the module does successfully complete its own processing, a second read will be attempted after control is returned to the overall control module. This attempt to read another record will most definitely generate an error condition, because we cannot read another record when end-of-file has already been reached.

In most of our examples, we have placed an IFTHENELSE statement just prior to the main processing loop to test for this condition—that is, to detect when EOF is reached on the first READ attempt. Figures 8–17 and

Figure 8–16
DOWHILE vs. DOUNTIL—
No Initial Test

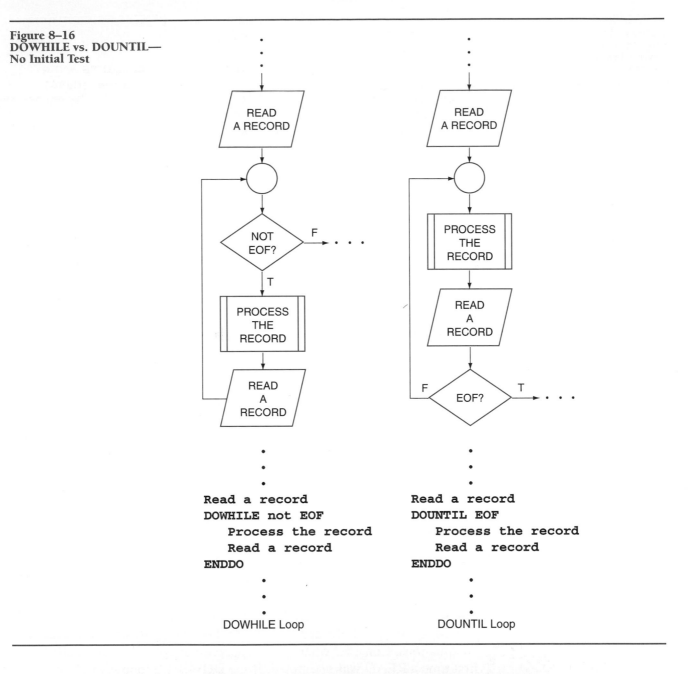

```
Read a record
DOWHILE not EOF
    Process the record
    Read a record
ENDDO
         .
         .
         .
    DOWHILE Loop
```

```
Read a record
DOUNTIL EOF
    Process the record
    Read a record
ENDDO
         .
         .
         .
    DOUNTIL Loop
```

8–18 show the two flowchart and pseudocode representations again, with the inclusion of this IFTHENELSE statement.

Although this extra step is not necessary when a DOWHILE loop is used, it does provide a convenient way to output a message indicating that no data was found. It is very important, however, to see that this additional IFTHENELSE statement is not merely helpful when it precedes a DOUNTIL loop: It is a requirement. Now if EOF is reached on the first READ attempt, the true path of the IFTHENELSE will be executed, and the loop steps will be bypassed entirely (as they should be). Make sure that you understand the limitations as well as the conveniences of using a DOUNTIL loop.

Figure 8–17
DOWHILE vs. DOUNTIL—
Initial Test (Flowchart)

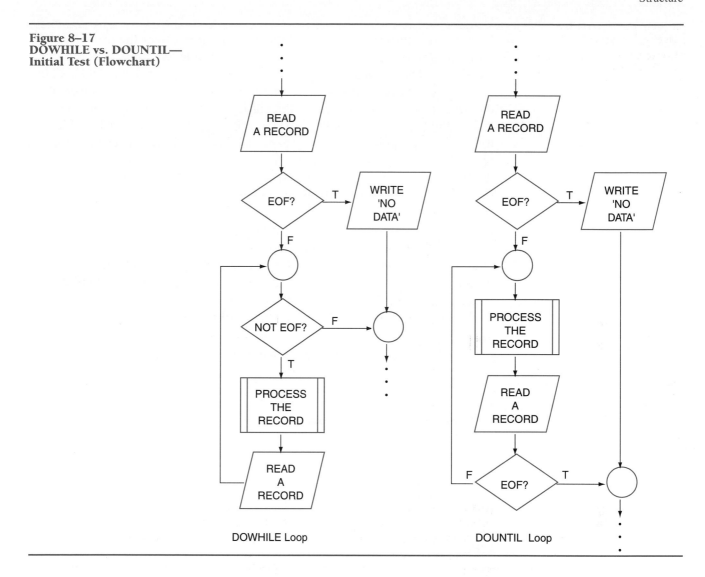

DOWHILE Loop   DOUNTIL Loop

Figure 8–18
DOWHILE vs. DOUNTIL—
Initial Test (Pseudocode)

```
.
.
.
Read a record
IF EOF THEN
    Write 'No data'
ELSE
    DOWHILE not EOF
        Process the record
        Read a record
    ENDDO
ENDIF
.
.
.
```

```
.
.
.
Read a record
IF EOF THEN
    Write 'No data'
ELSE
    DOUNTIL EOF
        Process the record
        Read a record
    ENDDO
ENDIF
.
.
.
```

DOWHILE Loop   DOUNTIL Loop

Enrichment (Basic)

Figure 8–19 illustrates a listing of the program that solves the property problem (see Figures 8–5 and 8–6). In this example we implement the DOUNTIL loop with a construct in Basic that is very similar to DOUNTIL pseudocode. Notice, however, that the loop test is placed at the bottom of the loop in Basic. This improves the clarity of the program by more clearly reflecting the logic of a DOUNTIL loop. Remember that the DOUNTIL loop is a trailing-decision loop. Again, the keyword LOOP is used in Basic instead of the keyword ENDDO. The same type of indentation is used in the Basic program, and the logic of the Basic DOUNTIL loop is identical to the logic of the DOUNTIL loop in pseudocode. In addition, most of the statements in the Basic program parallel the pseudocode. We did not need to include either a Read statement or an Input statement since the property values are generated by the loop steps. The Print statements are used to output the headings prior to entering the loop. Two Print statements are used to print the headings on separate lines. A final property value of 2500 is used instead of 20,000 to limit the output to one page.

Figure 8–20 illustrates the output that will be produced when the program is executed. Note that there is one line of output (a property value and a storage cost) for every property value that is processed.

Figure 8–19
Property Problem (Basic List)

```
PROP = 1000
PRINT "Property", "Storage"
PRINT "Value", "Cost"
PRINT
DO
        STOR = PROP * .05
        PRINT PROP, STOR
        PROP = PROP + 100
LOOP UNTIL PROP > 2500
END
```

Figure 8–20
Property Problem (Basic Run)

Property Value	Storage Cost
1000	50
1100	55
1200	60
1300	65
1400	70
1500	75
1600	80
1700	85
1800	90
1900	95
2000	100
2100	105
2200	110
2300	115
2400	120
2500	125

Enrichment (Visual Basic)

Figure 8–21 illustrates the graphical interface for the property problem (see Figures 8–5 and 8–6). In this example, only two command buttons are created. When the user clicks the End button, the program execution stops. When the user clicks the Begin button, an input box is presented, as illustrated in Figure 8–22. We have modified this example to allow the user to input the property values. Remember, the input box is another way (in addition to the text box control) to accept input from the user. At this point the user should enter a property value, as shown in Figure 8–23, and click the OK button. Once the user enters a property value and clicks OK, a message box displaying the property value and corresponding storage cost is presented, as shown in Figure 8–24.

When the user clicks the OK button shown in Figure 8–24, another message box is presented as shown in Figure 8–25. This message box displays a message asking if the user wishes to enter another property value. The user can choose Yes or No at this point. Notice that this message box contains a Yes button and a No button, not an OK button. The contents of a message box, as well as what types of buttons are displayed, are specified by the programmer. Note that we have modified this example to allow the user to determine how many property values will be processed. (Remember: The original algorithm actually generated the property values and stopped at the first property value greater than 20,000.)

Figure 8–21
Property Problem (Visual Basic—Screen 1)

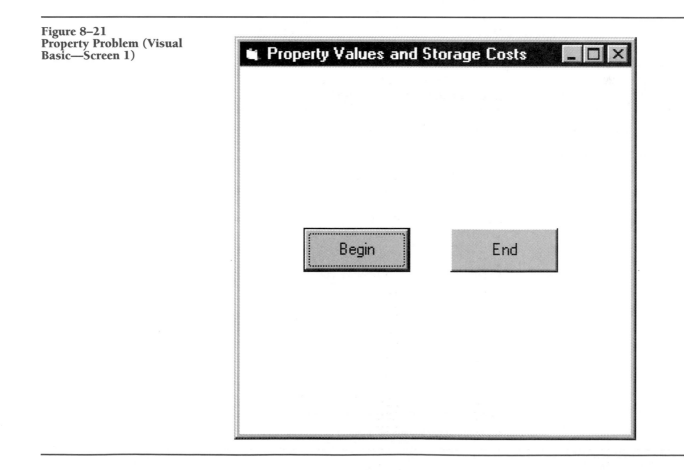

Figure 8–22
Property Problem (Visual Basic—Screen 2)

Figure 8–23
Property Problem (Visual Basic—Screen 3)

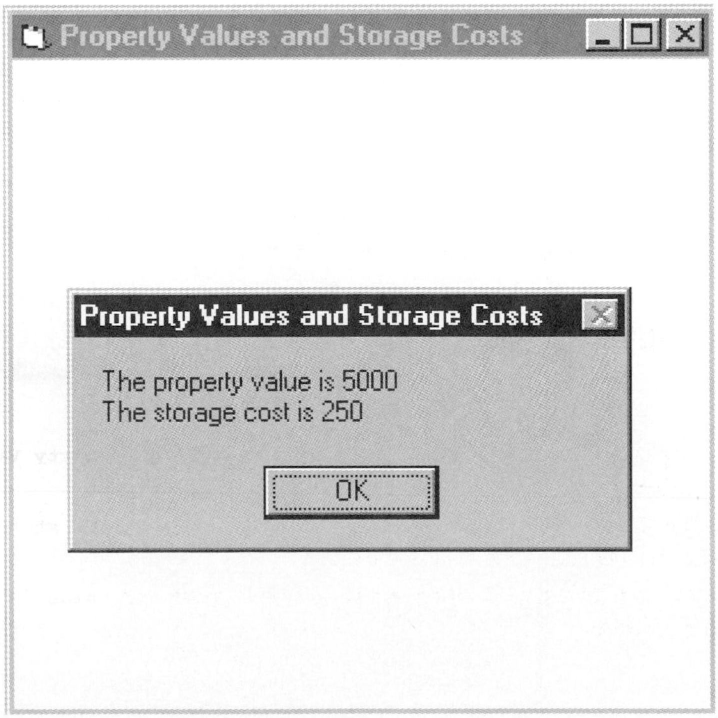

Figure 8–24
Property Problem (Visual Basic—Screen 4)

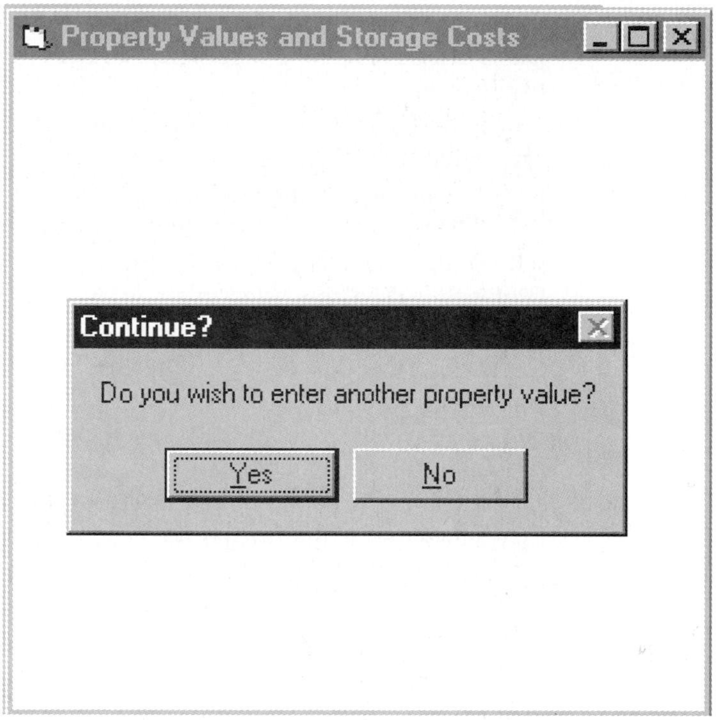

Figure 8–25
Property Problem (Visual
Basic—Screen 5)

Figure 8–26 illustrates the program that is associated with the click event of the Begin button. In this example a DOUNTIL loop is used to control the processing of each property value. We implement the DOUNTIL loop with a construct in Visual Basic that is very similar to DOUNTIL pseudocode. Notice, however, that the loop test is placed at the bottom of the loop in Visual Basic. This improves the clarity of the program by more clearly reflecting the logic of a DOUNTIL loop. Remember that the DOUNTIL loop is a trailing-decision loop. Again, the keyword LOOP is

Figure 8–26
Property Problem (Visual
Basic—cmd_BEGIN_Click)

```
Private Sub cmd_BEGIN_Click()


Do

        Property = InputBox$("Please input a property value", "Property Value Input Box")

        Storage = Val(Property) * 0.05

        MsgBox "The property value is " & Property & Chr$(13) & "The storage cost is " & Storage,,_
               "Property Values and Storage Costs"

        Response = MsgBox("Do you wish to enter another property value?",vbYesNo, "Continue?")

Loop Until Response = vbNo


End Sub
```

used in Visual Basic instead of the keyword ENDDO. The same type of indentation is used in the Visual Basic program, and the logic of the Visual Basic DOUNTIL loop is identical to the logic of the DOUNTIL loop in pseudocode.

The first statement in the loop causes an input box that requests a property value from the user to be displayed. Remember, the programmer determines both the message to be displayed and the title of the input box window. When a user enters a value into the input box and clicks the OK button, that value is assigned to the variable named Property. The value of Property is then converted to a numeric value by the Val function (Input-Box$ returns a string), multiplied by 0.05, and assigned to the variable named Storage. If the user clicks the Cancel button instead of the OK button, a null string (" ") will be assigned to Property. The null string will be evaluated to 0 by the Val function, resulting in a computed storage cost of 0. Although this situation will not cause an error to occur, the output will be worthless. A similar situation may occur if the user enters a non-numeric property value. Visual Basic includes additional functions and capabilities that can aid a programmer in ensuring good data is provided as input. These functions are included in any book that describes all Visual Basic capabilities, but are beyond the scope of this book.

Once the storage cost is computed, a message box displaying the property value and corresponding storage cost is presented. The message box displays the output as one long string. The string is composed of several parts—identifying labels, the variable names representing the property value and storage cost, and a special function, *Chr$(13)*. The concatenation operator (&) is used to join the parts of the string. Chr$(13) is a special function that converts the internal code of 13 to a character. A 13 represents a carriage return and is used to output the string on two separate lines. The underline character (_) is not part of the string, but rather a Visual Basic code to indicate continuation. In this case, two lines were needed to represent the message box statement in the code.

After the property value and storage cost are output, another message box is presented to the user. This time a message box function is being used. The message box function will return a value and store it in the variable named Response. This value indicates which button the user clicked—Yes or No. A special variable—*vbYesNo*—is included in the message box function to cause the buttons Yes and No to be displayed in the message box instead of the default button (the OK button). After the user clicks one of the buttons (Yes or No), the DOUNTIL loop test is made. This test determines whether or not the user wishes to enter another property value. Another special variable—*vbNo*—represents a No answer and is tested against the value of the variable Response. If the user has chosen No, the loop test will evaluate to true and the loop will be exited. If the user has chosen Yes, the loop test will evaluate to false and the loop steps will be re-executed. Note that the programmer does not need to define the special variables vbYesNo and vbNo. These variables are predefined within the Visual Basic programming system.

Figure 8–27 illustrates the program that is associated with the click event of the End button. The single statement End is used to stop program execution.

**Figure 8–27
Property Problem (Visual
Basic—cmd_END_Click)**

```
Private Sub cmd_END_Click()

End

End Sub
```

Key Terms

DOUNTIL control structure	leading-decision program loop	trailing-decision program loop

---

Exercises

1. (a) What kind of program loop is formed in a DOWHILE pattern?
   (b) What kind of program loop is formed in a DOUNTIL pattern?
   (c) When is the difference between DOWHILE and DOUNTIL particularly important? Why?

2. (a) If condition q is "X is less than or equal to Y," what is condition $\bar{q}$?
   (b) If condition q is "$A + B < C$," what is condition $\bar{q}$?
   (c) If condition q is "C is greater than or equal to $A + B$," what is condition $\bar{q}$?
   (d) If condition q is "$X - Y > A + B$," what is condition $\bar{q}$?
   (e) If condition q is "$X * Y = Z$," what is condition $\bar{q}$?
   (f) If condition q is "Z not equal to A / B," what is condition $\bar{q}$?

For the remainder of the exercises, include report and column headings, as well as a page number on each page, with 55 detail lines per page unless directed otherwise. Output an appropriate message if the input contains no records; include descriptive messages in the total lines; and use a modular design. Construct flowcharts and pseudocode for each module.

3. Modify the program flowchart and pseudocode shown in Figures 8–3 and 8–4 to make a more general-purpose algorithm. The revised algorithm should describe how to read and add *any number* of data values (not just six). A count of the number of values added should be printed along with the sum of the values. Try this with both the header record logic approach and the trailer record logic approach. Report and column headings and page numbers are not required.

4. Redo Sample Problems 8.2 and 8.3 to output the following totals:

   ■ Total number of records processed

   ■ Total property value of all records processed

   ■ Total storage cost for all property records processed

5. (a) Construct an algorithm to solve the following problem: One data item, MULT, is to be read as input. The sum of the following operations is to be computed and printed: 1, 1 + 1 * MULT, 1 + 2 * MULT, . . . , 1 + 9 * MULT. Assume that only one value of MULT will be

input; that is, there will be one and only one input record. Headings, page numbers, the no-data test, and modules are not required.

(b) Perform a procedure execution (trace) of the solution algorithm that you constructed in your response to Exercise 5(a) to answer the following questions:

- What value is provided as output if MULT is 5?
- What value is provided as output if MULT is 10?
- What value is provided as output if MULT is 0?

(c) Redo Exercise 5(a) to process any number of input values, that is, one or more values of MULT. Use both the header record logic approach and the trailer record logic approach.

6. Construct an algorithm to show the processing steps needed to solve the following problem. Data values to be added are provided as input. The number of items to be added will be indicated by the first value provided as input (which is not to be included in the sum). A count of the number of data items whose value exceeds 30,000 is also to be accumulated. The sum of the values added and the count of those exceeding 30,000 are to be provided as output. Headings and page numbers are not required.

7. Simulate the exception of your solution to Exercise 6, assuming the following input values:

10	Header Value
2000	
10000	
45000	
31000	Other
30000	data
30001	values
29000	
5000	
50000	
1000	

What values will the computer provide as output?

8. (a) Construct an algorithm to show the processing steps needed to solve the following problem: Two data items, LOWER and UPPER, are to be read as input. The sum of all odd numbers from LOWER to UPPER, inclusive, is to be computed and printed. Assume that LOWER is less than UPPER and that only one input record exists. Headings, page numbers, the no-data test, and modules are not required.

(b) Redo Exercise 8(a) to process any number of input values; that is, many records containing LOWER and UPPER limits may be input. Use both the header record logic approach and the trailer record logic approach.

9. To answer the following questions, perform a procedure execution (trace) of the solution algorithm that you constructed in response to Exercise 8(a):
    (a) What value is provided as output by the algorithm if LOWER is 3 and UPPER is 18?
    (b) Is the output of the algorithm correct? If the output is not correct examine both your solution algorithm and your procedure execution to determine where errors have occurred.

10. Repeat Exercise 9, but assume that LOWER is 3 and UPPER is 4.

11. Assume that you are given a file of records, each containing two fields, NAME and QUANTITY. The last record is a trailer record containing DUMMY in the name field and 999 in the QUANTITY field. Two algorithms follow. (At least one is incorrect—possibly both are.)
    (a) Will algorithm A work?
    (b) Will algorithm B work?
    (c) If your answer to (a), (b), or both was no, explain why.
    (d) If either algorithm A or algorithm B is incorrect, write the pseudocode for either or both, as the case may be, to correct.

```
A.  Start                              B.  Start
      ACCUM = 0                              ACCUM = 0
      DOWHILE NAME not = 'DUMMY'            DOUNTIL NAME = 'DUMMY'
        Read NAME, QUANTITY                    Read NAME, QUANTITY
        ACCUM = ACCUM + QUANTITY               ACCUM = ACCUM + QUANTITY
      ENDDO                                  ENDDO
      Write ACCUM                            Write ACCUM
      Stop                                   Stop
```

12. Redo Exercise 12 from Chapter 6 using DOUNTIL loops instead of DOWHILE loops in your solution. You will need to add some additional tests in this solution.

Introduction to Arrays

Objectives

Upon completion of this chapter you should be able to

- Distinguish between a simple variable and a subscripted variable.
- Input, output, and manipulate values stored in a list, or one-dimensional array.
- Input, output, and manipulate values stored in a table, or two-dimensional array.
- Distinguish between row-major ordering and column-major ordering and explain how each ordering affects the efficiency of processing.

Introduction

In preceding chapters we have directed our attention to the reading, processing, and writing of single values. Each value has been stored in a particular location and referred to as a **single** (or **simple**) **variable**. For each such variable, we selected and used a particular variable name (for example, COUNT, N, A, ITEM, NUM). This need not always be the case.

Suppose a list of 10 input values is to be read into 10 consecutive storage locations. We could assign a unique name to each of the 10 locations—say, INPUT1, INPUT2, and so on. We could also use the names to refer to the values throughout the program.

But what if the list contained 100 or even 1,000 values? The same approach might work, but it would not be convenient. Writing the program would be a tedious, time-consuming chore. With so many different values and corresponding variable names to keep track of, errors would be apt to occur.

List Structures

Suppose that, instead of treating the 10 input values as 10 similar but separate data items, we treat them as a group of data items. We reserve a storage area large enough for all the values and then assign a name to the area. Various terms are used to describe data items stored and identified in this way. In COBOL this kind of data group is usually known as a **single-level table**. In Basic, C++, and Java it may be called a **list**, **vector**, or **one-dimensional array**.

Figure 9–1
One-Dimensional Array
Example

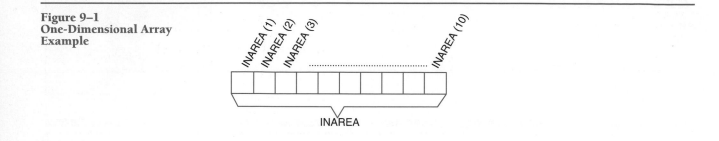

Only the group of data items is given a name. An individual item in the group is referred to by stating its relative position (on a left-to-right basis). This position is specified by means of a **subscript** or **index** following the group name.[1]

As an example, then, let us assume that a storage area is reserved for the 10 input values mentioned above, and that the name INAREA is chosen for the area (see Figure 9–1). Use of the **unsubscripted variable name** INAREA, if permitted in the programming language, is interpreted as a reference to all the items in the group. Individual items in the group are referred to by means of the **subscripted variable names** INAREA(1), INAREA(2), and so on.

When arrays are used in a program, a storage area must be reserved to hold the values of the array. Although the assignment of storage differs in each programming language, there is usually some sort of statement at the beginning of the program that defines both the name of the array and the maximum size of the array. Many language use a *dimension statement (DIM statement)* that includes the array name followed, in parentheses, by the maximum size of the array. For example, the programmer would need to include the statement DIM INAREA(10) in a program that processes an array with 10 storage locations. The DIM statement would reserve 10 storage locations that could be used to hold the data in the array INAREA. The DIM statement, or its equivalent, is not an executable statement as are the other statements in an algorithm. Thus, it would be inappropriate to include this statement as a step within the program flowchart or pseudocode representation of an algorithm. You just need to be aware that this type of statement exists and will be needed in the programming language representation of an algorithm.

Grouping capabilities are available in numerous programming languages. Because it is to our advantage to use them, we will examine the program logic needed to effectively manage group data items in this chapter.

List Examples

Figures 9–2, 9–3, and 9–4 illustrate three common procedures used in the processing of one-dimensional arrays. Figure 9–2 shows the processing steps required to initialize an array. Figure 9–3 shows the steps to input values into an array. Figure 9–4 shows the steps to output values from an

---

[1]In COBOL a distinction exists between a *subscript*, which is an integer value (occurrence number) that may range from 1 to the total number of entries in a table, and an *index*, which is a storage displacement value from the beginning of the table. In other programming languages, the terms *subscript* and *index* are used interchangeably with respect to the identification of data items in a group. In this discussion the term *subscript* can be understood to mean either.

Figure 9–2
Initialize One-Dimensional Array

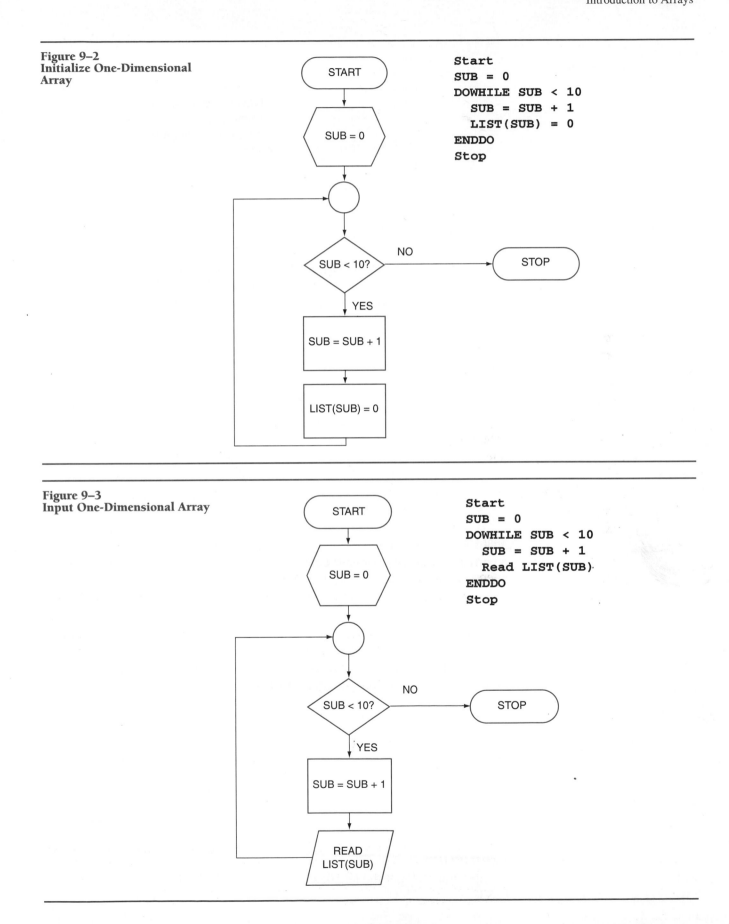

```
Start
SUB = 0
DOWHILE SUB < 10
  SUB = SUB + 1
  LIST(SUB) = 0
ENDDO
Stop
```

Figure 9–3
Input One-Dimensional Array

```
Start
SUB = 0
DOWHILE SUB < 10
  SUB = SUB + 1
  Read LIST(SUB)·
ENDDO
Stop
```

Figure 9–4
Output One-Dimensional
Array

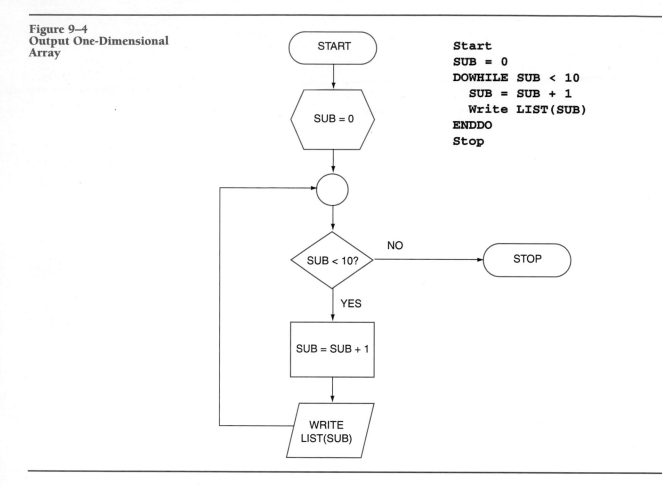

```
Start
SUB = 0
DOWHILE SUB < 10
   SUB = SUB + 1
   Write LIST(SUB)
ENDDO
Stop
```

array. In all three examples the variable SUB is used as a subscript to point to a particular **member** (or **element**) of an array called LIST. When SUB represents a value of 1, the subscripted variable LIST(1) refers to the first element of LIST; when SUB represents a value of 2, the subscripted variable LIST(2) points to the second element; and so on. A DOWHILE loop is executed repetitively, with the value represented by SUB increased by 1 on each execution. The loop is exited when the value represented by SUB is equal to 10. Notice that all three algorithms contain the same processing steps to control the loop. The subscript SUB plays two roles: It is used both as a counter to control the DOWHILE loop and as a pointer into the array LIST. Note also that the READ statement in Figure 9–3 indicates the input of one data value each time through the loop. Think of that READ step as "Read a new input value into the current position of LIST." The WRITE statement in Figure 9–4 can be understood in a similar manner. Think of the WRITE step as "Write out the value stored in the current position of the array LIST."

Sample Problem 9.1 (Finding the Smallest Number)

Problem:

Compute and output the smallest number in a 10-element array called LIST.

Solution

Figure 9–5 shows a structure chart consisting of three modules: A000 (overall control), B000 (input array), and B010 (compute and output small). Figure 9–6 shows the logic within the overall control module A000. A000 calls upon two lower-level modules: one to read 10 values into 10 consecutive storage locations as a group, or array (B000); one to find and output the smallest value in the array (B010).

Module B000 contains the steps to read the 10 values into LIST. This is essentially the same algorithm shown in Figure 9–3; the START and STOP steps are replaced with, respectively, an indication of entry into B000 and a RETURN to A000 when the module completes its processing.

Figure 9–7 shows the actual steps to find and output the smallest of the 10 values. We begin our search for the smallest value by assuming arbitrarily that the first member of LIST, which is LIST(1), holds the smallest value. We store this value in SMALL, the location set aside for the smallest value. We also set the variable SUB to 1.

The DOWHILE control structure in Figure 9–7 controls subsequent processing. All succeeding members of LIST are compared with the value of SMALL by means of the IFTHENELSE nested inside the loop. Any value

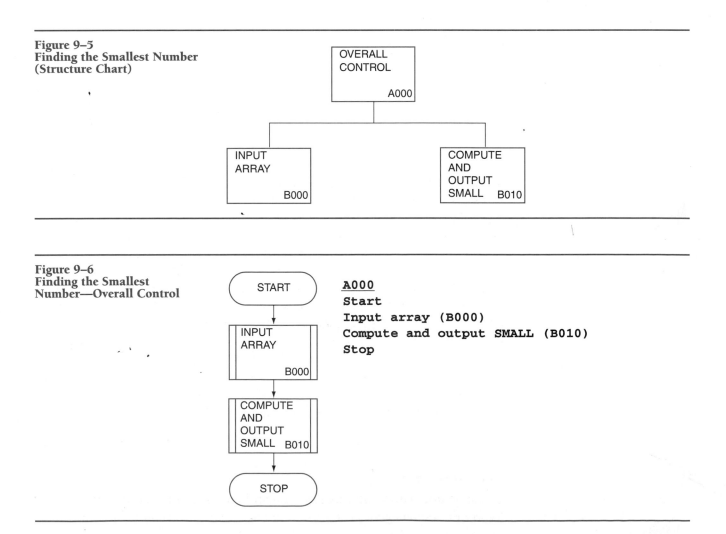

Figure 9–5
Finding the Smallest Number
(Structure Chart)

Figure 9–6
Finding the Smallest
Number—Overall Control

Figure 9–7
Finding the Smallest
Number—Compute and
Output SMALL

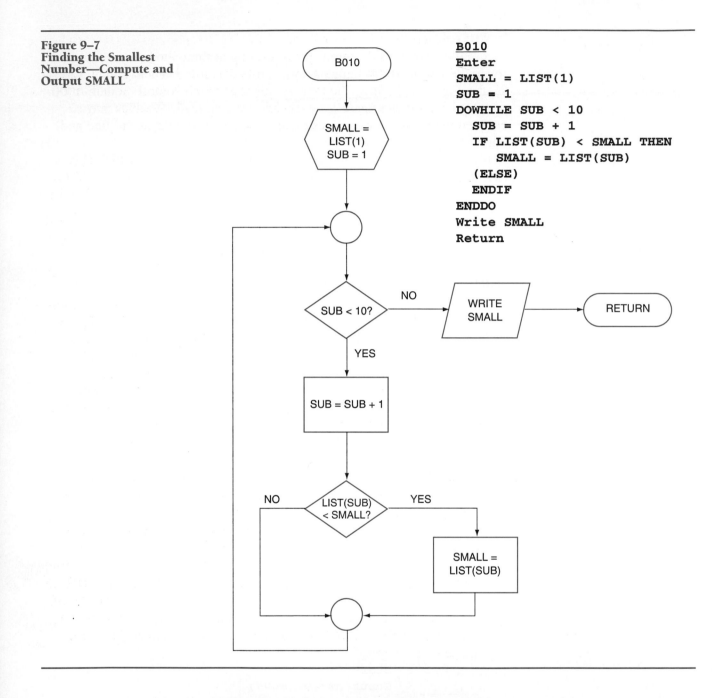

```
B010
Enter
SMALL = LIST(1)
SUB = 1
DOWHILE SUB < 10
  SUB = SUB + 1
  IF LIST(SUB) < SMALL THEN
     SMALL = LIST(SUB)
  (ELSE)
  ENDIF
ENDDO
Write SMALL
Return
```

smaller than the current content of SMALL becomes, itself, the current content of SMALL, via the THEN clause (YES path) of the IFTHENELSE. The DOWHILE loop is exited when SUB is known to be not less than 10—at which time the content of SMALL is the smallest value. This value is written as output and control is returned to A000, which terminates processing.

The value 10 used in this algorithm could easily be replaced by another value. This would permit us to handle a different number of values, and therefore to find the smallest value in a different-sized group of data items or array. Changes to the program flowchart or pseudocode and to the programming-language representation of the algorithm would be minimal.

Even greater flexibility would be available if the number of values to be processed was indicated by means of the first input value (remember header record logic from Chapter 4). This value could be assigned to a variable, say N, for use in loop control. The program could be used to find the smallest of any number of values without any program modification; only the first input value would have to be changed. (See Exercise 4.)

Sample Problem 9.2 (Finding the Average)

Problem:

Compute and output the average number in a 10-element array called LIST. Output the values in LIST as well.

Solution:

Figure 9–8 shows a structure chart consisting of four modules: A000 (overall control), B000 (input array), B010 (output array), and B020 (compute and output average). A000 calls upon three lower-level modules: one to read 10 values into 10 consecutive storage locations as a group, or array (B000); one to write 10 values from 10 consecutive storage locations (B010); and one to find and output the average of the values in an array (B020). The module A000 is much like A000 in the previous example (see Figure 9–6) and is not shown here.

Modules B000 and B010 contain the steps to read the 10 values into LIST and write the 10 values from LIST. Again, these algorithms contain the same processing steps as shown in Figures 9–3 and 9–4. Only the beginning and ending steps need be changed to indicate the proper module entry and exit points.

In Figure 9–9 we see the actual steps to find and output the average of the 10 values. We begin by initializing the variable ACCUM to 0. ACCUM will be used to sum the values in LIST. We also set the variable POS to 0. Notice that in this example we use the name POS, not SUB, to represent the subscript. Any name can generally be used, but you should try to pick names that are descriptive. For example, POS (for position), SUB (for subscript), and IND (for index) are better choices than names like I, J, and K.

Figure 9–8
Finding the Average
(Structure Chart)

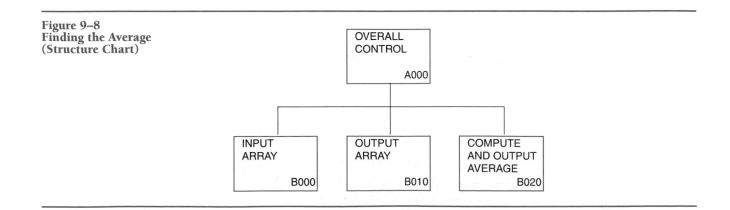

Figure 9–9
Finding the Average—
Compute and Output Average

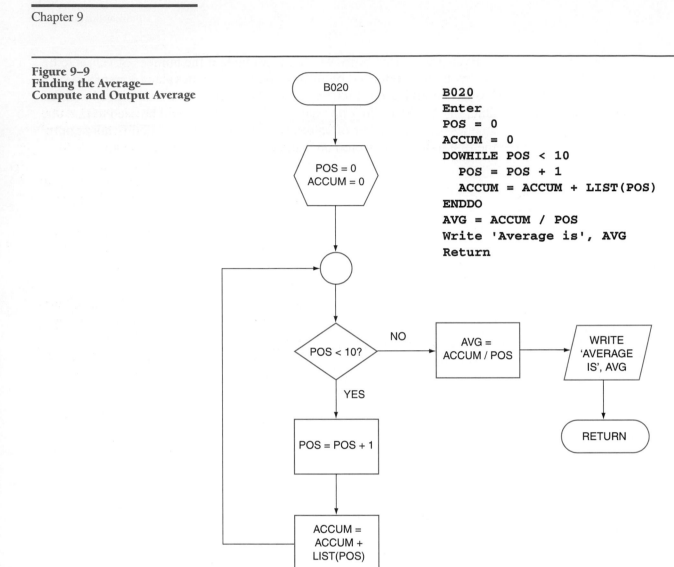

```
B020
Enter
POS = 0
ACCUM = 0
DOWHILE POS < 10
  POS = POS + 1
  ACCUM = ACCUM + LIST(POS)
ENDDO
AVG = ACCUM / POS
Write 'Average is', AVG
Return
```

The DOWHILE control structure in Figure 9–9 controls subsequent processing. Each member of LIST is added to ACCUM during one iteration of the loop. The DOWHILE loop is exited when the value of POS is not less than 10—at which time the average is computed. Notice that the current value of POS is used in the computation of the average, not the constant 10. This makes the algorithm somewhat more flexible. (Fewer program changes would be necessary, then, if we decided to compute the average of a different number of values.) The average is written as output and control is returned to A000, which terminates processing.

Sample Problem 9.3 (Counting Words)

Problem:

Compute and output the number of occurrences of the word "CAT" in a 100-element array called BOOK.

Solution:

Figure 9–10 shows a structure chart consisting of three modules: A000 (overall control), B000 (input array), and B010 (count words and output). Figure 9–11 shows the logic within the overall control module A000. A000 calls on two lower-level modules: one to read 100 values into 100 consecutive storage locations as a group, or array (B000); another to compute the number of occurrences of the word "CAT" in the array BOOK (B010).

Module B000 (see Figure 9–12) contains the steps to read the 100 values into BOOK. Notice again that this module contains the same general processing steps shown in Figure 9–3.

Figures 9–13 (flowchart) and 9–14 (pseudocode) show the actual steps to compute the number of times "CAT" is found in the array BOOK. We use a counter (COUNT) to keep track of each word "CAT" that we find. The variables COUNT and POS are set to 0 as we begin.

The DOWHILE control structure in Figures 9–13 and 9–14 controls subsequent processing. All members of BOOK are compared with "CAT" by means of the IFTHENELSE nested inside the loop. Any value equal to "CAT" will cause the counter to be incremented by 1 via the THEN clause (YES path) of the IFTHENELSE. The DOWHILE loop is exited when

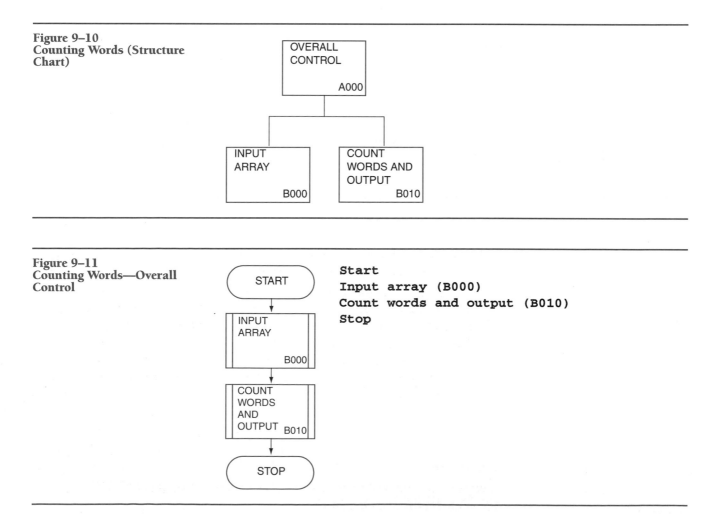

Figure 9–10
Counting Words (Structure Chart)

Figure 9–11
Counting Words—Overall Control

```
Start
Input array (B000)
Count words and output (B010)
Stop
```

**Figure 9–12
Counting Words—Input
Array**

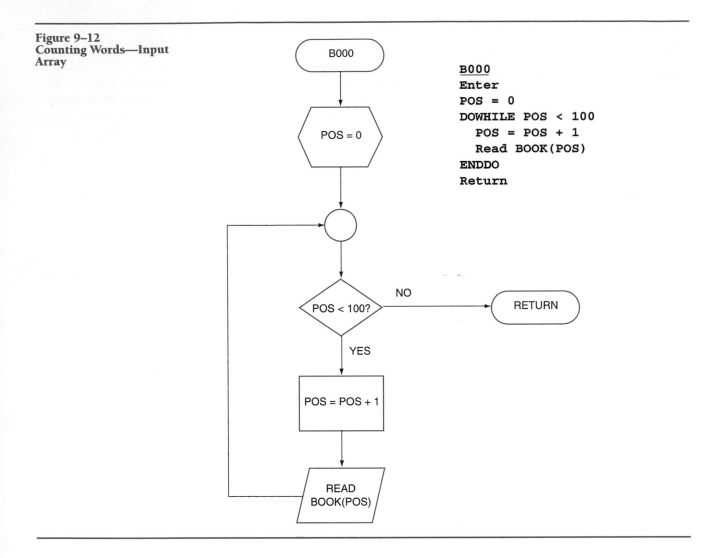

```
B000
Enter
POS = 0
DOWHILE POS < 100
  POS = POS + 1
  Read BOOK(POS)
ENDDO
Return
```

POS is not less than 100. At that time the counter is tested to determine whether the word "CAT" was found. An appropriate message is output and control is returned to A000, which terminates processing.

Sample Problem 9.4 (Doubling an Array)

Problem:

Input 50 values into an array called SINGLE. Create an array called DOUBLE that will contain the values in the array SINGLE, doubled. Output both arrays, SINGLE and DOUBLE.

Solution:

Figure 9–15 shows a structure chart consisting of six modules: A000 (overall control), B000 (input array SINGLE), B010 (create array DOUBLE), B020 (output arrays), C000 (output array SINGLE), and C010 (output array DOUBLE).

**Figure 9–13
Counting Words—Count
Words and Output
(Flowchart)**

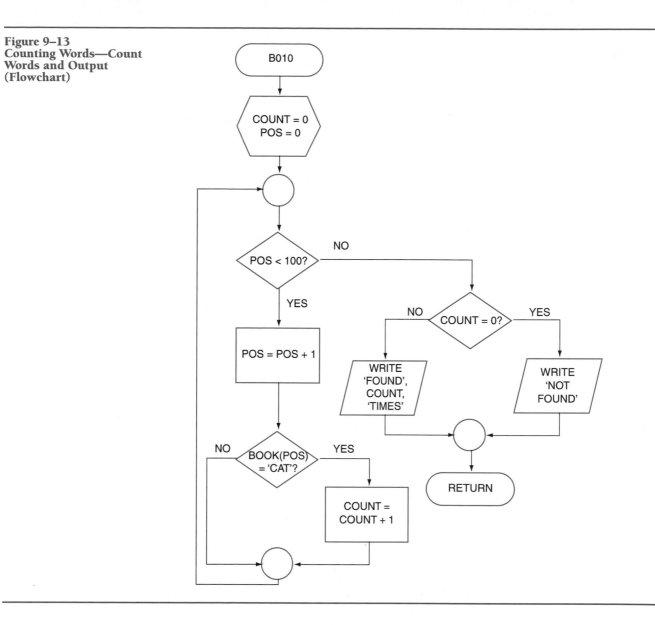

**Figure 9–14
Counting Words—Count
Words and Output
(Pseudocode)**

```
B010
Enter
COUNT = 0
POS = 0
DOWHILE POS < 100
    POS = POS + 1
    IF BOOK(POS) = 'CAT' THEN
        COUNT = COUNT + 1
    (ELSE)
    ENDIF
ENDDO
IF COUNT = 0 THEN
    Write 'Not found'
ELSE
    Write 'Found', COUNT, 'times'
ENDIF
Return
```

Figure 9–15
Doubling an Array (Structure Chart)

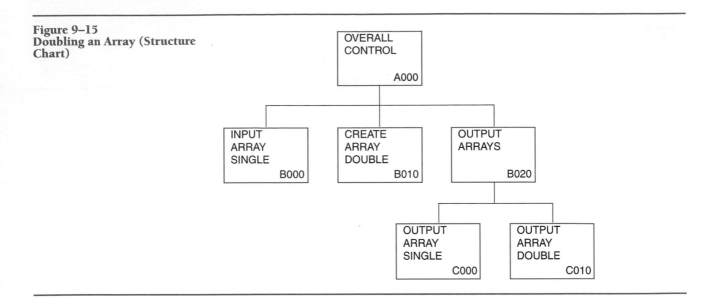

Module B000 contains the steps to read the 50 values into the array SINGLE. Modules C000 and C010 contain the steps to write the 50 values from the arrays SINGLE and DOUBLE. Again, these modules contain processing steps equivalent to the algorithms shown in Figures 9–3 and 9–4. Module B020 simply calls on C000 and C010 to output each array separately; it is therefore not shown here.

Figure 9–16 shows the actual steps to create the array DOUBLE. Notice that the values for DOUBLE do not need to be input; rather, they are generated from the values already input into the array SINGLE.

Sample Problem 9.5 (Squaring and Cubing an Array)

Problem:

Create and output two arrays containing the squares and cubes of the numbers from 1 to 100.

Solution

Figure 9–17 shows a structure chart consisting of five modules: A000 (overall control), B000 (create arrays SQUARE and CUBE), B010 (output arrays), C000 (output array SQUARE), and C010 (output array CUBE).

Modules C000 and C010 contain the steps to write the 100 values from the arrays SQUARE and CUBE. Again, these modules contain processing steps equivalent to the algorithm shown in Figure 9–4. Module B010 simply calls upon C000 and C010 to output each array separately, as module B020 did in the previous example.

Figure 9–18 shows the actual steps to create the arrays SQUARE and CUBE. Notice that the values for SQUARE and CUBE do not need to be input. Rather, they are generated from the value of the subscript represented by the variable POS.

The DOWHILE control structure in Figure 9–18 controls subsequent processing, as it has in all the previous examples. It might be worth

**Figure 9–16
Doubling an Array—Create
Array DOUBLE**

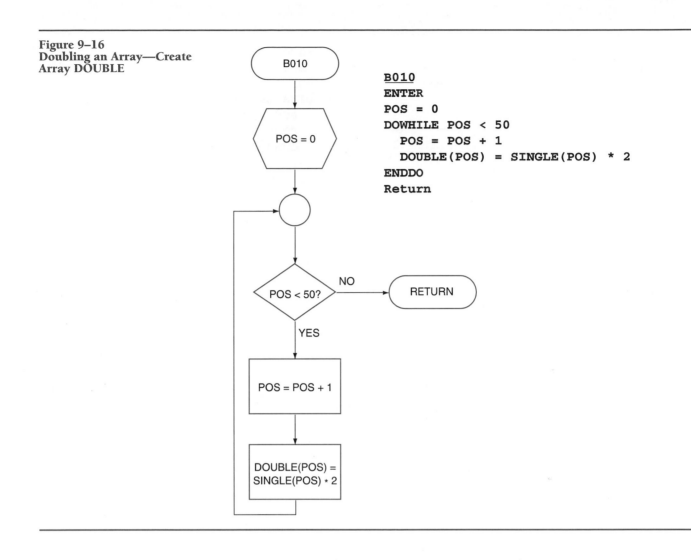

```
B010
ENTER
POS = 0
DOWHILE POS < 50
  POS = POS + 1
  DOUBLE(POS) = SINGLE(POS) * 2
ENDDO
Return
```

**Figure 9–17
Squaring and Cubing an
Array (Structure Chart)**

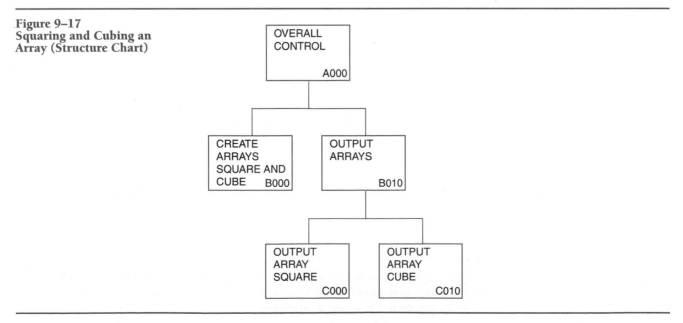

Figure 9–18
Squaring and Cubing an
Array—Create Arrays
SQUARE and CUBE

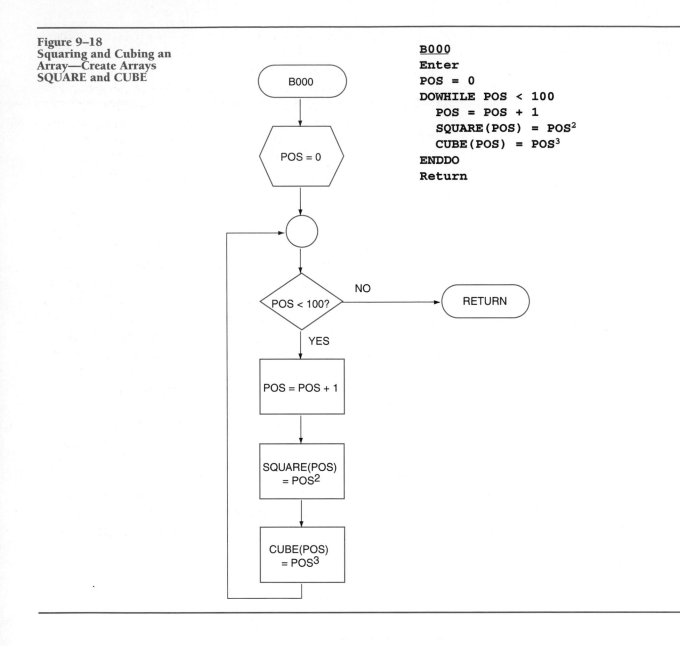

```
B000
Enter
POS = 0
DOWHILE POS < 100
  POS = POS + 1
    SQUARE(POS) = POS²
    CUBE(POS) = POS³
ENDDO
Return
```

mentioning at this point that a DOUNTIL loop also could have been used to control the processing in any of our examples. You may wish to convert the DOWHILE loops in some of the modules to DOUNTIL loops.

Table Structures

To refer to a specific element in any of the data groups discussed thus far, we have used a subscripted variable name containing only a single subscript [INAREA(1), INAREA(2), LIST(POS), where POS = 1,2, . . . , n].

One subscript was sufficient to identify a particular element because each group had a very simple structure; it could be treated as a list. We say that such a group has one **dimension**.

In some problem situations it is convenient to treat a group of values as though it has more than one dimension. For example, instead of storing values as a list, we can store them as a **table**. Such a data group has two dimensions: **rows** and **columns**. A specific member of the group is identified by a subscripted variable name having two subscripts. The first subscript identifies a particular row; the second identifies a particular column. Here again, the terminology used for the data group varies. In COBOL, this kind of group is called a **two-level** or **(multilevel) table**. In Basic, C++, and Java it may be called a **table**, **matrix**, **two-dimensional array**, or **multidimensional array**.

Tables are widely used in problem solving. To determine how much sales tax is owed on items purchased at a local supermarket, a checkout clerk may refer to sales tax tables. When preparing annual income tax returns, we refer to other types of tax tables. An insurance representative refers to rating tables to determine the premiums to be charged for insurance policies. Postal clerks refer to tables showing weights and distances to determine mailing costs. The results of market research and statistical analyses are often displayed in tabular form. To convert temperatures from Fahrenheit to Celsius, or to convert English measurements (inches, feet, yards, etc.) to metric units, we use tables. It should not surprise us, then, that tables are used widely in the computer-program representations of solution algorithms.

The table in Figure 9–19 contains numerical values arranged in four rows and five columns. M is the name of the table. Double-subscripted variables are used to refer to specific members of the table. For example, M(2,3) refers to the value in the second row and third column of the table, or 8. Similarly, M(4,1) refers to the value in the fourth row and first column, or 9. Multiplying the number of rows by the number of columns (4 × 5, in this example) gives the total number of elements in the table.

Figure 9–19
Two-Dimensional Array
Example

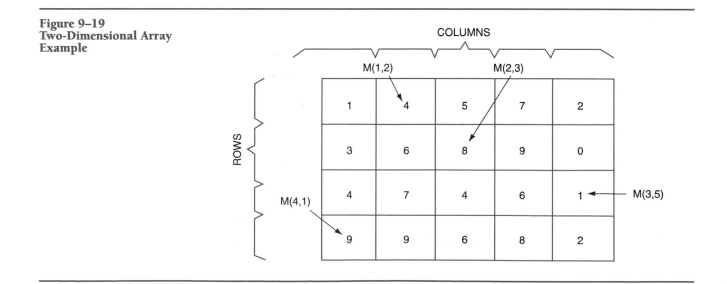

Table Examples

Figures 9–20, 9–21, and 9–22 illustrate three common procedures used in the processing of two-dimensional arrays. Figure 9–20 shows the processing steps required to initialize an array; Figure 9–21 shows the steps to input values into an array; and Figure 9–22 shows the steps to output values from an array. In all three examples the variables ROW and COL are used as subscripts to point to a particular member (or element) of an array called GRID. These algorithms use **row-processing logic**: All

Figure 9–20
Initialize Two-Dimensional Array

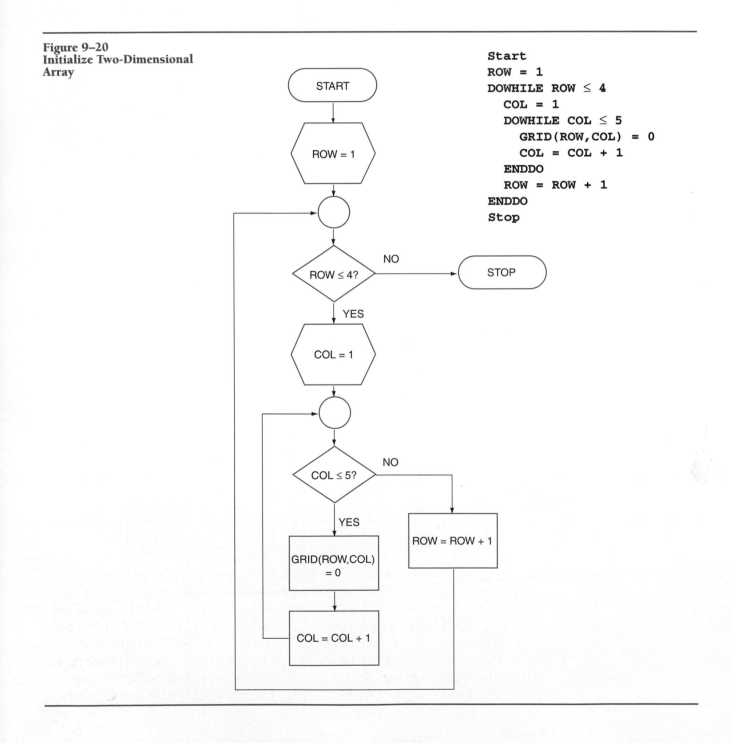

```
Start
ROW = 1
DOWHILE ROW ≤ 4
   COL = 1
   DOWHILE COL ≤ 5
      GRID(ROW,COL) = 0
      COL = COL + 1
   ENDDO
   ROW = ROW + 1
ENDDO
Stop
```

Figure 9–21
Input Two-Dimensional Array

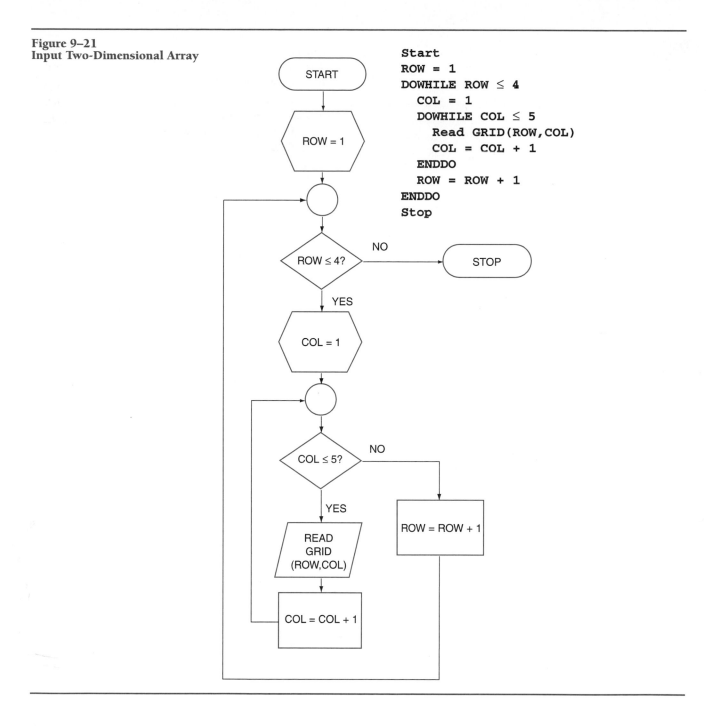

```
Start
ROW = 1
DOWHILE ROW ≤ 4
   COL = 1
   DOWHILE COL ≤ 5
     Read GRID(ROW,COL)
     COL = COL + 1
   ENDDO
   ROW = ROW + 1
ENDDO
Stop
```

members of row 1 are accessed first, starting with the value in the first column and proceeding to the value in the last (fifth) column. Then all members of row 2 are accessed, and so on. Nested DOWHILE loops are used to set up the processing logic required. The inner DOWHILE loop controls the initializing, reading, or writing of values into or from specific columns; the outer DOWHILE loop controls the row in which these columns are located. Notice that all three algorithms contain the same processing steps to control the loop. The subscripts ROW and COL play

Figure 9–22
Output Two-Dimensional
Array

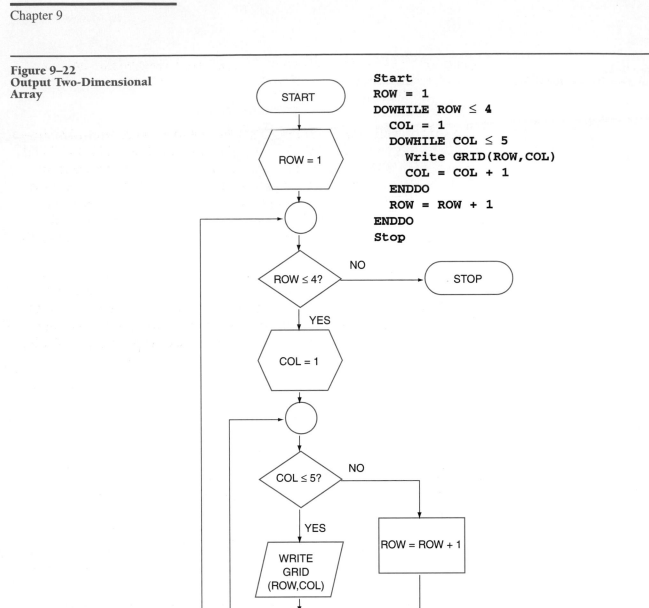

```
Start
ROW = 1
DOWHILE ROW ≤ 4
  COL = 1
  DOWHILE COL ≤ 5
    Write GRID(ROW,COL)
    COL = COL + 1
  ENDDO
  ROW = ROW + 1
ENDDO
Stop
```

two roles. They are used both as counters to control the DOWHILE loops and as pointers into the array GRID.

Sample Problem 9.6 (Seating Chart Problem)

Problem:

An instructor at Parkdale High School wants to create an electronic version of her seating chart. Each student's name will be entered into the seating chart, which will then be output for future reference. The instructor also wants to be able to find and output the row, and the seat within that row, of any given student in the class.

Solution:

The structure chart in Figure 9–23 shows that this algorithm will be implemented as four modules. The overall control module (A000) will call upon module B000 to input the student names into a two-dimensional array. The rows in the array will represent the rows of seats in the class and the columns in the array will represent the seats within each row. Another module (B010) will be used to output the seating chart and a third module (B020) will handle the processing that locates the student's seat in the class.

Figure 9–24 shows the logic within the overall control module A000. Module B000 contains the steps to read 25 names into a two-dimensional array called SEAT. The array is organized into five rows and five columns to parallel the layout of the classroom. B000 is essentially the same algorithm shown in Figure 9–21; the START and STOP steps are replaced with,

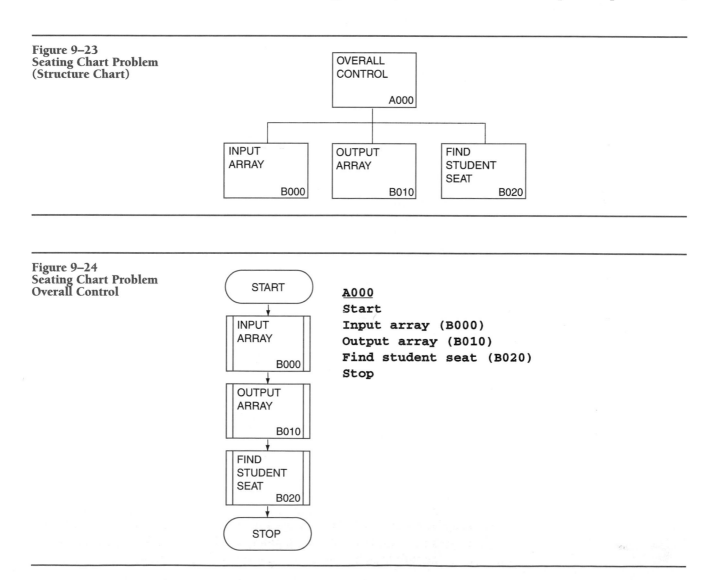

Figure 9–23
Seating Chart Problem
(Structure Chart)

Figure 9–24
Seating Chart Problem
Overall Control

```
A000
Start
Input array (B000)
Output array (B010)
Find student seat (B020)
Stop
```

respectively, an indication of entry into B000 and a return to A000 when the module completes its processing. The array name GRID is replaced by the array name SEAT and the number of rows is now 5, not 4. B010 is essentially the same algorithm shown in Figure 9–22; again, the START and STOP steps are replaced with, respectively, an indication of entry into B010 and a return to A000 when the module completes its processing. The array name GRID is again replaced by the array name SEAT and the number of rows is 5, not 4.

Figures 9–25 (flowchart) and 9–26 (pseudocode) show the actual steps to locate the row and seat number of a student in the class. The first step in B020 reads in the student name to be located in the seating chart. The variables ROW and COL are used as subscripts to point to a particular element of the array SEAT. The variables SAVEROW and SAVECOL are used to save, respectively, the row number and the seat within the row, where the student name is located. Note that SAVEROW is initially set to zero as an indication, later in the algorithm, that the student name could not be located in the seating chart.

Nested DOWHILE loops are used to set up the processing logic to search the array for the student name. All members of SEAT are compared with NAME by means of the IFTHENELSE nested inside the inner DOWHILE loop. If the name is found in the array, then the row number, represented by ROW, is saved in SAVEROW and the column or seat number, represented by COL, is saved in SAVECOL. After the entire array is searched, the outer DOWHILE loop is exited and an IFTHENELSE is used to test the value of SAVEROW. If SAVEROW is still zero, then we know that the student name was not found in the array and an appropriate message is output. If SAVEROW is greater than zero, then we know it represents the row in which the student name is located. At the same time, we know that SAVECOL represents the seat number within the row where the student name is located. A message is then output specifying the student name with the corresponding seat location. Control is then returned to A000, which terminates processing. Note that we did not initialize SAVECOL to zero as we did SAVEROW. Do you see why?

Sample Problem 9.7 (Finding the Highest Average)

Problem:

An instructor from Lehigh High School wants a computer program to input and store the names and averages of all students in his class. Each student name and corresponding average will be input as a separate record. He would also like the program to compute and print the name and average of the student with the highest average in the class.

Solution:

The structure chart in Figure 9–27 shows that this algorithm will be implemented as three modules. The overall control module (A000) will call upon module B000 to input the student names and averages into a two-dimensional array. The array will accommodate 25 students. Each row in the array will contain the student name in column 1 and the corresponding

**Figure 9–25
Seating Chart Problem—Find
Student Seat (Flowchart)**

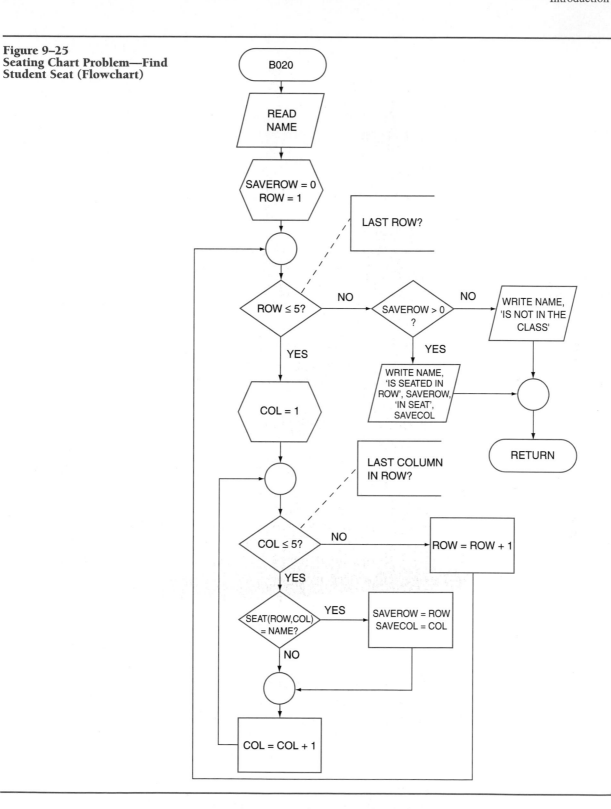

student average in column 2. Thus the size of the array will be defined with
25 rows and 2 columns (25 × 2). Another module (B010) will be used to
compute and output the name and average of the student with the highest
average in the class.

Figure 9–26
Seating Chart Problem—Find Student Seat (Pseudocode)

```
B020
Enter
Read NAME
SAVEROW = 0
ROW = 1
DOWHILE ROW ≤ 5
     COL = 1
     DOWHILE COL ≤ 5
          IF SEAT(ROW,COL) = NAME THEN
                    SAVEROW = ROW
                    SAVECOL = COL
          (ELSE)
          ENDIF
          COL = COL + 1
     ENDDO
     ROW = ROW + 1
ENDDO
IF SAVEROW > 0 THEN
     Write NAME, 'is seated in row', SAVEROW, 'in seat', SAVECOL
ELSE
     Write NAME, 'is not in the class'
ENDIF
Return
```

Figure 9–27
Finding the Highest Average (Structure Chart)

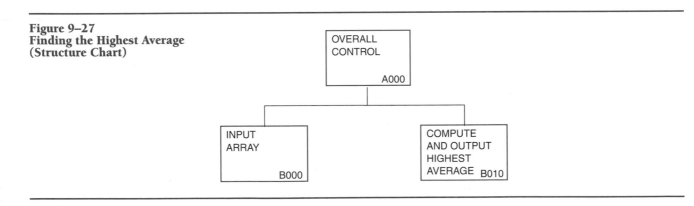

Figure 9–28 shows the logic within the overall control module A000. Figure 9–29 shows the logic within module B000. Module B000 contains the steps to read the 25 student names and averages into a two-dimensional array called STUDENT. Note that in B000 only one DOWHILE loop is used to control the process that inputs the data into the array STUDENT. The loop executes 25 times, once for each student record. Each execution of the READ statement actually inputs two values into the array STUDENT: the student name in column 1 and the student average in column 2. The variable SUB represents the current record being processed and is used to point to the corresponding row in the array STUDENT. Note how the steps in this algorithm differ from the steps used to input values into the array GRID shown in Figure 9–21. When all 25 student records have been

Figure 9–28
Finding the Highest
Average—Overall Control

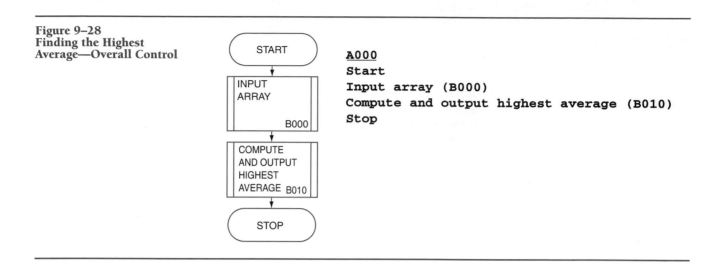

```
A000
Start
Input array (B000)
Compute and output highest average (B010)
Stop
```

Figure 9–29
Finding the Highest
Average—Input Array

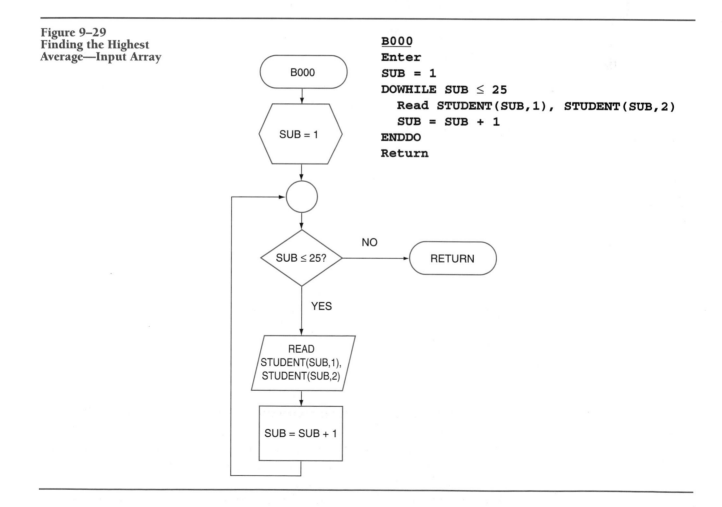

```
B000
Enter
SUB = 1
DOWHILE SUB ≤ 25
   Read STUDENT(SUB,1), STUDENT(SUB,2)
   SUB = SUB + 1
ENDDO
Return
```

input, the DOWHILE loop is exited and control returns to A000. Module B010 is then given control. Figure 9–30 shows the flowchart and corresponding pseudocode for module B010. B010 contains the processing logic to compute and output the name and average of the student with the highest average in the class.

**Figure 9–30
Finding the Highest
Average—Compute and
Output Highest Average**

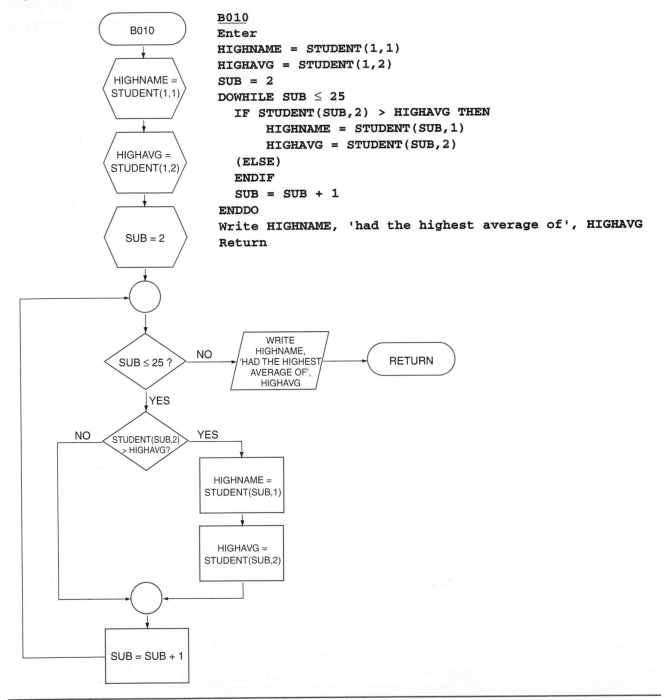

```
B010
Enter
HIGHNAME = STUDENT(1,1)
HIGHAVG = STUDENT(1,2)
SUB = 2
DOWHILE SUB ≤ 25
  IF STUDENT(SUB,2) > HIGHAVG THEN
      HIGHNAME = STUDENT(SUB,1)
      HIGHAVG = STUDENT(SUB,2)
  (ELSE)
  ENDIF
  SUB = SUB + 1
ENDDO
Write HIGHNAME, 'had the highest average of', HIGHAVG
Return
```

We begin our search for the highest average by assuming arbitrarily that the first student named in the array has the highest average. We store the first student name (from row 1: (STUDENT(1,1)) in HIGHNAME, the variable that is to hold the name of the student with the highest average.

We also store the first student average (from row 1: (STUDENT(1,2)) in HIGHAVG, the variable that is to hold the highest average. We then set the variable SUB to 2. Do you see why we set SUB to 2 rather than 1? We have used the values in row 1 of the array STUDENT. SUB will represent each row in the array STUDENT beginning with row 2. The DOWHILE control structure in Figure 9–30 controls subsequent processing. All averages, beginning with the value in row 2 (STUDENT(SUB,2)), are compared with the current content of HIGHAVG by means of the nested IFTHENELSE inside the loop. Any average greater than the current content of HIGHAVG becomes, itself, the current content of HIGHAVG, via the THEN clause (YES path) of the IFTHENELSE. Similarly, the corresponding student name (STUDENT(SUB,1)) becomes the current content of HIGHNAME. The DOWHILE loop is exited when SUB is known to be not less than or equal to (therefore, greater than) 25—at which time the current content of HIGHAVG is the highest average and the current content of HIGHNAME is the name of the student with the highest average. A message containing both the name and average of the student with the highest average is written as output. Control is returned to A000, which terminates processing.

Sample Problem 9.8 (Two-Dimensional Array Computation)

Problem:

Input values into a two-dimensional array called A. A header record will contain two numbers representing the number of rows (M) and the number of columns (N) in the array A. Also, input N values into a one-dimensional array called V. Create and output an array called T that will be generated by multiplying each element in every row of A by the list element in the corresponding position in V.

Solution:

The structure chart shown in Figure 9–31 consists of four modules to do array processing. Figure 9–32 shows the logic within the overall control module.

The first module that is executed reads values into consecutive storage locations as a two-dimensional array, Table A. The flowchart and pseudocode representations of the algorithm are shown in Figure 9–33. The first pair of values submitted as input indicates the number of rows (M) and

Figure 9–31
Two-Dimensional Array Computation (Structure Chart)

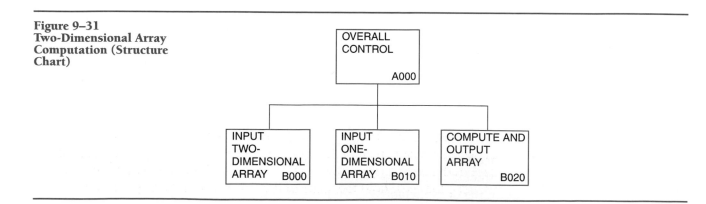

**Figure 9–32
Two-Dimensional Array
Computation—Overall
Control**

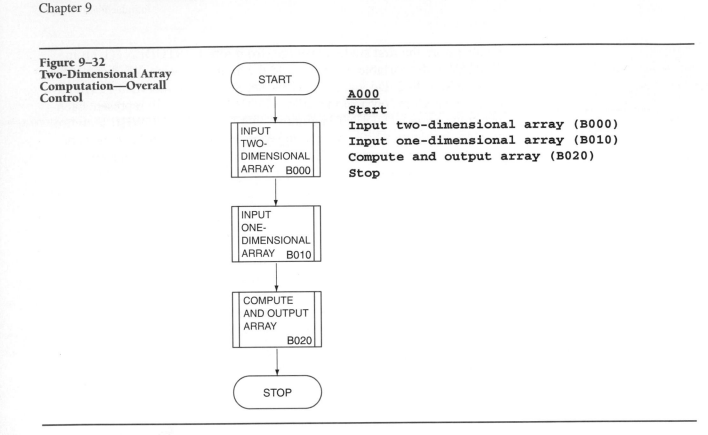

```
A000
Start
Input two-dimensional array (B000)
Input one-dimensional array (B010)
Compute and output array (B020)
Stop
```

the number of columns (N) to be read. Specific members (elements) of the table are entered row by row, one member at a time. The algorithm uses row-processing logic and contains the same processing steps as shown in Figure 9–21.

After all members of the table have been read into storage (i.e., when both the inner DOWHILE and the outer DOWHILE have been exited), control is returned to the overall control module. Specific values from the table can now be used in subsequent processing.

The next module that is executed reads a second group of data items—a list having N elements. Figure 9–34 shows the reading of input into the list V in flowchart and pseudocode forms. Note the similarities and differences between this part of the algorithm and the logic in Figure 9–3.

We are now to multiply each element in every row of the table by the list element in the corresponding position in V. First we multiply each element of the first row by each corresponding element of the list; then we multiply each element of the second row by each corresponding element of the list; and so on. The number of elements in the list must be the same as the number of columns in the table. In our example, with A representing the table and V representing the list, we find

$$T(ROW,COL) = A(ROW,COL) * V(COL)$$

repeatedly, with ROW ranging from 1 to the total number of rows (M) in the table, and COL ranging from 1 to the total number of columns (N). In effect, a new two-dimensional group, which we have called T, is constructed element by element. Its dimensions are the same as the dimensions of the original table, A, but its contents differ. Each value in T is the result

Figure 9–33
Two-Dimensional Array
Computation—Input Two-
Dimensional Array

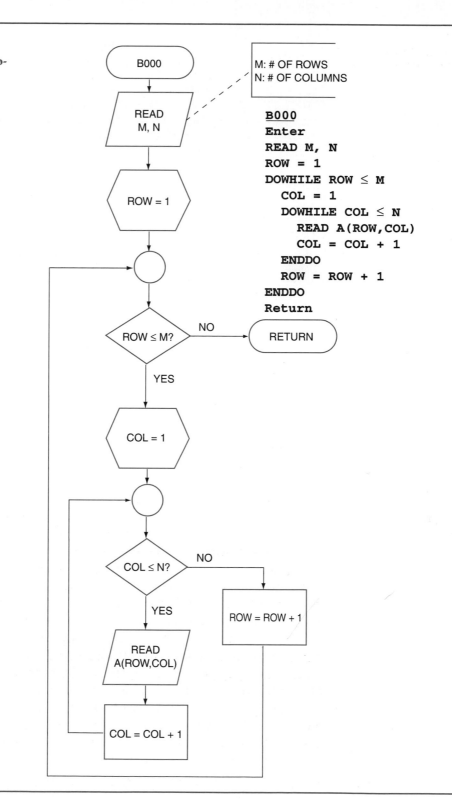

M: # OF ROWS
N: # OF COLUMNS

```
B000
Enter
READ M, N
ROW = 1
DOWHILE ROW ≤ M
  COL = 1
  DOWHILE COL ≤ N
    READ A(ROW,COL)
    COL = COL + 1
  ENDDO
  ROW = ROW + 1
ENDDO
Return
```

of a multiplication operation. Figure 9–35 shows module B020 in this example; it does the required multiplication and the printing of the new table, T. Documentation within the annotation symbols helps to explain some of the steps.

Figure 9–34
Two-Dimensional Array
Computation—Input One-
Dimensional Array

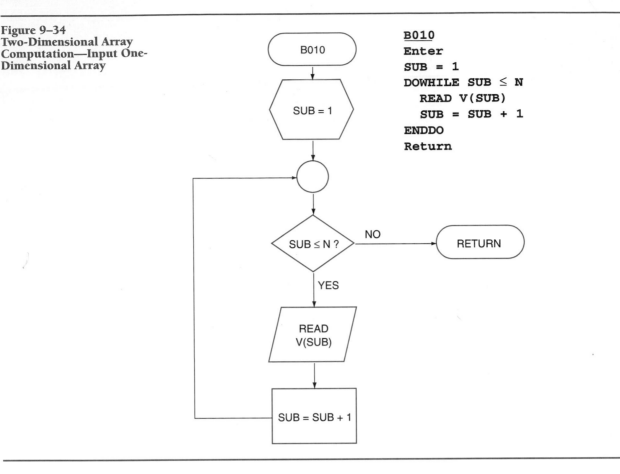

```
B010
Enter
SUB = 1
DOWHILE SUB ≤ N
  READ V(SUB)
  SUB = SUB + 1
ENDDO
Return
```

Multidimensional Structures

Some programming languages allow specification of data groups with 200 or more dimensions. In all cases, the number of a group's dimensions determines the number of subscripts needed to refer to a particular element of the group. Although groups with many dimensions are difficult or impossible for us to visualize, they may be extremely useful in computation. The same kind of program logic—loops within loops—can be used repetitively for all data groups.

The system designer or programmer who develops an algorithm involving multidimensional data groups should be aware of the programming language that will be used to express the algorithm in computer-program form. Some programming languages read, write, and store the data items in a group in **row-major order** (with the first subscript varying least rapidly, and the last subscript varying most rapidly). We assumed such logic was used in our previous examples. However, some programming languages read, write, and store the data items in a group in **column-major order** (with the first subscript varying most rapidly, and the last subscript varying least rapidly). The logic within the solution algorithm must be set up according to the specifications of the language being used—otherwise, the computer cannot process the data efficiently. To understand why, consider a simple analogy. Assume you are shopping at a local supermarket. Do you try to select all the items you want from aisle 1, then all from aisle 2, then all from aisle 3, and so on? Or do you prefer to select oranges from the

**Figure 9–35
Two-Dimensional Array
Computation—Compute and
Output Array**

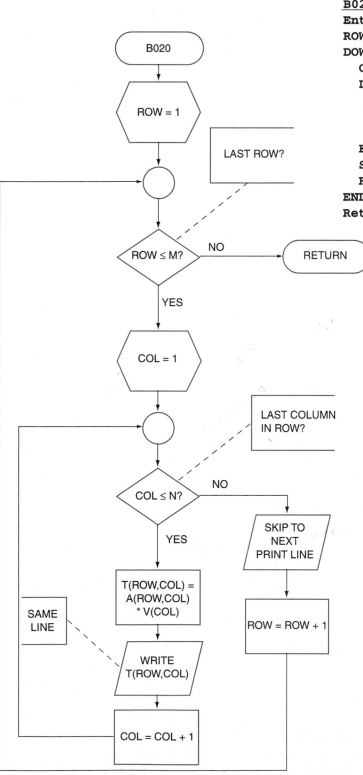

```
B020
Enter
ROW = 1
DOWHILE ROW ≤ M
  COL = 1
  DOWHILE COL ≤ N
    T(ROW,COL) = A(ROW,COL) * V(COL)
    Write T(ROW,COL)
    COL = COL + 1
  ENDDO
  Skip to next print line
  ROW = ROW + 1
ENDDO
Return
```

**Figure 9–36
Two-Dimensional Array
Ordering**

ROWS

COLUMNS

```
M(1,1)  =  1
M(1,2)  =  4
M(1,3)  =  5
M(1,4)  =  7
M(1,5)  =  2

M(2,1)  =  3
M(2,2)  =  6
M(2,3)  =  8
M(2,4)  =  9
M(2,5)  =  0

M(3,1)  =  4
M(3,2)  =  7
M(3,3)  =  4
M(3,4)  =  6
M(3,5)  =  1


M(4,1)  =  9
M(4,2)  =  9
M(4,3)  =  6
M(4,4)  =  8
M(4,5)  =  2
```

```
M(1,1)  =  1
M(2,1)  =  3
M(3,1)  =  4
M(4,1)  =  9

M(1,2)  =  4
M(2,2)  =  6
M(3,2)  =  7
M(4,2)  =  9

M(1,3)  =  5
M(2,3)  =  8
M(3,3)  =  4
M(4,3)  =  6

M(1,4)  =  7
M(2,4)  =  9
M(3,4)  =  6
M(4,4)  =  8

M(1,5)  =  2
M(2,5)  =  0
M(3,5)  =  1
M(4,5)  =  2
```

(a) Row-major ordering
of Figure 9–19

(b) Column-major ordering
of Figure 9–19

fruits/vegetables section, then syrup from aisle 3, then bacon from the meat counter, then onions (which are, in fact, positioned right next to the oranges), and then return to aisle 3 for strawberry jam?

Now look at Figure 9–36. If row-major ordering is used, the table shown in Figure 9–19 is processed as shown in the first column of Figure 9–36. In contrast, the second column of Figure 9–36 shows column-major ordering of the same table. Basic, COBOL, C++, and Java are designed to store and process data groups in row-major order as in Figure 9–36a. Some implementations of Basic, however, may store and process data groupings in column-major order. It is important to find out which of the two orderings has been used in the Basic implementation available to you. FORTRAN uses column-major ordering as in Figure 9–36b. In either approach, the total numbers of elements in the various dimensions of the data groups are often specified as loop controls.

Enrichment (Basic)

Figure 9–37 illustrates a listing of the program for the finding the smallest number problem (see Figures 9–6 and 9–7). We have modified this example to include an additional module—OutputArray.

The program begins with three declaration statements specifying the name of each subprogram that will be invoked during program execution. The overall control module is very similar to the pseudocode in Figure 9–6. In this example, however, we invoke an additional module—OutputArray—to output the values of the array. The first statement in the overall control

Figure 9–37
Finding the Smallest Number
(Basic List)

```
DECLARE SUB ComputeAndOutputSmall()
DECLARE SUB OutputArray()
DECLARE SUB InputArray()

REM A000 - Overall Control Module
DIM SHARED LISTOFVALS(10)
CALL InputArray
CALL OutputArray
CALL ComputeAndOutputSmall
END

REM B000
SUB InputArray
SUBVAL = 0
PRINT "Enter ten numbers, one per line"
DO WHILE SUBVAL < 10
        SUBVAL = SUBVAL + 1
        INPUT LISTOFVALS(SUBVAL)
LOOP
END SUB

REM B010
SUB ComputeAndOutputSmall
SMALL = LISTOFVALS(1)
SUBVAL = 1
DO WHILE SUBVAL < 10
        SUBVAL = SUBVAL + 1
        IF LISTOFVALS(SUBVAL)< SMALL THEN
                SMALL = LISTOFVALS(SUBVAL)
        END IF
LOOP
PRINT
PRINT "The smallest number in the list is "; SMALL
END SUB

REM B020
SUB OutputArray
SUBVAL = 0
PRINT
PRINT "The numbers in the list are:"
DO WHILE SUBVAL < 10
        SUBVAL = SUBVAL + 1
        PRINT LISTOFVALS(SUBVAL);
LOOP
END SUB
```

module is a *Dim* statement that specifies the size of the array. In this case, 10 memory locations are specified to hold the values of the array named LISTOF-VALS. In addition, since the array is global in scope, (all modules have access to its values) the keyword *Shared* also must be included in the Dim statement. In Basic we must include a Shared statement for each global variable.

The program statements for each module or subprogram are listed after the overall control module. The statements in InputArray, ComputeAndOutputSmall, and OutputArray are very similar to the pseudocode in Figures 9–3, 9–4, and 9–7. Note that the variable SUBVAL is used in the three subprograms. SUBVAL holds the value of the index, or subscript, of the array LISTOFVALS. There is no Shared statement in-cluded for SUBVAL because SUBVAL is not a global variable. SUBVAL is specific or local to each subprogram. We could just as easily have used three

| Figure 9–38 Finding the Smallest Number (Basic Run) | `Enter ten numbers, one per line`
`?6`
`?8`
`?5`
`?3`
`?9`
`?7`
`?6`
`?9`
`?4`
`?8`

`The numbers in the list are:`
`6 8 5 3 9 7 6 9 4 8`
`The smallest number in the list is  3` |

different variables to represent the index of the array LISTOFVALS in each of our three subprograms.

Figure 9–38 illustrates the output that will be produced when the program is executed. The user is prompted to enter the 10 numbers, one per line. Each of these numbers is stored in one location in the array LISTOFVALS. After all 10 numbers are input, the contents of the array are output. Note that all the numbers are printed on one line. Placing a semicolon at the end of the Print statement in module OutputArray prevents a linefeed from occurring each time the Print statement is executed. Finally, the smallest value in the list is output with an identifying label.

Enrichment (Visual Basic)

Arrays can be used to represent data in a program as we saw in this chapter. They also can be used to group controls created in a Visual Basic program. An array that is used to group controls is called a *control array*. The following example (shape problem) illustrates the concept of a control array. The graphical interface for the shape problem is shown in Figure 9–39.

In this example, four command buttons are created and grouped together into a control array. A control array is created for a group of controls by giving all controls the same name. Each command button is given the name cmd_SHAPE. Each individual array item is actually one of the command buttons and is uniquely identified by an index. This index is similar to the index or subscript used with an array of data. The index is automatically assigned by Visual Basic when each control in a control array is created. In this example the index is assigned as follows:

Command Button	Index Value
Rectangle	0
Square	1
Oval	2
Circle	3

Figure 9–39
Shape Problem (Visual
Basic—Screen 1)

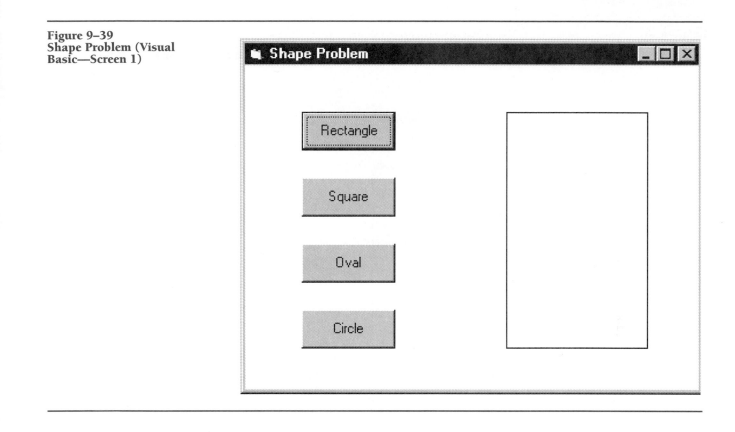

The Rectangle command button can be referenced in code as cmd_SHAPE(0); the Square command button can be referenced in code as cmd_SHAPE(1); and so on. Note that Visual Basic assigns a value of 0 to the first index in each control array. Zero is the default value and can be changed by the programmer. Although the names of the command buttons in the control array must be identical, the caption of each command button can be unique. For example, the caption of the first command button is "Rectangle." The caption is simply a way to label the control on the interface.

When the user clicks the first command button (Rectangle), a rectangle shape appears on the right of the interface, as shown in Figure 9–39. Similarly, when the user clicks any of the other command buttons, the corresponding shape is displayed, as shown in Figures 9–40, 9–41, and 9–42.

Figure 9–43 illustrates the program associated with the click event of each command button. Since the four command buttons constitute a control array, only one click event is written. The first statement in this event includes a parameter (Index As Integer), an integer variable automatically generated by Visual Basic. This variable identifies which command button was actually clicked by the user. If the user clicked the Rectangle button, the value of Index will be 0 when the event code is executed. If the user clicked the Square button, the value of Index will be 1 when the event code is executed. If the user clicked the Oval button, the value of Index will be 2 when the event code is executed. Finally, if the user clicked the Circle button, the value of Index will be 3 when the event code is executed. The programmer can then refer to the value of Index. In this example, the value of

Figure 9–40
Shape Problem (Visual Basic—Screen 2)

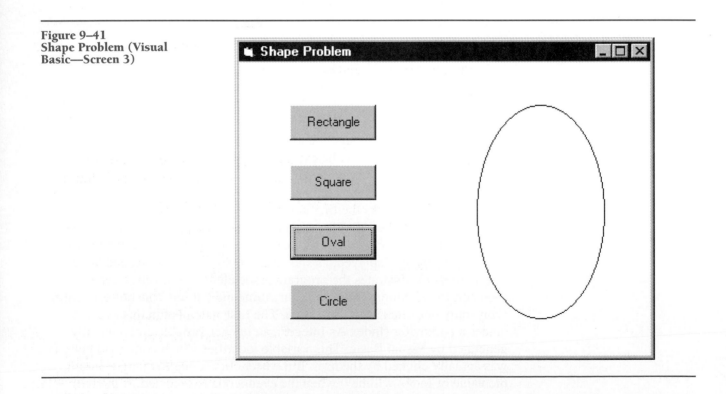

Figure 9–41
Shape Problem (Visual Basic—Screen 3)

Index is assigned to the *shape* property of a *shape* control named shp_SHAPE.

A shape control is used in Visual Basic to display various shapes. The name of a shape control begins with *shp* as specified by standard naming conventions. The shape control in this example is named shp_SHAPE and

Figure 9–42
Shape Problem (Visual Basic—Screen 4)

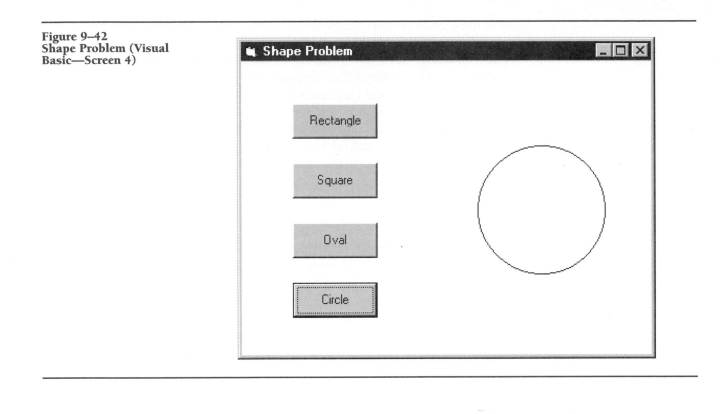

Figure 9–43
Shape Problem (Visual Basic—cmd_SHAPE_Click)

```
Private Sub cmd_SHAPE_Click(Index As Integer)

shp_SHAPE.Shape = Index

End Sub
```

is used to display one of four different shapes. The shape property of a shape control defines the type of shape that will be displayed as follows:

Shape Property Value	Type of Shape
0	Rectangle
1	Square
2	Oval
3	Circle
4	Rounded Rectangle
5	Rounded Square

The program in Figure 9–43, although simple, is quite powerful. The command buttons were created in a specific sequence to ensure that the index of each command button would match the shape property value of the shape control corresponding to the command button's caption. For example, the first command button (caption: Rectangle) has an index of 0 because it was the first button created. Remember, indices begin at 0. At the same time, a shape property value of 0 corresponds to a rectangle shape. The second command button (caption: Square) has an index of 1 because it was the second button created. At the same time, a shape property value of 1 corresponds to a square shape. The third command button (caption: Oval) has an index of 2 because it was the third button created. At the same time, a shape property value of 2 corresponds to an oval shape. Finally, the fourth command button (caption: Circle) has an index of 3 because it was the fourth button created. At the same time, a shape property value of 3 corresponds to a circle shape. Clearly, our choice to use a control array for the command buttons and relate the index to the shape property of the shape control greatly simplified the program. A well-thought-out interface is essential in the design of a Visual Basic program.

Key Terms

single (simple) variable
single-level table
list
vector
one-dimensional array
subscript
index
unsubscripted variable
   name

subscripted variable
   name
member (of an array)
element (of an array)
dimension
table
row
column

two-level table
multilevel table
matrix
two-dimensional array
multidimensional array .
row-processing logic
row-major order
column-major order

Exercises

1. (a) What is a data group?
   (b) Give some examples of data for which grouping capabilities are likely to be appropriate.

2. Using the term *COLOR(I)*, distinguish between an array name, a subscript, and a subscripted variable name.

3. E is an eight-member one-dimensional array. Its contents are shown as follows:

| 57 | 12 | 03 | 48 | 34 | 16 | 50 | 22 |

   (a) What is the content of E(4)?
   (b) Write the subscripted variable name that should be used to refer to the location containing 57.
   (c) Write the subscripted variable name that should be used to refer to the Rth position in the array E.
   (d) The variable R, in Exercise 3(c), must have a value within a range between what two numbers?

4. Refer to Figures 9–3 through 9–7. Modify either the program flowchart or the pseudocode as suggested in this chapter to show how to find the smallest of any number of values.

5. Modify either the program flowchart or the pseudocode in Figure 9–7 to keep track of which position in LIST contains the smallest value. The position of the smallest value and the value itself should be provided as output.

6. Refer to Figures 9–10 through 9–14. Modify either the program flowchart or the pseudocode to
   (a) Output the position of the first occurrence of the word "CAT."
   (b) Output the position of the last occurrence of the word "CAT."
   (c) Output the position of each occurrence of the word "CAT."
   (d) Compute and output the number of occurrences of any word in which the value of the word is input.
   (e) Modify Exercise 6(d) to process any number of words. Each word will be input as a separate record. The word "EOF" will signal the end of the input.

7. Construct a program flowchart and corresponding pseudocode to show the logic required in the following problem: Read numbers into a 10-element one-dimensional array called LIST. Create another 10-element one-dimensional array called REVERSE that is to contain the same elements as LIST but in reverse order. For example, the first element in LIST will be placed in the last position in REVERSE; the second element in LIST will be placed in the second-to-last position in REVERSE; and so on. After REVERSE has been created, output the contents of both arrays. Be sure to plan a well-structured, modular program.

8. A, B, and C are one-dimensional arrays of sizes 100, 50, and 200, respectively. Construct a program flowchart and corresponding pseudocode for an algorithm to store the first 100 numbers (1,2,3,4,5, . . . ,100) into array A, the first 50 positive odd numbers (1,3,5,7, . . .) into array B, and the reciprocal of each position [C(5) = 1/5] into array C. After all the arrays have been defined, output each array. Notice that no input is required. Be sure to plan a well-structured, modular program.

9. Construct a program flowchart and corresponding pseudocode to solve the following problem: Read the items for two 10-element arrays named I1 and I2. Each item can be assumed to be on a separate record. Multiply the items in array I1 by the items in corresponding positions in array I2. For example, I1(1) is multiplied by I2(1); I1(2) is multiplied by I2(2); and so on. The resulting products are to be stored in the corresponding item positions in a 10-element array called I3. When the multiplication process is complete, output the items for each array. Be sure to plan a well-structured, modular program.

10. Construct a program flowchart and corresponding pseudocode to solve the following problem: Input 25 numbers, one per record, into an array called NUMBER. For each of these numbers, print out the number itself, and, beside it, the value of the difference between the number and the average of all the numbers. Be sure to plan a well-structured, modular program.

11. A is a 30-member two-dimensional array with six rows and five columns. S is a simple variable. The contents of these locations have been set to 0; they appear as follows:

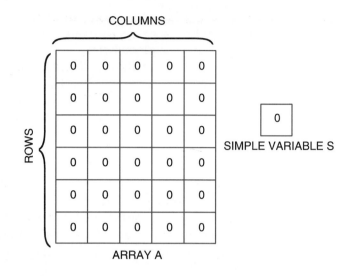

Show the contents of these locations after the following statements have been executed:

$$S = 5$$
$$A(2,4) = 40$$
$$A(1,2) = 67$$
$$A(6,2) = A(4,3) + A(2,4)$$
$$A(S,1) = 99$$
$$A(5,4) = A(1,2) - A(2,S - 1)$$

12. Modify the solution algorithm shown in Figures 9–25 and 9–26 to include the following logic:
  (a) Process a variable number of student names received as input. Use automatic EOF to signal the end of the input. Include an initial IF and a modular design. You may assume that one input is fully processed before another student name is read as input.
  (b) Terminate the execution of both DOWHILE loops as soon as a student name is located in the array SEAT. Note that the current algorithm continues to search the entire array even after a name is found. Be sure to adhere to structured programming techniques.

13. (a) Construct a program flowchart and corresponding pseudocode to show the logic required in the following problem situation: Read a two-level table called SALES into storage. The table has three rows and each row contains 18 values. Compute the total value of the elements in each row. Provide three computed totals and a grand total, which is the sum of all values, as output. Be sure to plan a well-structured, modular program.
  (b) Modify your solution to the problem in Exercise 13(a) to total the values of the elements in each column. Provide the column totals as well as the grand total as output. Note that only one of

the two solutions to this exercise will be efficient. The other solution will be very inefficient, depending on which programming language is used.

14. Construct a program flowchart and corresponding pseudocode to show the logic required in the following problem: Read into storage a two-level table, called MATRIX, that has five rows and five columns. Compute and output the sum of the elements on each of the two diagonals in the table MATRIX. Be sure to plan a well-structured, modular program. Remember to keep efficiency in mind when designing the order in which the elements will be processed.

15. Construct a program flowchart and corresponding pseudocode for an algorithm that searches a 4×5 table named M (to be input) and counts all the occurrences of the items that match the value of MVAL, which also will be input. Output table M in row-order fashion and output a line at the end identifying how many times MVAL was found in table M. Be sure to plan a well-structured, modular program.

16. Construct a program flowchart and corresponding pseudocode to solve the following problem: Assume there are 10 records, each containing two temperature values. The first temperature value on each record represents a high temperature from a weather station, and the second temperature on each record represents a low temperature from a weather station. Input these temperatures into a table TEMP(10,2) with the high temperatures from the 10 stations in column 1 of the table, and the low temperatures from the 10 stations in column 2 of the table. After reading in all the temperatures, find and output the average high temperature and the average low temperature for the 10 weather stations. Be sure to plan a well-structured, modular program.

Part 2

Introduction to Object-Oriented Design

10

Objectives

Upon completion of this chapter you should be able to

- Distinguish between procedure-oriented design and object-oriented design.
- Define the terms *class* and *object* and give examples of each.
- Understand what an abstraction is and show why dealing with abstractions is sometimes useful.
- Distinguish between a data member and a method and use examples of both in the creation of a class definition.
- Explain why encapsulation and data hiding are good programming practices, distinguish between them, and apply both in program design.
- Explain the purpose of a driver program.
- Define the term *instantiation*.
- Explain the role of messages in object-oriented design and programming.
- Explain the purpose of a constructor and of a destructor.
- Distinguish between parameters and arguments and use examples of both in program design.
- Define the term *overloading* and apply overloading in program design.
- Explain what a return value is, suggest situations where return values are useful, and use return values in program design.

Introduction

In Chapter 1 of this book, we looked at the system development life cycle and how program design fits within it. We introduced the basic control structures of structured programming, which can be used to develop the program logic within any program. In succeeding chapters, we applied a procedure-oriented approach to program design. Our emphasis was on the actions to be performed and on the sequence of those actions. We focused on the algorithms needed to solve a problem.

As the problems that we solve with computer help increase in complexity, the programs that we create to solve the problems tend to increase in complexity also. That means the programs are harder to design, harder to understand, and thus harder and more expensive to maintain.

The limitations of our own memory—how much we can remember and how long—impose restrictions on the number of details we are able to receive, process, and remember. In addition, as graphical user interfaces have gained popularity, programming issues are changing. Programs must be designed to allow users to interact frequently with computers. The programs must be easy to use—the sequence of steps, or operations, to be followed, often should be chosen by the user, not controlled by the author of the program.

One approach to program simplification is the development of program modules, which can be called from other program modules, as we saw in Chapter 6. That approach helps because we can focus on specific modules of an application at a time. We do not have to remember or check the names, data types, and uses of variables throughout the application if we just need to change one part of it. Nevertheless, developing a top-down procedure-oriented solution to a complex real-world problem may prove to be a serious challenge. Some development teams choose an alternative approach. The approach preferred by many developers uses **objected-oriented analysis and design (OOAD)**, followed by **object-oriented programming (OOP)**.

Objects and Classes

What is an **object**? The world around us consists of objects: people, cars, buildings, trees, sweaters, books, and so on. In the everyday business world, there are objects such as employers, employees, customers, invoices, orders, accounts, and the like. The names of objects are usually nouns. That means objects are usually persons, places, or things. Each object has certain characteristics and exhibits certain behaviors. We learn about objects by studying their characteristics and their behaviors.

When we recognize that there are many similarities among certain objects, we tend to organize them into groups, or **classes**. For example, your friend's white Accord, your neighbor's dark blue Acura, and your father's gray ("pearl mist") Toyota Camry are all objects. They are also all cars. That is, they are all members of the car class. Cars have certain characteristics such as manufacturer, model, color, weight, wheels, doors, top speed, and so on. They also have certain behaviors: acceleration, deceleration, idling, turning, needing fuel or fuels of some kind (even electricity can be regarded as a fuel), and so on. Your friend's car is a specific example, or **instance**, of the car class. It possesses all the characteristics and behaviors of any car, but it is a specific example of one. Its manufacturer is Honda; its color is white; and so on. A Schwinn bicycle is not a car. A Boeing C17 airplane is not a car. When someone tells us that a certain "thing" is a car, we know some facts about it right away; that is, we have a generalized idea—a **model**, or an **abstraction**—of it. The abstraction is a good abstraction if it includes essential facts that distinguish the car, for example, from other types of vehicles. We can learn more about a car by studying the details of its characteristics and behaviors. The characteristics of the car are **data** about the car. The behaviors of the car are **operations** it performs.

Object-Oriented Design

Object-oriented analysis and design is based on organizing a program around a collection of classes. Each class represents a template from which any number of objects can be created. For example, if we were designing a program that allows students to view their course grades on a display screen, we might design a general Window class. Then individual objects of the Window class could be created. Two pieces of data associated with any object of the Window class might be its size and its position. Some operations that could be performed on any object of the Window class might be to move or reposition the window on the screen, to resize it, to minimize it, to maximize it, and so on. The specific objects of the Window class are the things that will be manipulated in the program, not the Window class itself, just as your father's Toyota Camry or your Ford Mustang is the thing you drive, not your generalized idea of a car.

A class definition contains both data and operations. As we saw previously, the data describes what an object looks like and the operations describe what an object does. The data that defines a class also may be referred to as **attributes**, **properties**, **instance variables**, **variables**, or **data members** (of the class). The behaviors also can be referred to as **methods**, **method members** (of the class), or **member functions**. We will use the terms *data members* and *methods* from this point forward when discussing classes and objects. We will sometimes use the term *variable* when referring to a certain data member in specific pseudocode.

As we have just explained, data members of a car class could include manufacturer, model, color, weight, and so on. Data members of a student class could include student name, student number, student permanent address, student local address, and so on. An example of a method associated with the car class might be to retrieve the color of the car for inclusion on a car registration document. One method associated with the student class might be to retrieve a student's name, given the student's number.

Developers of object-oriented applications often document their designs using the **Unified Modeling Language**, or **UML**. We will learn more about UML later; it is an industry-standard language for specifying, visualizing, and documenting object-oriented designs. Figure 10–1 shows two alternative icons that could be used in UML notation to represent the car class on a UML class diagram. Some developers always use the rectangle. Others prefer the cloud with the dashed lines as edges to emphasize that the class definition is an abstraction. Remember that program statements operate on specific objects, not on classes. The class definition in Figure 10–1

Figure 10–1
Car Class

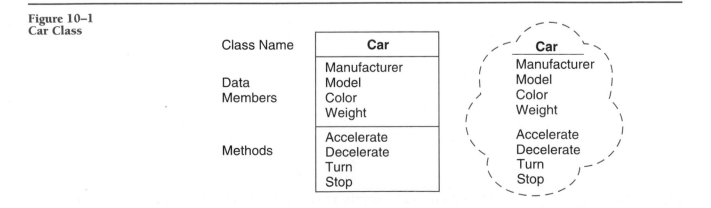

is not complete. Classes can be described at various levels of abstraction, precision, and completeness during the design process. In early stages of design, the diagram captures the basic aspects of the problem solution. In later stages, the developer may include design decisions and implementation details.

GradeBook Class

Now let us look at an example of a class in more detail. We define a class called GradeBook. It has five data members: Name, Grade1, Grade2, Grade3, and Average. The class has three methods: a method to input a student's name and three grades (GetStudent), a method to compute the average of the three grades (ComputeAverage), and a method to output the student's name (ShowStudent). Figure 10–2 shows the class GradeBook.

Encapsulation and Data Hiding

When we are actually implementing a class, we have some design decisions to make. How much information about the class do we want to be known only within the class? What information about the class needs to be known by other parts of the application? Suppose, for example, that you were designing and implementing a class to operate on bank account balances. You would need to create data members that were the account number, account balance, and so on. You would need to create methods to change the balance based on deposits, withdrawals, interest earned, and so on—methods written according to the policies of the bank. Would you want another developer to be able to write other methods to access the account balance and to operate on it differently—say, to direct funds within the account to another person's bank account? Obviously not!

The term **encapsulation** means to package data members and methods into a single well-defined programming unit (in our case, a class or an object created of that class). Encapsulation is most often achieved through **data hiding**, which is the process of hiding all the facts, or secrets, of a class that do not contribute directly to others' ability to use it. Generally, it means that the structure of a class and the code that implements its methods are hidden, or not revealed to other developers who are users of the class. It is good programming practice to design an application with these

Figure 10–2
GradeBook Class

	GradeBook
Class Name	**GradeBook**
Data Members	Name Grade1 Grade2 Grade3 Average
Methods	GetStudent ComputeAverage ShowStudent

two concepts in mind. In doing so, you will help to protect your code and
the data controlled by it from inadvertent or deliberate but unauthorized access. You also will make it possible for other developers to use the classes
you provide without having to know how the classes are implemented. At
some later time you may want to change some part of a class implementation. You can do so without having to worry about "side effects" elsewhere,
provided that you do not change the external boundary, or **interface**, of the
class that you have made available to other developers.

We can specify which parts of a class we want to be known only within
the class and which parts can be known outside the class by means of certain modifiers. Generally, we declare all data members to be **private**. A
data member that is private cannot be referred to by name from anywhere
outside the class within which it is declared. That is, only methods within
the class can access such a data member. Look, for example, at the
pseudocode representing the GradeBook class in Figure 10–3. The data
members Name, Grade1, Grade2, Grade3, and Average are all declared to
be private. Therefore, they can be accessed only by methods within our
GradeBook class.

Most methods are declared to be **public**. Such methods form the external
interface to the class. They can be accessed, or called, from outside the
class to perform the operations, or services, that the class provides. Remember that if the methods are defined wisely, the users of the class still
have no idea how the methods are really implemented. Our GetStudent,
ComputeAverage, and ShowStudent methods in Figure 10–3 are all declared to be public.

Methods also can be declared to be **private**. Such methods are only accessible within the class and cannot be called outside the class. They may
perform housekeeping functions, such as setting the values of certain data
members to zero.

A third modifier—**protected**—is sometimes useful. A protected data
member or method can be accessed (1) from within the class itself, (2)
from within subclasses of the class (we will learn more about subclasses

---

**Figure 10–3
GradeBook Class
Implementation 1**

```
Class GradeBook
        private Name
        private Grade1
        private Grade2
        private Grade3
        private Average
        public GetStudent
                Input Name, Grade1, Grade2, Grade3
        End
        public ComputeAverage
                Average = (Grade1 + Grade2 + Grade3) / 3
        End
        public ShowStudent
                Display Name, Average
        End
End Class
```

later), and (3) from other classes included with the class containing the protected data member or method in a grouping of related classes, known as a **package**. Although the concept of protected elements is not supported in all languages used for object-oriented programming, it is supported in C++ and Java, so we shall assume its availability in our discussions.

GradeBook Class Pseudocode

The title statement (Class GradeBook) in Figure 10–3 identifies and names the class we are creating. The End Class statement simply defines the end of the statements that make up the class. All statements inside the class are indented for readability. As in the GradeBook class definition, the data members are listed first in the pseudocode. The next group of statements shows the implementation plan for the three methods defined for the class. The code within each method is simple and should be familiar to you at this point. You may be thinking that listing the data members in our pseudocode is something we did not do when we used a procedure-oriented approach to design. You are right. That approach pays most attention to algorithms and the processing steps within them. An object-oriented approach emphasizes data and, once having identified the data, the operations on that data.

Once a class is defined, it can be used when building a program. A developer usually begins by creating an object, or instance, of the class. An object is a tangible representation of the data members and methods of the class to which it belongs. (Remember, just like a Toyota Camry is a specific implementation of the car class.) For example, a developer could create two grade books, that is, two objects belonging to the GradeBook class. Each object has its own set of data members and methods as defined in the GradeBook class. This concept is illustrated in Figure 10–4. Only a method in GradeBookM101 can access a data member in GradeBookM101. That is, a method in GradeBookS204 cannot access a data member in GradeBookM101. Even though GradeBookM101 and GradeBookS204 are objects of the same class, they are different objects. The class is simply a template. It is not until the developer creates objects of the class that representations of real grade books exist. We can assume here that Mary Jones is one of the students in Class M101. John Davis is a student in Class S204.

Driver Program— GradeBook Example

Now let us take a look at pseudocode (see Figure 10–5) for a program that actually creates and interfaces with the GradeBookM101 and GradeBookS204 objects we have discussed. This type of program is sometimes referred to as a **driver program**. It is the program that gets processing going. Our example illustrates one type of driver program, a relatively simple one. Driver programs that are just this simple often are used by developers to test the methods they create, to ensure the methods really behave as they expect. On a larger scale, driver programs can be fairly complex and are used in a wide range of contexts.

Figure 10–4
GradeBook Class Objects

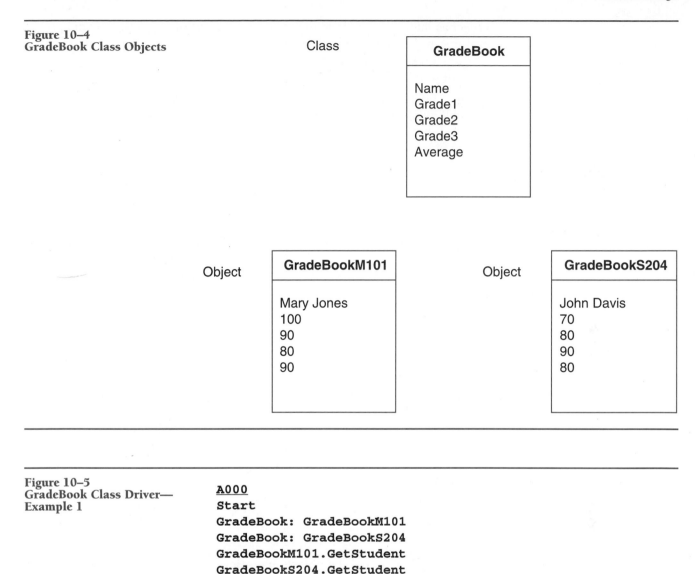

In this example, we are labeling our driver program A000. The first two statements following the Start statement cause two objects from the GradeBook class to be created. The objects are GradeBookM101 and GradeBookS204, which we have been discussing. Each statement consists of a class name and a colon followed by an object name:

class name: object name

The process of creating an object, or more precisely, of creating an instance of a class, is called **instantiation**. Most people treat the terms *instance* and *object* as interchangeable in the context of object-oriented design and programming. Any number of objects (instances) may be created from a class. The objects have the same behavioral capabilities, which are defined by the operations defined for the class. The objects differ in the particular data values stored in their data members, or variables. The access modifiers (private, public, protected) that we learned about earlier in this chapter apply to classes (all instances of a class) rather than to instances (a particular instance of a class).

Subsequent statements in Figure 10–5 show how methods within these two objects are called, or invoked. The general form of these statements is

object name.method name

This notation expresses the notion that we are sending a **message** to an object. Stated simply, a message is a call to an object method. As the term *message* implies, it is the means of communicating with the object. The object responds by doing the operations within the particular method identified by the method name. The method will send back one or more return values to the caller if doing so is a part of the processing logic within the method that is called. We might say that an object's methods are services that it offers to its callers. The set of all public methods of a class forms the external boundary, or interface, of the class, as we mentioned earlier.

The statement GradeBookM101.GetStudent sends a message to the object GradeBookM101 to execute the code in the method GetStudent. We know from our earlier look at the GetStudent method that, when it executes, the user of the application containing the GradeBookM101 object will be prompted to enter a name and three grades. The entries will be stored in the variables Name, Grade1, Grade2, and Grade3 of the object GradeBookM101 by the Input statement within the GetStudent method (see Figure 10–3). Similarly, the GradeBookS204.GetStudent statement calls the GetStudent method of object GradeBookS204. This method executes, causing four input values to be stored in the variables Name, Grade1, Grade2, and Grade3 of the object GradeBookS204.

You might wonder how two different values can be stored in what appears to be the same variable—for example, Name. Remember, each object has its own set of data members. Thus, the user may have input "Mary Jones" the first time and "John Davis" the second time. In that case, we could use a similar dot notation to say GradeBookM101. Name is now equal to "Mary Jones" and GradeBookS204.Name is now equal to "John Davis." Note, however, that we would get an error message if we included the statement GradeBookM101.Name in our driver program. Name is a private variable and cannot be referred to outside the object. Indeed, we (the caller, in this case) may not even know the names of the variables within the GetStudent method (remember data hiding). All we need to know is the name of the method. The methods are public so we can refer to them by name from outside the object, as we do in our driver program.

After the data for each object is input, the ComputeAverage method for each object is invoked. When the ShowStudent method is invoked, Name and Average are displayed (see the Display statement in the ShowStudent method in Figure 10–3), but which name and average? The method invocation determines which set of data will be accessed.

When GradeBookM101.ShowStudent is invoked, GradeBookM101.Name and GradeBookM101.Average are displayed. When GradeBookS204.-ShowStudent is invoked, GradeBookS204.Name and GradeBookS204.-Average are displayed.

Constructors

Whereas a class is an abstraction, an object exists in time and space. It has a life span. When we create an object of a certain class, we often want certain initializing actions performed. Examples of such actions are allocating memory for the data members of the class and setting their initial values. A **constructor** is a special kind of method that is designed to perform such initializations.

If you are certain you are not going to refer to a variable within an object until a value has been set for it explicitly by a statement in a method in the object, you may choose to let the memory allocation be done by a **default constructor**. Such a constructor is commonly provided by the C++ or Java language-processor program (interpreter or compiler) that will convert your program to an executable form. However, the default constructor may not (generally, will not) initialize any of the data members. So far, we have assumed the use of a default constructor in our driver program (Figure 10–5). But, if the ComputeAverage method were invoked by a caller before the GetStudent method, the data members representing the three grades would be undefined. There is nothing we include in our GradeBook class definition that prevents a caller from doing just that. When the computer attempted to add the three grades, an error would occur. To avoid such errors, it is good programming practice to define your own constructors in any classes you design.

We write a constructor method in much the same way as we write other methods. However, a unique characteristic of a constructor method is that it must have the same name as the class that contains it. This name identifies it to a language-processor program as a constructor. Also, unlike other methods, a constructor is executed automatically when the object that contains it is created. That is, constructors do not need to be called explicitly like other methods do.

Figure 10–6 is a simple constructor for the GradeBook class. This constructor sets all data members to their initial values. It ensures that all of the data members will be defined as soon as the object is created.

Although this constructor addresses the problem of undefined data, it is not very interesting. Consider the constructor in Figure 10–7. This constructor method makes use of a list of **parameters** placed in parentheses

Figure 10–6
GradeBook Class Constructor (No Parameters)

```
public GradeBook
        Name = "Any Student"
        Grade1 = 0
        Grade2 = 0
        Grade3 = 0
        Average = 0
End
```

| **Figure 10–7**<br>**GradeBook Class Constructor**<br>**(Four Parameters)** | ```
public GradeBook(AnyName, AnyGr1, AnyGr2, AnyGr3)
        Name = AnyName
        Grade1 = AnyGr1
        Grade2 = AnyGr2
        Grade3 = AnyGr3
        Average = 0
End
``` |

| **Figure 10–8**<br>**GradeBook Class Driver—**<br>**Example 2** | ```
A000
Start
GradeBook: GradeBookM101
GradeBook: GradeBookS204
GradeBook: GradeBookG200("Sally Marsh", 100, 90, 80)
GradeBookM101.GetStudent
GradeBookS204.ComputeAverage
GradeBookG200.ComputeAverage
GradeBookM101.ShowStudent
GradeBookS204.ShowStudent
GradeBookG200.ShowStudent
Stop
``` |

after the constructor name. The parameters are a list of four variable names that serve as placeholders. The placeholders will be replaced by some actual values when the method is called.

Now let us modify our driver program so that, when it causes a third new object of the class GradeBook to be created, it provides initial values for the first four data members in the list of data members in the new GradeBook object. Let us name the new object GradeBookG200 (see Figure 10–8). The new statement looks like this:

```
GradeBook: GradeBookG200("Sally Marsh", 100, 90, 80)
```

This statement causes a new object called GradeBookG200 to be created. When the constructor in Figure 10–7 is executed, the four values (called **arguments**) in the parentheses are assigned to the four variables in the constructor parameter list as follows:

AnyName	= "Sally Marsh"
AnyGr1	= 100
AnyGr2	= 90
AnyGr3	= 80

This constructor provides flexibility because the initial values of the first four data members can be determined when a GradeBook object is created. Of course, the caller can still invoke the GetStudent method to input new values for the name and three grades during program execution. Note that the data member Average is still initialized to 0 by the constructor in Figure 10–7, just as it is by the constructor in Figure 10–6, because there is no variable name in the parameter list that corresponds to it.

Although we have introduced the use of parameters and arguments in discussion of a constructor method, their use is not confined only to constructor methods. You can define parameters as part of any method you create. A general form of the **method heading**, or **method signature**, which is the first statement within the method, is

public method-name(parameter1, parameter2, . . . , parametern)

The general form of a statement to invoke such a method is

object-name.method-name(argument1, argument2, . . . , argumentn)

When this call is executed, the value specified as the first argument is assigned to the variable named in the first parameter of the method's parameter list; the value specified as the second argument is assigned to the variable named in the second parameter; and so on. If no parameters are specified in the method signature, then no arguments can be specified in an invocation of the method. The signature of a method is the external boundary, or **interface**, between the method and its callers.

Overloading

Figure 10–9 shows the complete class definition of GradeBook with both constructors included. You will note that both constructors have the same name—the name of the class. How, then, does the computer know which constructor method to execute?

When an object is created, the number and type of arguments specified in the statement that causes the object to be created determine which constructor will be called. For example, the statement GradeBook: MyGradeBook, which does not include any arguments, will cause the first constructor in Figure 10–9 to be executed. The statement GradeBook: YourGradeBook("Tom Nelson", 70, 80, 80), which has four arguments, will cause the second constructor in Figure 10–9 to be executed. The ability to use the same method name to invoke different methods that perform different actions based on the number or type of arguments in the method invocation is called **overloading**. We say that the method is an **overloaded method**. Like the use of parameters and arguments, the use of overloading is not restricted only to constructor methods. Within one class, you can have two or more definitions of any single method name. When you overload a method name, either (1) any two definitions of the same method name must have different numbers of parameters or (2) some parameter position within any two definitions of the same method name must be of different data types.

Look again at the driver program shown in Figure 10–8. This program creates three objects of the GradeBook class. When objects GradeBookM101 and GradeBookS204 are created, the first constructor for each object is called because no arguments are included in the object creation statements. When the object GradeBookG200 is created, the second constructor is called because there are four arguments in the object creation statement. Note that in this example the GetStudent method is invoked only for GradeBookM101. The ComputeAverage method is invoked only for GradeBookS204 and GradeBookG200.

Figure 10–9
GradeBook Class
Implementation 2

```
Class GradeBook
        private Name
        private Grade1
        private Grade2
        private Grade3
        private Average
        public GradeBook
                Name = "Any Student"
                Grade1 = 0
                Grade2 = 0
                Grade3 = 0
                Average = 0
        End
        public GradeBook(AnyName, AnyGr1, AnyGr2, AnyGr3)
                Name = AnyName
                Grade1 = AnyGr1
                Grade2 = AnyGr2
                Grade3 = AnyGr3
                Average = 0
        End
        public GetStudent
                Input Name, Grade1, Grade2, Grade3
        End
        public ComputeAverage
                Average = (Grade1 + Grade2 + Grade3) / 3
        End
        public ShowStudent
                Display Name, Average
        End
End Class
```

Figure 10–10
GradeBook Class Driver
Output

```
Pat Morgan          0

Any Student         0

Sally Marsh         90
```

Let us assume that when the GetStudent method of the GradeBookM101 object is invoked, the user enters the data values "Pat Morgan", 100, 100, 100. Figure 10–10 shows the output when the ShowStudent method of each object is invoked. The data members of GradeBookM101 are initialized by the first constructor and then reset to the specific values entered by the user when the GetStudent method of GradeBookM101 is called. Note, however, that the average is never computed for GradeBookM101 because the ComputeAverage method of GradeBookM101 is not called. Therefore, an average of 0 is displayed on the first line. Although the ComputeAverage method of GradeBookS204 is invoked, the values for Grade1, Grade 2, and Grade 3 are still 0. (Remember they were set to 0 by the first constructor.) Therefore, an average of 0 is displayed on the second line of output. Can you see why the name "Any Student" is displayed? The third object GradeBookG200 is created using the second constructor. Because no additional data is input, the ComputeAverage method of GradeBookG200 uses the values for Grade1 (100), Grade2 (90), and Grade3 (80) that were set up when the second constructor was executed. The average is computed to be 90 by the ComputeAverage method of GradeBookG200. That value and the name "Sally Marsh" as set up by the constructor are displayed. Make sure that you understand how each object manipulates its own data. Write some variations of the driver program and see what output results. (See Exercises 12 and 13.)

Destructors

Just as constructors may be executed when objects are created, **destructors** may be executed when objects are destroyed. The destructor method of an object may, for example, release the memory allocated for the object's data members. In our examples so far, we assume a **default destructor** is executed for each object. As with constructors, we can write our own destructors. Because destructors are less commonly specified explicitly, they will not be covered in this book.

Sample Problem 10.1 (Determining the Perimeter)

Problem:

The developer of several homes within a housing project has decided to build a redwood fence to enclose each plot of land on which a home is built. Define a class that will accept the dimensions (in feet) of a plot of land as input and calculate the perimeter of the plot as one measure of how much redwood will have to be purchased to complete the project. The result of the perimeter calculation, together with the plot number of the plot, should be displayed as output.

Solution:

Figure 10–11 shows one way to define a class that can be used in addressing this problem. You should see readily the data members of the class, the constructor method for the class, and the methods that perform the required operations. Do you see any ways the class might be improved? (See Exercises 14 and 15.)

Sample Problem 10.2 (Determining the Perimeter and Area with Parameters and Return Values)

Problem:

Now suppose that the developer wants to know not only the perimeter of each plot of land but also the area of the plot. Instead of allowing the measurements of length and width to be entered as input by the user, assume they are passed from a calling object or program to an object of the class that provides the perimeter and area calculations. Assume that the computed values for perimeter and area are to be returned to the calling object or program rather than displayed as output.

Figure 10–11
PerimeterCalc Class

```
Class PerimeterCalc
        private PlotNumber
        private Length
        private Width
        private Perimeter
        public PerimeterCalc
                PlotNumber = "None"
                Length = 0.0
                Width = 0.0
                Perimeter = 0.0
        End
        public GetMeasurements
                Input PlotNumber, Length, Width
        End
        public ComputePerimeter
                Perimeter = (2.0 * Length) + (2.0 * Width)
        End
        public ShowResults
                Display PlotNumber, Perimeter
        End
End Class
```

Figure 10–12
PlotCalc Class—
Implementation 1

```
Class PlotCalc
        private Length
        private Width
        private PlotCalc
                Length = 0.0
                Width = 0.0
        End
        public PlotCalc(PlotLength, PlotWidth)
                Length = PlotLength
                Width = PlotWidth
        End
        public float ComputePerimeter
                 Return 2.0 * (Length + Width)
        End
        public float ComputeArea
                Return Length * Width
        End
End Class
```

Solution:

Figure 10–12 shows one way to define a class that can be used in addressing this problem. Note that length and width are listed as parameters in the second constructor method heading. That means that values for length and width can be passed as arguments to an object of this class when the object is created.

As we said earlier, a method can return a value to its caller; that is, it can provide a **return value**. That is exactly what the ComputePerimeter and

ComputeArea methods of the PlotCalc class in Figure 10–12 are designed to do.

A general form of the method heading for a method that returns a value is

public data-type method-name(parameterl,
parameter2, . . . , parametern)

The phrase *data-type* in this general form indicates that we must tell the data type of the value to be returned by the method as a part of the method heading. In our example, we have specified *float*, which stands for floating-point data; this specification allows data values with decimal points such as 210.2 and 350.75 to be returned. If we had specified *integer* as the data type, only whole numbers such as 210 and 350 could be returned. We include the data type, even in our pseudocode, because any caller of the method must know the data type of the return value. The data type of the return value determines the legal uses of the method in any calling object or driver program. In our example, the ComputePerimeter and ComputeArea methods can be invoked any place in a calling object or program that it is legal to use a value that has a data type of float.

Figure 10–13 shows a driver program to test our class definition in Figure 10–12. Since we want to pass the values for two arguments in our instantiation of the PlotCalc class, we initialize two variables (MyEastSide and MySouthSide) that will contain the argument values in our driver program. Two additional variables—MyPerimeter and MyArea—will be used to hold return values. The statement PlotCalc: BigPlot(MyEastSide, MySouthSide) causes an object named BigPlot of the PlotCalc class to be instantiated and passes the current values of MyEastSide and MySouthSide (300 and 125, in our case) to the BigPlot object. The statement MyPerimeter = BigPlot.ComputePerimeter causes the ComputePerimeter method of the BigPlot object to be executed, and the result of that execution ($2.0 \times (300 + 125)$, or 850) to be returned to our driver program and assigned to the variable MyPerimeter. Similarly, the statement MyArea = BigPlot.ComputeArea causes the ComputeArea method of the BigPlot object to be executed and the result of that execution (300×125, or 37,500) to be returned to our driver program and assigned to the variable MyArea.

A method can be designed to return more than one value. If more than one one value is returned, all of the values must be of the same data type.

**Figure 10–13
PlotCalc Class Driver—
Example 1**

```
A000
Start
MyEastSide = 300.0
MySouthSide = 125.0
MyPerimeter = 0.0
MyArea = 0.0
PlotCalc: BigPlot(MyEastSide, MySouthSide)
MyPerimeter = BigPlot.ComputePerimeter
MyArea = BigPlot.ComputeArea
Write MyPerimeter, MyArea
Stop
```

Figure 10–14
PlotCalc Class—
Implementation 2

```
Class PlotCalc
        private Length
        private Width
        public PlotCalc
                Length = 0.0
                Width = 0.0
        End
        public PlotCalc (float PlotLength, float PlotWidth)
                Length = PlotLength
                Width = PlotWidth
        End
        public float ComputePerimeter(float Length, float Width)
                Return 2.0 * (Length + Width)
        End
        public float ComputeArea(float Length, float Width)
                Return Length * Width
        End
End Class
```

Most likely, the values are part of an array structure. The concepts and use of arrays are covered in Chapters 9 and 13 of this book. We will not deal with them in this chapter.

Look again at Figure 10–12. Do you see what might be a drawback of the program? Only the values provided to the PlotCalc object at object instantiation time can be used in the computations. We might want to design the class so that a user of either of the computational methods could include the values to be used in the computations in any number of different calls to the methods (see Figure 10–14). To do so, we include the length and width as parameters in the method headings. We go a step further than we did in our pseudocode previously by also indicating the data type of each parameter in the method headings. That is, the general form of the method heading becomes

public data-type method-name(data-type parameter1,
data-type parameter2,. . .)

You should include the data types of parameters as well as of return values in the method headings of all methods you create that provide return values and/or have parameters. Why? Because not only the number of arguments but also the data types of arguments passed in a method call must match, or be compatible with, those in the heading of the method that is being called. If you are designing with data hiding in mind, you can make your method headings available to any other developers who may use your classes; the method headings will contain the information needed by those developers to use them effectively. In our example here, since we have included two parameters of data type float in the ComputePerimeter and ComputeArea method headings, any call to either of these methods must include two arguments whose values are floating-point data. If not, an error will occur.

Key Terms

object-oriented analysis
  and design (OOAD)
object-oriented
  programming (OOP)
object
class
instance
model (abstraction)
data
operation
attribute
property
instance variable
variable

data member
method
method member
member function
Unified Modeling
  Language (UML)
encapsulation
data hiding
interface (of a class)
private member
public member
protected member
package
driver program

instantiation
message
constructor
default constructor
parameter
argument
method heading
  (method signature)
interface (of a method)
overloading
overloaded method
destructor
default destructor
return value

Exercises

1. Distinguish between procedure-oriented design and object-oriented design.

2. Distinguish between a class and an object.

3. Name and define the two parts of a class definition.

4. What is an abstraction? How do we use abstractions in object-oriented design?

5. What is Unified Modeling Language (UML)?

6. What is meant by encapsulation, and why is it useful?

7. What is meant by data hiding, and why is it useful?

8. Why might you specify that a certain data member is private when constructing a class definition?

9. What is one purpose of a driver program?

10. What is meant by instantiation?

11. What is meant by overloading?

12. Given the following driver program and the class definition shown in Figure 10–9, what output will result?

```
A000
Start
GradeBook: MyGradeBook("Jane Rynn", 80, 70, 100)
GradeBook: YourGradeBook
YourGradeBook.ComputeAverage
MyGradeBook.ShowStudent
YourGradeBook.ShowStudent
Stop
```

13. Given the following driver program and the class definition shown in Figure 10–9, what output will result?

```
A000
Start
GradeBook: MyGradeBook("Jane Rynn", 80, 70, 100)
GradeBook: YourGradeBook
MyGradeBook.ComputeAverage
MyGradeBook.ShowStudent
YourGradeBook.ShowStudent
Stop
```

14. Look at the PerimeterCalc class definition shown in Figure 10–11. Modify the class so that it has a second constructor that provides for plot number, length, and width to be provided as arguments of the constructor. What difference does that make on the potential use of the class?

15. Now suppose the PerimeterCalc class shown in Figure 10–11 is to be used, but a caller who has a plot that is a square should be able to provide only one argument value for length and width (rather than two that are just the same) in the method invocation. Modify the PerimeterCalc class to accommodate this requirement.

16. Assume that a plot of land for which perimeter and area are to be calculated is 100 feet in width across the front and 180 feet in depth (front to back). Modify the driver program in Figure 10–13 to instantiate a new object named CornerPlot and use it to compute the perimeter and area. What values should be provided as output by the Display statement in your driver program?

17. Construct a driver program to test the ComputePerimeter and ComputeArea methods in Figure 10–14. If the length of the plot of land you are dealing with is 230 feet and the width is 330 feet, what value should be determined for perimeter? What value should be determined for area?

18. Define a class that captures the logic of the temperature conversion problem shown in Figure 2–9.

19. Construct a driver program to test the class that you created in response to Exercise 18.
    (a) What are the input values?
    (b) What is the expected output?

    Check to verify that the output received is what you expected. If not, find where errors have occurred.

Inheritance

Objectives

Upon completion of this chapter you should be able to

- Explain what is meant by a generalization/specialization relationship and give examples.
- Explain inheritance and use it in program design.
- Distinguish between a base class and a derived class.
- Distinguish between overloading and overriding and use either or both in program design.
- Distinguish between a polymorphic method and an overloaded method and use either or both in program design.
- Define the term *polymorphism*.
- Illustrate a class hierarchy by constructing a class diagram.
- Understand how constructors of different classes in a family relate to one another at execution time.
- Identify, and use in program design, an abstract class.
- Explain the purpose and use of the Unified Modeling Language, or UML.
- Show why inheritance is useful in object-oriented design and programming.

Introduction

In the preceding chapter, you were introduced to the ideas of classes and objects. You learned about data members and methods, about instantiating objects and invoking an object's methods. In this chapter, we introduce additional concepts of object-oriented design.

Classes can have certain relationships with other classes. One type of relationship between classes is a generalization/specialization relationship. An inheritance mechanism applied to such a relationship enables sharing of identifications and descriptions. Developers often use inheritance to simplify and shorten their implementation tasks. In fact, some people argue that a programming language that does not support inheritance is not an object-oriented language and a program that does not employ inheritance is not an object-oriented program.

You will also be introduced to the concept of polymorphism. As we will see, the capabilities related to polymorphism are also useful to developers of classes and objects, to other developers who write calls to the objects, and even to users of the applications built using object-oriented design and programming.

Generalization/ Specialization

Think again of our discussion of cars in Chapter 10. Suppose we had really begun our discussion by talking about vehicles. We might have listed manufacturer, owner, color, weight, and maximum capacity as data members of the vehicle class. Not only cars, but also airplanes, buses, carriages, even bicycles, might belong to the vehicle class. Now suppose we want to deal with objects at a less abstract level. We want to distinguish among cars, airplanes, buses, carriages, and so forth. We can say that the vehicle class and the car class have a **generalization/specialization relationship**. We also can say that the vehicle class and the airplane class have a generalization/ specialization relationship. The ways cars and airplanes are alike are facts that we use to discern that both cars and airplanes belong to the general class of vehicles. But there are also facts that are true about cars that are not true about airplanes. These are the specializations. We might ask ourselves: What makes a car different from an airplane, or an airplane different from a bus, or a carriage different from a bicycle?

Perhaps we are writing an application for a transportation firm and we want to recognize some important characteristics of buses that do not apply to all of the other vehicles. Examples might be the city authorized vehicle license number, assigned company garage location, cubic feet of luggage space underneath the passenger area, cubic feet of luggage space in overhead racks, and the like. These details are specializations, relative to the more general vehicle class.

When two classes have a generalization/specialization relationship, the more specific class is always fully consistent with the more general class and contains additional information. An instance of the more specific class may be used wherever the more general class is allowed.

Inheritance

In object-oriented design, **inheritance** is the mechanism that allows us to define a very general class and then later define more specialized classes by adding some new details to the existing, more general class definition. We take advantage of the generalization/specialization relationship. Doing so saves work because the more specialized classes inherit all the properties of the general class, and we only have to define the new details. In our previous example, we have our existing vehicle class. Now we define a bus class that is a subclass of the vehicle class (see Figure 11–1). Our bus class can inherit all the characteristics, or data members, of the vehicle class and all of its methods. We only have to add the data members and methods that are unique to buses. For another application, we might define a car class that is another subclass of the vehicle class, a truck class that is a subclass of the vehicle class, and so on.

Figure 11–1
A Generalization/
Specialization Relationship

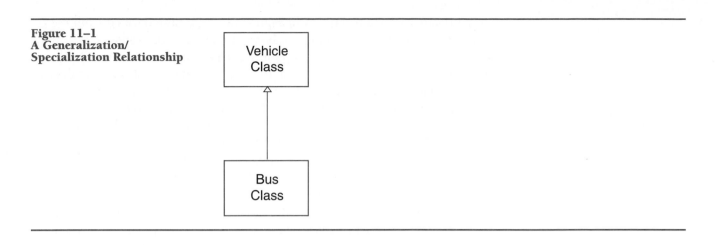

Class Hierarchy

The concept of inheritance is illustrated more fully in the UML **class diagram** in Figure 11–2. We see an original class named Employee and two subclasses named Faculty and Staff. The Employee class is the **base class** containing data members and methods that apply to all employees. For example, all employees have a Name and an Id. Employee, GetEmployee, ShowEmployee, and ZapEmployee are the methods of the Employee class. A base class may also be called a **parent class** or a **superclass**.

Figure 11–2
Class Diagram—Employee
Example

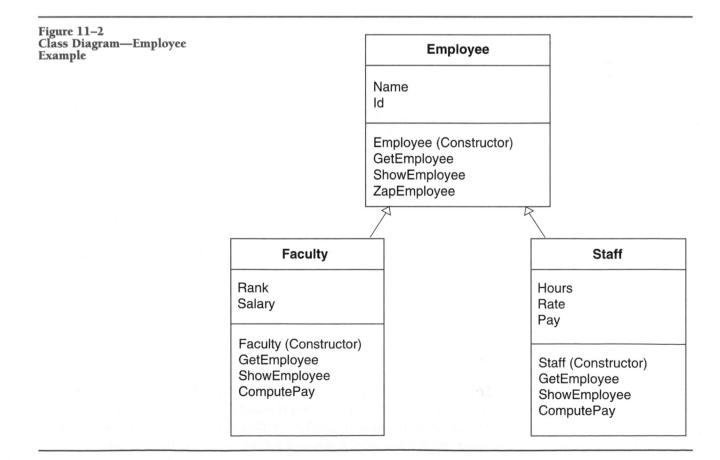

The Faculty and Staff classes in Figure 11–2 are both derived from the base class Employee. We call them **derived classes**, **child classes**, or **subclasses**. The unshaded arrows pointing from the derived classes to the base class are used in UML notation to denote a generalization/specialization relationship. This relationship is also sometimes referred to as an **"is-a" relationship** (perhaps because that is easier for us to remember). A faculty member *is a* kind of employee. A staff member *is a* kind of employee. Just as a car *is a* kind of vehicle, a bus *is a* kind of vehicle, and so on. An unshaded arrow always points from the more specific class to the more general, or more abstract, class. A family of classes that are related in this manner is referred to as a **class hierarchy**.

The Faculty class includes two additional data members that are not in the base class: Rank and Salary. The Staff class includes three additional data members that are unique to the Staff class: Hours, Rate, and Pay. Name and Id are inherited from the Employee class and do not have to be restated in the subclass definitions.

Notice also that each class has a constructor method, identified by the name of the class, as well as several other methods. All three classes also have two methods with the same name—GetEmployee and ShowEmployee. Even though a subclass inherits all the methods from its base class, that subclass can contain a new design and new implementation of any inherited method. The process of reimplementing in a subclass a method inherited from a base class is called **overriding**. In this example, each subclass has its own implementation of the two methods GetEmployee and ShowEmployee. In contrast, the method ZapEmployee has not been reimplemented in either subclass. Therefore, the original implementation specified in the Employee class is reused in each subclass.

Polymorphism

Both the Faculty and Staff subclasses include a ComputePay method that was not included in the design and implementation of the base class Employee. This method was not inherited from the Employee class. It was added to each subclass as an additional method. A subclass cannot inherit a method from another subclass at its same level (any more than you can inherit characteristics, or traits, from a sister or brother). Therefore, we can conclude that each implementation of the ComputePay method is unique. That makes sense because the pay for a faculty member is likely to be calculated differently than the pay for a staff member. We say that the ComputePay method in the Faculty subclass and the ComputePay method in the Staff subclass are polymorphic methods. A **polymorphic method** is a method that has the same name as a method in another class (or even another method in each of several classes) within the same class hierarchy.

Do you see how this technique is different from overloading, which we studied in Chapter 10? An overloaded method is one that has different implementations within the same class, only one of which will be invoked by any message that is sent to an object of that class. Which method implementation is executed depends on the data types and/or number of arguments in the invocation. As we have explained, a polymorphic method is a method that has the same name as a method in one or more other classes of the same family but has different implementations for the various classes. Which method is wanted is stated explicitly in the method invocation. To call the ComputePay method in the Faculty subclass, we might write

`Faculty.ComputePay`

To call the ComputePay method in the Staff subclass, we might write

`Staff.ComputePay`

From a conceptual point of view, both overloaded methods and polymorphic methods are implementations of **polymorphism**. The term *polymorphism* comes from a Greek word meaning "many forms." Polymorphism is one of the important ease-of-use features of object-oriented design and programming. It would be inconvenient, for example, if multiple developers were creating methods in various subclasses within a class hierarchy, and they had to keep checking to ensure they were not unintentionally naming their methods the same. Conversely, if method implementations perform roughly the same functions but with certain unique differences in different subclasses, it is easier for callers of the methods if the methods have the same name. Programming errors are less likely to occur.

Polymorphism is used extensively in commercial programming. Think of the many times you use a double-click action to cause something to happen when you are using a computer. Each double-click causes a double-click event to be raised. It, in turn, calls upon a polymorphic method. If you double-click on the name of a file folder, a listing of the items in that folder is shown. If you double-click on an application icon on the display, a program is executed. If you double-click on the name of a Microsoft Word document, the Microsoft Word program is activated if you have not activated it previously, and the document is opened (shown on your display screen). Do you see how easy polymorphism makes things from the user's point of view?

Employee Class

Figure 11–3 shows the pseudocode for the Employee class. The makeup of this class is straightforward. A single constructor (the Employee method) initializes the data members Name and Id. The methods GetEmployee and ShowEmployee input and display the Name and Id. The method ZapEmployee displays a message that a particular employee is no longer employed. This method is included to illustrate a method that is inherited and not reimplemented in a subclass.

Faculty Subclass

Figure 11–4 shows the pseudocode of the Faculty subclass. The first statement defines the class name, as before, but indicates in the parentheses that this class is a subclass of the base class Employee. A single constructor (the Faculty method) initializes the data members Rank and Salary. The methods GetEmployee and ShowEmployee are reimplemented in this subclass to input Rank and to display Rank and Salary. Notice that both of these methods invoke the original GetEmployee and ShowEmployee methods defined in the base class Employee. This technique often is used to take advantage of previously written code. The method ComputePay is a new

Figure 11–3
Employee Class Pseudocode

```
Class Employee
        private Name
        private Id
        public Employee
                Name = "Any Employee"
                Id = 9999
        End
        public GetEmployee
                Display "Please enter employee name and id"
                Input Name, Id
        End
        public ShowEmployee
                Display Name, Id
        End
        public ZapEmployee
                Display Name, "no longer works here"
                Name = "Former Employee"
                Id = 9999
        End
End Class
```

Figure 11–4
Faculty Subclass Pseudocode

```
Class Faculty (base Employee)
        private Rank
        private Salary
        public Faculty
                Rank = "No rank assigned"
                Salary = 0
        End
        public GetEmployee
                Employee.GetEmployee
                Display "Please enter rank"
                Input Rank
        End
        public ShowEmployee
                Employee.ShowEmployee
                Display Rank, Salary
        End
        public ComputePay
                CASENTRY Rank
                        CASE "Instructor"
                                Salary = 25000
                        CASE "Assistant Professor"
                                Salary = 30000
                        CASE "Associate Professor"
                                Salary = 40000
                        CASE "Professor"
                                Salary = 50000
                        CASE other
                                Salary = 0
                                Display "Invalid rank"
                ENDCASE
        End
End Class
```

method included in the Faculty subclass to assign a specific salary based on the rank of the faculty member. A simple CASE control structure implements this design.

Staff Subclass

Figure 11–5 shows the implementation of the Staff subclass. The first statement defines the class name and, again, indicates in the parentheses that this class is a subclass of the base class Employee. A single constructor initializes the data members Hours, Rate, and Pay. The methods GetEmployee and ShowEmployee are reimplemented in this subclass as well to input Hours and Rate and to display Pay. Notice, again, that both of these methods invoke the original GetEmployee and ShowEmployee methods defined in the base class Employee. The method ComputePay is a new method included in the Staff subclass to compute the pay based on hours and rate. A simple IFTHENELSE control structure implements this design.

Figures 11–4 and 11–5 illustrate the two subclasses Faculty and Staff by listing all the data members and methods defined in each subclass. These figures do not, however, show the two data members—Name and Id—and the four methods—Employee, GetEmployee, ShowEmployee and ZapEmployee—that are inherited from the base class Employee. It can be

Figure 11–5
Staff Subclass Pseudocode

```
Class Staff (base Employee)
        private Hours
        private Rate
        private Pay
        public Staff
                Hours = 0
                Rate = 0
                Pay = 0
        End
        public GetEmployee
                Employee.GetEmployee
                Display "Please enter hours and rate"
                Input Hours, Rate
        End
        public ShowEmployee
                Employee.ShowEmployee
                Display Pay
        End
        public ComputePay
                IF Hours > 40 THEN
                        Pay = (40 * Rate) + ((Hours - 40) * (1.5 * Rate))
                ELSE
                        Pay = Hours * Rate
                ENDIF
        End
    End Class
```

difficult, especially in longer, more complex programs, to trace back up the inheritance chain to find the definition of an inherited method. We can alleviate this problem by providing a fully instantiated view of an object. This is done by showing not only the data members and methods of derived classes but also the data members and methods inherited from the base class. Figure 11–6 shows the fully instantiated view of the Faculty subclass. Figure 11–7 shows the fully instantiated view of the Staff subclass. At first glance, it may be confusing to see what appear to be two implementations of the GetEmployee method and two implementations of the ShowEmployee method in the Faculty subclass, for example. We will see how it all works in the following discussion of the driver program.

Driver Program— Employee Example

Figure 11–8 shows a driver program that creates two objects: Faculty1 and Staff1. Faculty1 is an object of the subclass Faculty. Staff1 is an object of the subclass Staff. When an object of a subclass is created, processing proceeds as if a call to the constructor of the base class were the first statement in the definition of the constructor for the subclass. Thus, when the object Faculty1 is created, the constructor for the Employee class executes first, initializing Name to "Any Employee" and Id to 9999. The statements within the constructor for the Faculty subclass execute next, initializing Rank to "No rank assigned" and Salary to 0. Similar actions occur when the object Staff1 is created.

The statement Faculty1.GetEmployee invokes the GetEmployee method in Faculty1. You may be wondering which GetEmployee method is invoked. The code in the GetEmployee method implemented especially for the Faculty subclass is executed first. It, in turn, invokes the GetEmployee method of the base class Employee. (See the statement Employee.GetEmployee within the GetEmployee method of Faculty.) The GetEmployee method from the Employee class requests the Name and Id from the user. When it finishes executing, control is returned to the GetEmployee method of Faculty1. The Display statement in that method requests the Rank of the employee (faculty member, in this case) from the user.

Refer again to the driver program in Figure 11–8. The Faculty1.ComputePay statement in the driver program invokes the ComputePay method of the Faculty1 object to compute the salary for a faculty member based on his or her rank. The Faculty1.ShowEmployee statement in the driver program invokes the ShowEmployee method implemented especially for the Faculty subclass. It, in turn, invokes the ShowEmployee method of the base class Employee. The ShowEmployee method from the Employee class displays the Name and Id of the employee for whom data is currently being processed. When it finishes executing, control is returned to the Display statement in the ShowEmployee method of Faculty1. The Display statement in that method displays the Rank and Salary of the employee (faculty member, remember).

A similar series of events occurs when the next three statements in the driver program are executed to invoke the GetEmployee, ComputePay, and ShowEmployee methods of the Staff1 object. We now know that the GetEmployee and ShowEmployee methods of Employee are, in turn, invoked during this processing.

Figure 11–6
Faculty Subclass
Pseudocode—Fully
Instantiated View

```
Class Faculty (base Employee)
        private Name (inherited from base class - Employee)
        private Id (inherited from base class - Employee)
        private Rank
        private Salary
        public Employee (inherited from base class - Employee)
                Name = "Any Employee"
                Id = 9999
        End
        public Faculty
                Rank = "No rank assigned"
                Salary = 0
        End
        public GetEmployee (inherited from base class - Employee)
                Display "Please enter employee name and id"
                Input Name, Id
        End
        public ShowEmployee (inherited from base class - Employee)
                Display Name, Id
        End
        public GetEmployee
                Employee.GetEmployee
                Display "Please enter rank"
                Input Rank
        End
        public ShowEmployee
                Employee.ShowEmployee
                Display Rank, Salary
        End
        public ZapEmployee (inherited from base class - Employee)
                Display Name, "no longer works here"
                Name = "Former Employee"
                Id = 9999
        End
        public ComputePay
                CASENTRY Rank
                        CASE "Instructor"
                                Salary = 25000
                        CASE "Assistant Professor"
                                Salary = 30000
                        CASE "Associate Professor"
                                Salary = 40000
                        CASE "Professor"
                                Salary = 50000
                        CASE other
                                Salary = 0
                                Display "Invalid rank"
                ENDCASE
        End
End Class
```

Figure 11–7
Staff Subclass Pseudocode—
Fully Instantiated View

```
Class Staff (base Employee)
        private Name (inherited from base class - Employee)
        private Id (inherited from base class - Employee)
        private Hours
        private Rate
        private Pay
        public Employee (inherited from base class - Employee)
                Name = "Any Employee"
                Id = 9999
        End
        public Staff
                Hours = 0
                Rate = 0
                Pay = 0
        End
        public GetEmployee (inherited from base class - Employee)
                Display "Please enter employee name and id"
                Input Name, Id
        End
        public ShowEmployee (inherited from base class - Employee)
                Display Name, Id
        End
        public GetEmployee
                Employee.GetEmployee
                Display "Please enter hours and rate"
                Input Hours, Rate
        End
        public ShowEmployee
                Employee.ShowEmployee
                Display Pay
        End
        public ZapEmployee (inherited from base class - Employee)
                Display Name, "no longer works here"
                Name = "Former Employee"
                Id = 9999
        End
        public ComputePay
                IF Hours > 40 THEN
                        Pay = (40 * Rate) + ((Hours - 40) * (1.5 * Rate))
                ELSE
                        Pay = Hours * Rate
                ENDIF
        End
End Class
```

Finally, the Faculty1.ZapEmployee statement in the driver program invokes the ZapEmployee method for the Faculty1 object. Because no ZapEmployee method was created especially for the Faculty subclass, the ZapEmployee method inherited from the base class Employee is executed.

Figure 11–8
Employee Class Driver

```
A000
Start
Faculty: Faculty1
Staff: Staff1
Faculty1.GetEmployee
Faculty1.ComputePay
Faculty1.ShowEmployee
Staff1.GetEmployee
Staff1.ComputePay
Staff1.ShowEmployee
Faculty1.ZapEmployee
Stop
```

Figure 11–9
Employee Class Driver
Output

```
Joe Black      1234    Professor    50000

Mary Dolan     5656    400

Joe Black no longer works here
```

Note that a subclass object can invoke a method from the base class because all methods from the base class really become part of the subclass. However, an object of a base class cannot invoke a method solely defined in a derived class. For example, if Employee1 were an object of the Employee class, then the statement Employee1.ComputePay would be illegal because the method ComputePay is not defined in the base class Employee.

Assume the driver program in Figure 11–8 is executed. Further assume the user enters "Joe Black", 1234, and "Professor" in response to the prompts for input in the GetEmployee method of the Faculty1 object. Assume the user enters "Mary Dolan", 5656, 40, and 10 in response to the prompts for input in the GetEmployee method for the Staff1 object. Figure 11–9 shows the output that would result. Make sure that you understand how each line of output is generated. Then write some variations of the driver program and see what output results.

Sample Problem 11.1 (A Structure for Bank Account Processing)

Problem:
Citizen's State Bank offers the following choices of checking and savings accounts to its customers: Standard Checking, Prime Checking, Access Checking, Student Checking, Regular Savings, and Maximize Interest Savings. Construct a high-level class diagram to suggest how classes might be organized to simplify the design and implementation of the required processing for these accounts.

Solution:
Figure 11–10 shows alternative UML class diagrams that could be established in architecting the solution to this problem. Figure 11–10a shows

**Figure 11–10
Class Diagram—Account
Example**

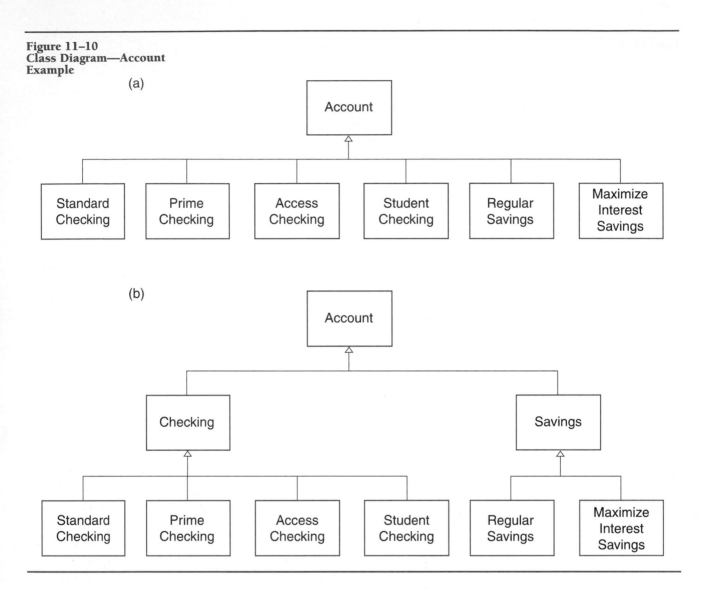

two levels of hierarchy in the family of account classes. Figure 11–10b shows three levels of hierarchy. Which one is correct? Well, there is no such thing as the "perfect class structure" or the "right set of objects." Some developers insist that identification of classes and objects is the hardest part of object-oriented analysis and design.

We know that the class at the highest level of abstraction in this diagram—the Account class—is the most general class. It contains information that applies to all classes in the family. If there are many data members and/or many methods that do not apply to all accounts, but do apply to all checking accounts, we may choose to have a level of abstraction that deals with checking accounts as a class (as in Figure 11–10b). That will save us the work of defining the same information for four different checking account classes. We will have another class at the same level as the class for checking accounts that contains data members and/or methods that apply to all savings accounts. On the other hand, if there is very little information that is common to all checking accounts but does not also apply to all savings accounts—or if there is very little information that is common to all savings accounts but does not also apply to all checking accounts—having the three levels of abstraction shown in Figure 11–10b will not prove to be very useful.

Sample Problem 11.2 (Implementing the Account Class)	**Problem:** Define an Account class that can be used in setting up an account-processing application that uses the family of account classes shown in Figure 11–10a. **Solution:** First, let us consider some data members that all types of accounts have in common: Account Type, Account Number, and Account Balance, to name a few. We are going to need at least some of these data members in our account processing (see Figure 11–11). There are also some operations that can be done on all types of accounts. We can deposit funds in any of these accounts. We can withdraw funds from any of these accounts, assuming the current balance in the account is large enough to cover the withdrawal. It is reasonable to expect that some of the data members and some of the methods can and should be specified in the Account class, which is the base class of the family. The pseudocode for the Account class is shown in Figure 11–11. The variables AccountNumber, NewBalance, and Message are defined to be protected members of the Account class. As we learned in Chapter 10, that means these variables are accessible not only to methods in this class but also to methods in any subclasses of this class. In contrast, methods of the Faculty subclass in the previous example are not able to access the variables Name and Id in the Employee superclass because these variables are defined to be private to the Employee superclass. Because a protected variable is accessible to methods in subclasses, the protected modifier should be used with care to ensure unintended modifications of the value of the variable cannot occur during processing. We will see why and how this capability is useful to us here shortly. Do you see why Message is needed? This variable will contain a status message that is returned to a calling object or program. There is a default constructor and a constructor that allows the instantiator of an Account object to specify values for AccountNumber and NewBalance when the Account object is created. As in some of our previous examples, these constructors are examples of an overloaded method. Which constructor is invoked at execution time depends on how many arguments are specified in the method invocation. Note that the data type of the parameter AccountNo is *string*. A string variable holds alphanumeric data referred to as a *character string* in preceding chapters. The Deposit method is pretty straightforward. Notice what happens in the Withdrawal method: If an attempted withdrawal causes the account balance (NewBalance) to be less than 0, the withdrawal is denied. Remember, a method can return one or more values, but all values returned by a single method must be of the same data type. That is why we need the other, separate methods RetAccountNumber and RetStatusMsg.

Figure 11–11
Account Class Pseudocode

```
Class Account
        protected AccountNumber
        protected NewBalance
        protected Message
        public Account
                AccountNumber = "ZZZ999999"
                NewBalance = 0.00
                Message = "Awaiting a transaction."
        End
        public Account(string AccountNo, float Balance)
                AccountNumber = AccountNo
                NewBalance = Balance
                Message = "Awaiting a transaction."
        End
        public integer RetAccountNumber
                Return AccountNumber
        End
        public string RetStatusMsg
                Return Message
        End
        public float Deposit(float Amount)
                NewBalance = NewBalance + Amount
                Message = "Deposit processed successfully."
                Return NewBalance
        End
        public float Withdrawal(float Amount)
                NewBalance = NewBalance - Amount
                IF NewBalance > 0 THEN
                   Message = "Withdrawal processed successfully."
                   Return NewBalance
                ELSE
                   NewBalance = NewBalance + Amount
                   Message = "Insufficient funds. Withdrawal denied."
                   Return NewBalance
                ENDIF
        End
End Class
```

A driver program to test this class is shown in Figure 11–12. We are passing constants as arguments to the instantiated object MJStandard. Alternatively, we could have assigned values to variables in our driver program and then specified those variables as arguments in our calls to methods of MJStandard. Either way works. If this were a long or complex program that was going to be reused extensively, assigning the values to variables in one section of the program would be advisable. You may wonder why we ask the MJStandard object to return the Account Number to us when we passed it to MJStandard in the first place. Well, when we see the Account Number displayed as output, we can check to ensure that we really typed what we think we typed. In an inherited method in a subclass of

Figure 11–12
Standard Checking Account
Class Driver

```
A000
Start
AccountNumber = "AAA654321"
MJBalance = 0.00
StatMessage = "OK"
Account: MJStandard("NHL123456", 623.00)
MJBalance = MJStandard.Deposit(125.00)
AccountNumber: MJStandard.RetAccountNumber
StatMessage = MJStandard.RetStatusMsg
Write AccountNumber, MJBalance, StatMessage
MJBalance = MJStandard.Withdrawal(150.00)
AccountNumber: MJStandard.RetAccountNumber
StatMessage = MJStandard.RetStatusMsg
Write AccountNumber, MJBalance, StatMessage
Stop
```

Figure 11–13
Standard Checking Account
Class Driver Output

```
NHL123456 748 Deposit processed successfully.
NHL123456 598 Withdrawal processed successfully.
```

the Account family, the Account Number may be used to find the matching Account Name, for example. We need to get it right. Do you see why any caller of the object MJStandard must invoke the RetStatusMsg after requesting a deposit or withdrawal? That is the way to check that the transaction happened successfully. The output of our driver program is shown in Figure 11–13.

Sample Problem 11.3 (Implementing the Standard Checking Account Class)

Problem:

Define a Standard Checking Account class to process deposits and withdrawals to a Standard Checking Account. Take advantage of the Account class you designed in response to Sample Problem 11.2. However, there are some specific rules that apply only to Standard Checking Accounts:

- If an attempt is made to withdraw more funds than are currently in the account, the withdrawal is refused, and an overdraft fee (currently $35.00) is deducted from the account.
- If, after a withdrawal is made from the account, the new balance is less than the minimum balance to avoid a fee per transaction (currently $100.00), a transaction fee (currently $3.00) is deducted from the account.

Solution:

The pseudocode for the Standard Checking Account class (StandardChecking) is shown in Figure 11–14. Remember, this class is really a subclass of the Account class in Figure 11–11. It inherits all of the data members and all of the methods from the Account class.

Figure 11–14
Standard Checking Account
Class Pseudocode

```
Class StandardChecking (base Account)
      private MinBaltoAvoidTranFee
      private TranFee
      private OverdraftFee
      public StandardChecking
            super("ZZZ999999", 0.00)
            MinBaltoAvoidTranFee = 100.00
            TranFee = 3.00
            OverdraftFee = 35.00
      End
      public StandardChecking(string AccountNo, float Balance)
            super(AccountNo, Balance)
            MinBaltoAvoidTranFee = 100.00
            TranFee = 3.00
            OverdraftFee = 35.00
      End
      public StandardChecking(string AccountNo, float Balance, float Minimum,
                              float Tran, float Overdraft)
            super(AccountNo, Balance)
            MinBaltoAvoidTranFee = Minimum
            TranFee = Tran
            OverdraftFee = Overdraft
      End
      public float Withdrawal(float Amount)
            NewBalance = NewBalance - Amount
            IF NewBalance > 0 THEN
               IF NewBalance < MinBaltoAvoidTranFee THEN
                  NewBalance = NewBalance - TranFee
                  Message = "Withdrawal processed. Transaction fee applied."
                  Return NewBalance
               ELSE
                  Message = "Withdrawal processed successfully."
                  Return NewBalance
               ENDIF
            ELSE
               NewBalance = NewBalance + Amount - OverdraftFee
               Message = "Insufficient funds. Overdraft fee applied."
               Return NewBalance
            ENDIF
      End
End Class
```

Look first at the default constructor and the two constructors that can be
used to initialize data members with values specified by the creator of a
StandardChecking object. Only one of the constructors will be executed in
any creation of a StandardChecking object, depending on how many

arguments are specified in the object activation statement in the calling object or program. Remember that, during execution, by default, the default constructor of the base class is invoked first, then the constructor for the subclass at the next level in this branch of the hierarchy, and so on. In this example, we do not want to invoke the Account default constructor. Instead, we want to use the Account constructor that accepts values for AccountNo and Balance. The keyword *super* in the first statement in the default constructor for this StandardChecking class

```
super("ZZZ999999", 0.00)
```

causes the values ZZZ123456 and 0.00 to be assigned to the variables AccountNo and Balance listed in the method heading of the second Account constructor. Thus, the default constructor of the base class Account will not be invoked. Alternatively, the keyword *super* in the statement

```
super(AccountNo, Balance)
```

in each of the other StandardChecking constructors causes values specified in the activation of a StandardChecking object to be passed to the second Account constructor before subsequent processing occurs.

We need not include a method for handling deposits in the subclass StandardChecking because the Deposit method in the base class Account works just fine for this subclass too. We do want to override the Withdrawal method in Account because of the unique rules that apply to Standard Checking Accounts (see the bulleted list in the problem statement for this sample problem). Study the processing within the Withdrawal method in StandardChecking to verify for yourself that it carries out the processing specified in the rules. Constructing a driver program to test the Withdrawal method is left as an exercise (see Exercise 17).

Abstract Class

Before we leave this problem, it is useful to consider one additional perspective. Suppose we decide to implement the base class Account as a way of grouping together all of the classes for the various types of accounts we may want to process. But we are convinced that the method for handling withdrawals, at least, will be unique to each type of account. We can implement the Withdrawal method in the base class Account and always override it. We can simply omit the Withdrawal method from the base class Account and implement it in each subclass (as we did the ComputePay method in our previous example). Another approach is to indicate that descendants of the Account class always support a withdrawal type of behavior but not to write the plan for a specific implementation of Withdrawal in the base class Account. That approach is used in Figure 11–15. We indicate in the pseudocode in Figure 11–15 that the Account class is abstract by placing the word *Abstract* at the beginning of the title statement that identifies the class Account.

A class in which one or more of the methods is identified but not defined is called an **abstract class**. Setting up some abstract classes can help you to simplify or to organize your thinking during the design stage. You or other developers responsible for designing new classes can then derive the new

Figure 11–15
Account Class as an Abstract
Class Pseudocode

```
Abstract Class Account
        protected AccountNumber
        protected NewBalance
        protected Message
        public Account
                AccountNumber = "ZZZ999999"
                NewBalance = 0.00
                Message = "Awaiting a transaction."
        End
        public Account(string AccountNo, float Balance)
                AccountNumber = AccountNo
                NewBalance = Balance
                Message = "Awaiting a transaction."
        End
        public integer RetAccountNumber
                Return AccountNumber
        End
        public string RetStatusMsg
                Return Message
        End
        public float Deposit(float Amount)
                NewBalance = NewBalance + Amount
                Message = "Deposit processed successfully."
                Return NewBalance
        End
        public float Withdrawal(float Amount)
End Class
```

classes using an abstract class as the base class for the new classes (i.e., create them as subclasses). However, you cannot create any objects of a base class that is an abstract class. Further, you must define a method for any method not defined in the abstract class in any subclass you create. Otherwise, your new subclass also will be an abstract class, and you will not be able to create any objects of it either. That is why this approach may help to ensure the completeness of class definitions. In discussion, classes that are not abstract classes are sometimes referred to as **concrete classes**, just for our own clarification purposes.

Think back to the class of vehicles discussed early in this chapter. If we were creating an application that deals with vehicles, we might very well decide to treat vehicles as an abstract class. All vehicles have some characteristics in common. Therefore, it is probably reasonable to establish a vehicle class as the base class. But we might want to deal at a finer level of abstraction—say, cars, bicycles, trucks, or whatever—before we define many of the specifics of implementation. If we are simply dealing with the buying and selling of vehicles, the actions (behavior) to buy or to sell may be definable at the vehicle level for all. On the other hand, the method for selling a multimillion-dollar airplane may differ significantly from the method for selling a $20,000 car, which may in turn differ significantly from the method for selling a scooter.

Unified Modeling Language (UML)

As discussed in Chapter 10, a model is a generalized idea, or abstraction. It is a pattern, or representation, of something. The "something" may already exist in the real world. Alternatively, the model may be an evolving plan for "something" that may or will exist at some point in the future. Engineers, architects, and, yes, even advertising agencies and fashion designers use models. In our study here, we are creating models for business applications.

The flowcharts we have used in planning the procedure-oriented logic within program modules in earlier chapters of this book are one kind of model. The pseudocode we are using in this chapter and in other chapters is another kind of model. The class diagrams we are using to plan object-oriented data structures are yet another kind of model. They are expressed using the **Unified Modeling Language** or **UML**.

UML is the industry-standard language for specifying, visualizing, and documenting object-oriented design. It is a very rich language and includes semantic concepts (i.e., meaning), notation, and guidelines for use of the language. We will use only a small portion of UML in our work here, but if you continue to work in information systems areas, you are likely to encounter UML, so you should know more about it.

As the term *unified* suggests, UML was created as an attempt to simplify and consolidate the large number of object-oriented development practices that had emerged in the 1960s, 1970s, 1980s, and early 1990s. The UML definition was led by Rational Software's Grady Booch, Ivar Jacobson, and Jim Rumbaugh, three of the early "fathers" of design methodologies.[1] UML was adopted unanimously as a standard by the Object Management Group, an industry consortium of many businesses and other organizations, in November 1997. Work on a major revision to the standard, UML 2.0, is underway and should be completed by 2005.

The definition of UML does not encompass the definition of a standard development process. It is intended to support iterative development—for example, the various types of UML diagrams may contain more and more levels of detail as a development effort proceeds. Developing an accurate, understandable model for an application is just as essential to software development as developing a blueprint is to the construction of a building. Good models foster effective communication among the users, analysts, designers, and programmers involved in the specification, implementation, and use of a system. CASE tools that can be used to create UML diagrams and even to generate programming-language statements from UML diagrams are widely available.

---

[1] Grady Booch, Ivar Jacobson, and James Rumbaugh are co-authors of *The Unified Modeling Language Reference Manual*, published by Addison-Wesley in 1999, as the definitive reference for UML. It also contains a useful bibliography.

Key Terms

generalization/ specialization relationship	superclass	polymorphic method
	derived class	polymorphism
	child class	abstract class
inheritance	subclass	concrete class
class diagram	"is-a" relationship	Unified Modeling
base class	class hierarchy	Language (UML)
parent class	overriding	

Exercises

1. What is meant by inheritance?

2. Distinguish between a base class and a derived class.

3. (a) Suggest some terms that mean the same as the term *base class*.
   (b) Suggest some terms that mean the same as the term *derived class*.

4. Suppose that rather than starting with Employee as the base class in Figure 11–2, we had started with Person, which is a higher level of abstraction. Some members of the Person class are employees. Others are students. Redraw the class diagram in Figure 11–2 to include this higher level of abstraction.

5. Now suppose the Faculty class can be subdivided further into college deans, full-time faculty members, and part-time faculty members. Show how the class diagram that you drew in response to Exercise 4 should be modified to reflect these additional classes.

6. Assume a superclass of Food has been defined. Suggest some generalization/specialization relationships with other items that might apply.

7. Assume a super class of Athlete.
   (a) Suggest some classes that might be defined at a second level of a class hierarchy.
   (b) Given the second level of classes that you defined in Exercise 7(a), suggest some classes at a third level in the hierarchy.
   (c) Suggest some classes at a fourth level in the hierarchy for at least one of the classes you have defined in Exercise 7(b).
   (d) Construct a class diagram of the class hierarchy you have defined.

   Check the class hierarchy you have defined by making certain that a generalization/specialization, or "is-a," relationship applies at every place you have shown such a relationship on your class diagram.

8. Distinguish between overloading and overriding.

9. Distinguish between an overloaded method and a polymorphic method.

10. Given the following driver program and the class definitions shown in Figures 11–3, 11–4, and 11–5, what output will result? Assume that the user entered "Harry Bell", 5567, and "Instructor" in response to the prompts for input in the GetEmployee method for the NewFaculty object and "Joan Mason", 7833, 50, and 15 in response to the prompts for input in the GetEmployee method for the NewStaff object.

```
A000
Start
Faculty: NewFaculty
Staff: NewStaff
NewFaculty.GetEmployee
NewFaculty.ComputePay
NewFaculty.ShowEmployee
NewStaff.GetEmployee
NewStaff.ComputePay
NewStaff.ShowEmployee
Stop
```

11. Given the following driver program and the class definitions shown in Figures 11–3 and 11–4, what output will result?

```
A000
Start
Faculty: AnotherFaculty
AnotherFaculty.ComputePay
AnotherFaculty.ShowEmployee
AnotherFaculty.ZapEmployee
Stop
```

12. Refer to the definition of the Account class in Figure 11–11. What is the meaning of the protected modifier preceding the names of the data members?

13. Refer to the method heading for the Deposit method in Figure 11–11.
  (a) What is the significance of the first term *float* in the method heading?
  (b) What is the significance of the second term *float* in the method heading?

14. (a) Expand the driver program in Figure 11–12 to test the Account class for correct handling of an attempted withdrawal from an account when there are not enough funds in the account to handle it.
  (b) Show the output you expect from your test execution.

Walk through the execution of your driver program. If the output you determine does not match the output you expected, find where an error has occurred.

15. Refer to the definition of the Standard Checking Account class in Figure 11–14.
  (a) What happens when the statement super("ZZZ999999", 0.00) is executed?
  (b) What happens when the statement super(AccountNo, Balance) is executed?

16. Refer to the definition of the Standard Checking Account class in Figure 11–14.
  (a) What advantage does the third constructor offer that the second constructor does not provide?
  (b) What advantage does the second constructor offer that the first constructor does not provide?

17. Assume that

- The rules for Sample Problem 11.3 remain in force as stated.
- The current balance for account MHR777888 is $316.30.
- An attempt is made to withdraw $120.00 from the account.

(a) Construct a driver program to test the Withdrawal method of the subclass StandardChecking in Figure 11–14 using this data.

(b) Show the output you expect when the driver program is executed.

Walk through the execution of your driver program. If the output you determine does not match the output you expected, find where an error has occurred.

18. Refer again to Figure 11–14. Assume that

- The rules for Sample Problem 11.3 remain in force as stated except that the overdraft fee is changed to $40.00.
- The current balance for account PLT444234 is $60.00.
- An attempt is made to withdraw $100.00 from the account.

(a) Modify the driver program you constructed in response to Exercise 17 to test the Withdrawal method using this data.

(b) Show the output you expect when the driver program is executed.

Walk through the execution of your driver program. If the output you determine does not match the output you expected, find where an error has occurred.

19. Refer to Figure 11–15.
(a) Why do we call the Account class in Figure 11–15 an abstract class?
(b) What restriction applies to it?

Other Class and Object Relationships

12

Objectives

Upon completion of this chapter you should be able to

- Understand the following kinds of relationships between classes (and objects of those classes) and give examples of each:
  - association
  - aggregation
  - composition
- Understand and explain relationships and accompanying notation on a UML class diagram.
- Suggest the classes and class relationships needed to solve a well-understood problem and show those classes and relationships on a UML class diagram.
- Explain what is meant by multiplicity and apply it in program design.
- Explain what an inner class is and use inner classes in program design.
- Explain what a program switch is and use program switches in program design.
- List the advantages of object-oriented design and programming.

Introduction

Think about the world around you. It is filled with objects—people, places, and things—and you often classify those objects without even realizing you are doing so. Do you see any flowers? Carnations? Daisies? Roses? Yellow roses? White roses? Rose petals? Thorns? Flower pots? Vases? A watering can?

We can make the following observations:

- A carnation is a kind of flower.
- A daisy is a kind of flower.
- A rose is a kind of flower.
- A yellow rose is a kind of rose.
- A white rose is a kind of rose.
- A rose petal is a part of a rose.
- A thorn is a part of a rose.

- A vase holds one or more flowers.
- A watering can is used to give water to one or more flowers.

Now let us think about the classes we see here. Flowers are a class. Carnations are a class. Daisies are a class. Roses are a class. In fact, a rose is a kind of flower. A daisy is a kind of flower. A carnation is a kind of flower. These are all examples of the generalization/specialization, or "is-a," relationship we saw in Chapter 11. We can say that roses, daisies, and carnations are all subclasses of flowers. What do they all have in common? Well, they all have petals, they all have colors, they all have fragrances, and so on. In what ways are the subclasses unique? What makes a rose different from a daisy? The thorns, for one thing! We have also noted a distinction between yellow roses and red roses—yellow roses and red roses can be another layer of subclasses in our family of flowers if we decide that is useful. Is a petal a kind of rose? No. Is a vase a kind of flower? No. Classes (and objects of those classes) have relationships with other classes (and objects) that are different than generalization/specialization relationships. (Stop to think—each of us is an object, and we certainly have relationships with other objects that are not generalization/specialization relationships.) We are going to learn about some of the relationships in this chapter.

Association

An **association** is a relationship among two or more specified classes that describes connections among objects (instances) of those classes. Associations provide the connections through which objects of different classes can interact. They are, in fact, the "glue" that holds a system together.

Look at the UML class diagram in Figure 12–1. It shows an employment relationship. A Company is an employer in this example. A Person is an employee. This is an example of the most common kind of association—a **binary association** between a pair of classes. The notation for a binary association is a line, or **path**, connecting the participating classes. We place the name of the association—WorksFor, in our example—near the path but away from either end to avoid confusion. We place a small filled triangle near the **association name** to show the ordering of the classes, or how to "read" the name. In Figure 12–1, we see "Person works for Company."

The terms *employee* and *employer* in Figure 12–1 are **role names**. The purposes of a role name are to identify an association end and to support navigation from one object to another using the association. For example, given an object Person1 of the Person class, the expression

Person1.employer

Figure 12–1
Association—Example 1: WorksFor

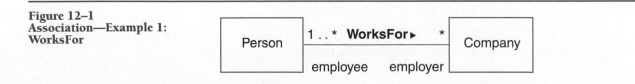

yields a set of objects of the Company class—the companies at which Person1 is employed. Within a **self-association** (an association involving the same class more than once—say, Person gave birth to Person), role names are necessary to identify the association ends attached to the same class (the role names might be Mother and Child, in our example). Otherwise, role names are optional because the class names can be used to identify the ends. They often are useful for documentation purposes.

Notice also the symbols containing asterisks in Figure 12–1. The single asterisk (*) near the Company class tells us that the number of companies a person may work for is zero or more, without limit. The symbol 1..* near the Person class tells us that the number of persons a company may employ is one or more, without limit. That is, a person may not work for any company or may work at any number of companies. A company must have at least one employee and may have any number of employees greater than one. If a person could work for at most one company, we would place the symbol 0..1 near the Company class instead of the asterisk (*) that is shown in Figure 12–1. We call this characteristic **multiplicity**. It tells how many instances of one class in the association relationship can be related to one instance of the other class in the relationship (see Figure 12–2).

An instance of an association is a **link** (just as an instance of a class is an object). As an application executes, links between objects are created and destroyed, just as objects themselves are created and destroyed. If an object that is an association end ceases to exist, the link between that object and the object that is the other end of the association ceases to exist as well. The link has no identity apart from the ordered list of objects in it. That is, you cannot refer explicitly to the link by means of a link name in a statement in a program.

No duplicate links may exist. That is, there cannot be two identical links between the same two objects. (In our example, a person cannot work for the same company more than once.) The number of appearances of an object in a set of links must be compatible with the multiplicity specified for each end of the association. (In our example, because both multiplicities are unlimited, the number of links in the set at any particular time may be hard to estimate. If, instead, the symbol near Company were 0..1, then for each person, there exists at most one person-employer pair. If we know the

Figure 12–2
Examples of Multiplicity

Symbol	Meaning
0..1	0 or 1
1..1	Exactly 1
0..*	0 or more, without limit
*	Same as 0..*
1..*	1 or more, without limit
0..3	0, 1, 2, or 3
3	Exactly 3
48..50	48, 49, or 50
12..*	12 or more, without limit
1..3,7,10	1, 2, 3, 7, or 10

number of persons we are dealing with, the total number of links is at most equal to that number.)

Sample Problem 12.1 (Vases and Flowers)

Problem:

Just for practice, let us apply what we have learned so far to our observation of a real-world relationship from earlier in the chapter: A vase holds one or more flowers. Assume we are tracking flowers in vases for a local nursery that sells floral arrangements to customers. Prepare a UML class diagram to show the classes needed, just for the tracking aspect of this situation, and their relationships.

Solution:

We might ask ourselves, is "vase" a general enough class for this association? Or do we want to allow for a broader class of containers—say, one that includes baskets or flower pots? Is "flower" a general enough class for the contents of the container? Or do we want to allow for a broader class of contents—say, one that includes plants as well? Well, for now we will stick with vases and flowers because that is the scope of the problem presented. A UML class diagram showing the association between these classes is shown in Figure 12–3.

The class diagram indicates that a vase holds one or more flowers. (A vase is not interesting to us, or relevant to our problem situation, if it is empty. The nursery does not sell vases without contents.) We may observe that there is a practical limit to the number of flowers that a single vase can contain. So far, none has been stated, but that is a refinement that may come at a later stage of design. A flower can be held by at most one vase. The 0..1 symbol indicates that a flower might not be contained in a vase at all. (Perhaps it has been picked but not yet included in a bouquet. The nursery might sell it without the vase so we need to track it—especially because it is perishable; that is, it has a "limited shelf life.")

Associations are first recognized at the analysis stage of problem solving. During design, associations capture design decisions about data structures as well as about the separation of responsibilities among classes. The directionality of an association is important and is reflected in its ordered list (meaning, for example, vase.contents when we are talking about the Holds relationship, not contents.vase or flower.container). The implementations of associations in various programming languages may be pointers, container classes embedded in other classes, or even table objects—that detail is beyond the scope of this text. The objective here is to introduce associations and facts about them that are important in object-oriented design.

**Figure 12–3
Association—Example 2:
Holds**

Aggregation

Recall our previous statements: A rose petal is a part of a rose. A thorn is a part of a rose. Here we see a special kind of association relationship—a **whole-part**, or an **aggregation, relationship** between an **aggregate** (a whole) and a **part**. A rose is an aggregate. A petal is a part. A thorn is a part. A clue to the existence of an aggregation relationship is the phrase "*is a part of.*" Some people also look for the phrase "has a" and refer to aggregation as a **"has-a" relationship**. (A rose *has a* rose petal; a rose *has a* thorn; and so on.) In any case, a whole-part situation must exist or the relationship is not aggregation.

Aggregation is not unique to object-oriented design. Any programming language that supports record-like structures containing fields, or data items, supports aggregation. Aggregation is not inheritance and does not imply inheritance. In our previous example, a petal does not inherit any attributes or methods from a rose. However, the combination of inheritance with aggregation is powerful and is not available in procedure-oriented design (see Exercise 9).

In a UML class diagram, one end of an association relationship is designated the aggregate by means of an **unfilled diamond**, as shown in Figure 12–4. The other end—the part—is unmarked. In Figure 12–4, the symbols for multiplicity indicate that one or more petals can be a part of one rose, and one or more thorns can be a part of one rose.

As Figure 12–4 suggests, an aggregate class can have multiple parts. Each relationship is separate from any other relationships. The drawing shown in the class diagram in Figure 12–4a always works. If two or more associations have the same multiplicity (as in our example—a rose has one or more petals, and a rose has one or more thorns), the aggregation associations may be drawn as a tree, as shown in Figure 12–4b.

Managing Assets Example

Now let us look at an example from the business world. The UML class diagram in Figure 12–5 shows an aggregate class InvestorAssets that has an aggregation relationship with a part class Stock, an aggregation

Figure 12–4
Aggregation—Example 1

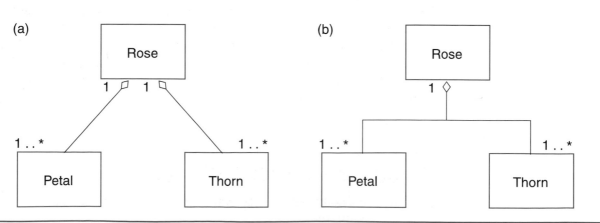

Figure 12–5
Aggregation—Example 2

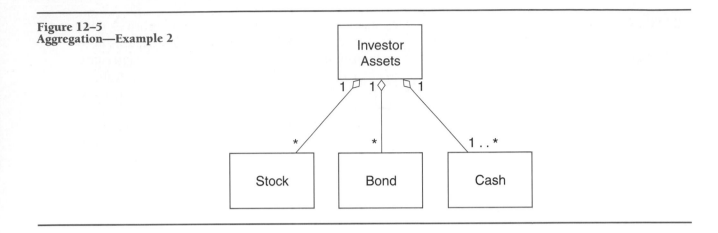

relationship with a part class Bond, and an aggregation relationship with a part class Cash. We understand from this diagram that an instance of the Investor Assets class is the whole, or aggregate. Zero or more instances of the Stock class is one of its parts. Zero or more instances of the Bond class is another of its parts. And one or more instances of the Cash class is another of its parts. That is, an investor may or may not own any stocks; an investor may or may not own any bonds; but an investor will have at least some cash, which may be kept all in one place or in multiple places.

We know, in the real world, individual stocks or bonds or supplies of cash may be acquired and disposed of independently. Similarly, an instance of the Stock class may be created or destroyed during processing. An instance of the Bond class may be created or destroyed during processing. And so on. But the class that is the whole, or aggregate—Investor Assets—will continue to exist. That is, the lifetimes of objects of these classes are completely independent, although there is still conceptually a whole-part relationship (any stock owned by the investor is always a part of the investor's assets during the time that it is owned, and so on).

A link instantiated from an aggregation association must conform to certain rules. A rule that applies to an aggregation relationship that does not apply to the more general association relationships described previously is that an object may not directly or indirectly be part of itself. That is, there can be no **cycles**, or circles back to a class (or to an object of that class), in the directed paths of aggregation links. In our real-world situation, an InvestorAssets can never be part of a Stock, or Bond, or Cash. A Stock can never be part of a Stock. A Bond can never be part of a Bond. A Cash can never be part of Cash. As the term *link* implies, an aggregate can communicate with its parts. That is, the aggregate may send messages to its parts during processing. Similarly, a part can communicate with its aggregate during processing.

Composition

Now look at the schematic in Figure 12–6. It shows a window as it might appear on a display screen—a familiar part of interactive applications in today's business world. We see a window that has two components: a header, or title, and a frame, or body. The frame in turn contains numerous icons (buttons, labels, and so on).

**Figure 12–6
A Frame for Measurement
Conversion**

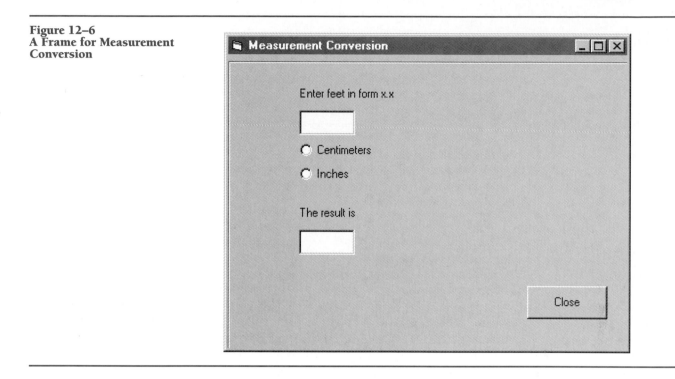

Figure 12–7 shows a UML class diagram of the relationships among the classes established to facilitate processing. This diagram looks a lot like the diagrams we saw earlier in the chapter, but the **solid-filled diamonds** indicate composition relationships. A **composition** is a form of aggregation relationship in which two additional rules apply: (1) An object may be a part of only one **composite** (the whole, in this relationship) at a time, and (2) a composite has sole responsibility for the disposition of its parts. Having responsibility for the disposition of its parts means that the composite is responsible for their creation and destruction. In actual implementation, that means it is responsible for their memory allocation and deallocation. As you will learn through experience, without adequate care, memory allocation and deallocation errors are all too likely to occur and can be very hard to find. The visible symptoms of memory errors, such as error messages that identify the effects of the memory errors but do not point to insufficient memory as the cause, may seemingly have no relation to memory conditions.

As shown in Figure 12–7, Window is a composite that has objects of the Header class and the Frame class as parts. Frame, in turn, is a composite that has any number of Button, Label, TextComponent, and CheckBoxGroup objects as its parts. TextArea and TextField are parts of the composite TextComponent. Any number of CheckBox objects are parts of the composite CheckBoxGroup. (In common usage, members of the CheckBox class grouped within a CheckBoxGroup are known as radio buttons.) Indirectly, the parts of TextComponent and CheckBoxGroup are also parts of the composite Frame and parts of the composite Window.

During its instantiation, a composite must ensure that all of its parts have been instantiated and correctly assigned to it. It can create a part or assume responsibility for an existing part. But, during the life of the composite, no

Figure 12–7
Composition—Example 1

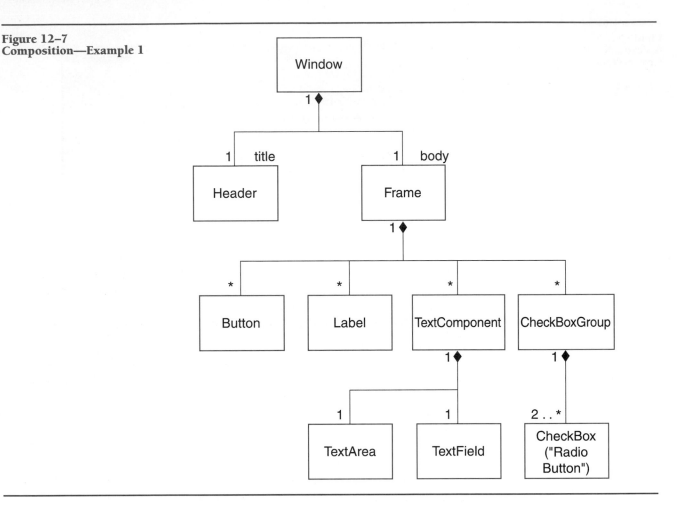

other composite may have responsibility for the part without the composite's concurrence. The behavior (methods) of a composite class can be designed knowing that no other class can destroy, or deallocate the memory for, any of its parts. A composite may add additional parts during its lifetime if the multiplicities permit (e.g., if only one of the parts exists and the part multiplicity is 1..3, not 1..1). It may destroy parts, provided the multiplicities permit. It may remove parts, provided the multiplicities permit and responsibility for the parts is assumed by other objects. The difference between destroying parts and removing parts is that when a part is simply removed it is not destroyed because responsibility for it is assumed by another composite. When a composite is destroyed it must either destroy all its parts or give responsibility for them to other composite objects.

Iterative Design

Let us look again at our earlier example of roses. We know that roses and petals have a whole-part, or aggregation, relationship. We also know that roses and thorns have a whole-part, or aggregation, relationship (see Figure 12–4). Suppose we are only interested in petals when they are part of a rose. Further suppose we are only interested in thorns when they are part of a rose. (Certainly, that is normally the case.) That means we can set up and manipulate them as composites. Figure 12–8 shows a refinement of our initial UML

**Figure 12–8
Composition—Example 2**

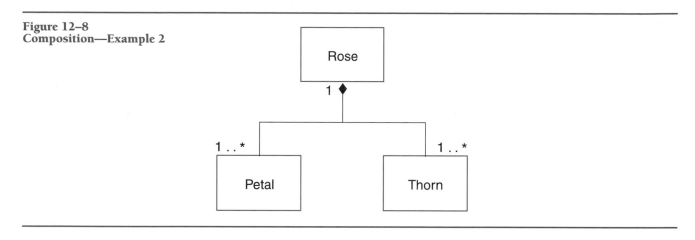

class diagram. We indicate that the relationships between the classes are aggregations, but we further define the aggregation relationships to be compositions (indicated by the solid-filled diamond). In doing so, we are not suggesting that we were wrong initially. We are just doing **iterative design**— looking again at our work in progress to improve it or add to it as our understanding increases or as we get "further into" a problem situation. Through iteration, we continue to evolve our class diagram as we more closely examine and understand the problem situation.

Sample Problem 12.2 (Planning a Group Event)

Problem:

One of the honor societies on campus is planning its annual spring dinner and initiation ceremony for new members. If you have been involved in planning for such an event, you know that pulling it off is a big job. Because this event and others similar to it are held each year, the committee responsible for the event is determined to use the computer as an aid to its planning and control. The committee's overall goal is to host a successful event. Their near-term objectives are to think of all the items involved; to track item-related decisions, tasks, and contacts; and to monitor item-related anticipated and actual costs. Construct a UML class diagram reflecting the classes and relationships that must be taken into account.

Solution:

A UML class diagram reflecting the classes and relationships identified by the committee is shown in Figure 12–9. Numerous classes are depicted; you can probably think of more.

At the top of the class diagram are the classes Organization and Event. Organization Sponsors Event is an association relationship. An organization may sponsor any number of events. An event has one and only one sponsoring organization. Location, Agenda, Decorations, OffsitePrinting, and Refreshments are all parts of the composite Event. Notice that even though Location is a part of the composition relationship with Event it is also an aggregate in relationships with other classes. In particular, the classes Building, Parking, and Furnishings are all parts of Location. Refreshments is also a part of a composition relationship with Event. Refreshments is a

Figure 12–9
Event Planning Problem
Class Diagram

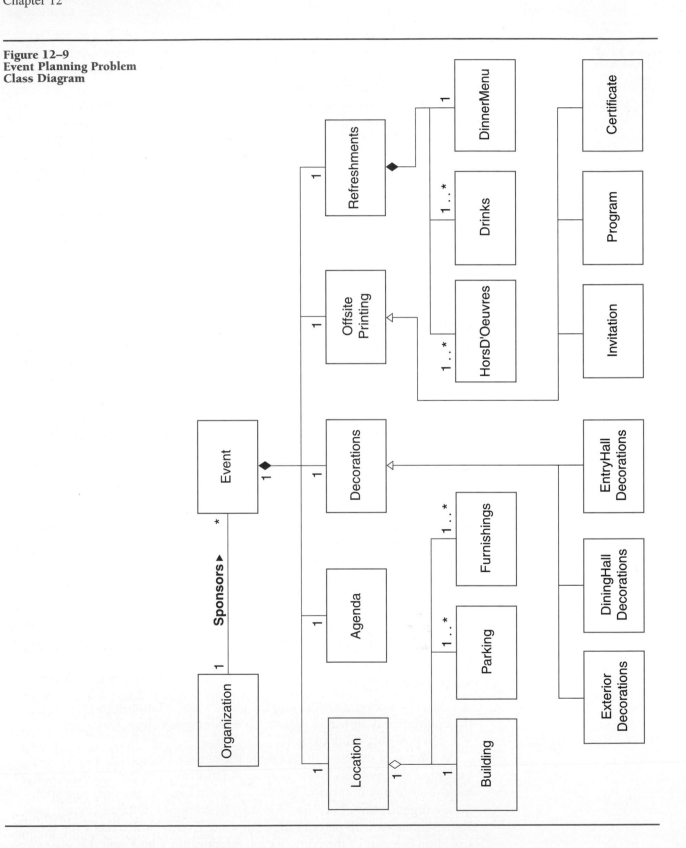

composite in a relationship with HorsD'Oeuvres, in a relationship with Drinks, and in a relationship with DinnerMenu. As this example illustrates, a single class may be a part in one relationship and a composite or an aggregate in another relationship at the same time.

Generalization/specialization relationships are also shown in Figure 12–9. The Decorations class is a parent; ExteriorDecorations, DiningHallDecorations, and EntryHallDecorations are subclasses of the Decorations class. The OffsitePrinting class is a parent; Invitation, Program, and Certificate are subclasses of the OffsitePrinting class. Why is it likely to be convenient to handle decorations and offsite printing in this manner?

Many additional facts are recorded on this UML class diagram (see Exercise 12). Through iterative design, additional classes are likely to be identified as the committee's work on the dinner and initiation ceremony proceeds (see Exercise 13).

Inner Classes

Now we are going to talk about a very different relationship between classes. If you spend time with application developers or if you explore object-oriented design and implementation more fully, you are likely to encounter the use of inner classes. As the term *inner class* implies, an **inner class** is a class defined within another class (the **enclosing class**). That is, one class is nested inside another one. Use of inner classes is especially common in Java because they are used in predefined classes provided as part of the Java language or in toolkits available with it.

When should you use an inner class? You should define a class within another class when the inner class makes sense only in the context of its enclosing class or when it relies on the enclosing class for its function. For example, a text cursor (which keeps track of the current, active position in text) makes sense only in the context of a particular document or other text component. Sometimes the block of logic within an inner class performs functions that are just generally different from other functions within the surrounding class. (In the following example, the inner class writes a list of exceptions.) Sometimes the block of logic is functionality that is used again and again throughout an application. Such logic is often found in event-driven, interactive programs that respond to the clicking of icons on a screen.

Sometimes the use of an inner class stems not so much from a design need as from a desire for programming convenience. When using Java, an inner class has access to all of the enclosing class's members, even if they are declared private. The modifiers *public, private*, and *protected* restrict access by classes *outside* of the enclosing class but not by classes *inside* of it. Inner classes are a particularly convenient way to set off some block of logic as a separate class, but not have to send messages to pass it the current values of a lot of variables that the enclosing class already knows about.

Look at Figure 12–10. It shows the definition of the MonthlyInventory class. You should see easily the default constructor and one alternate constructor for this class. The GetSales method obtains the item number, beginning quantity, and units sold during the month for a particular item in inventory.

Figure 12–10
MonthlyInventory Class with
Inner Class WriteExcep
Pseudocode

```
Class MonthlyInventory
        private Month
        private ItemNo
        private BeginQty
        private UnitsSold
        private PercentSold
        private EndQty
        private Flag
        public MonthlyInventory
                Month = "Not specified"
                ItemNo = "ZZZ999"
                BeginQty = 0
                UnitsSold = 0
                PercentSold = 0.0
                EndQty = 0
                Flag = 1
        End
        public MonthlyInventory (string CurMonth)
                Month = CurMonth
                ItemNo = "ZZZ999"
                BeginQty = 0
                UnitsSold = 0
                PercentSold = 0.0
                EndQty = 0
                Flag = 1
        End
        public GetSales
                Input ItemNo, BeginQty, UnitsSold
        End
        public float RetPercentSold
                IF UnitsSold > 0 THEN
                    PercentSold = UnitsSold / BeginQty
                ELSE
                    PercentSold = 0.0
                    IF Flag = 1 THEN
                        Exception: ExcepNow
                        Flag = 0
                    (ELSE)
                    ENDIF
                    ExcepNow.WriteExcep
                ENDIF
                Return PercentSold
        End
        public integer RetEndQty
                EndQty = BeginQty - UnitsSold
                Return EndQty
        End
```

Figure 12–10
(continued)

```
private Class Exception
        private MaxLines
        private LineCnt
        private PageCnt
        private Exception
            MaxLines = 55
            LineCnt = MaxLines
            PageCnt = 1
        End
        private WriteExcep
            IF LineCnt ≥ MaxLines THEN
                Write "Inventory Exception List:", Month, PageCnt
                Write "Item No.", "Begin Qty", "Units Sold"
                LineCnt = 0
                PageCnt = PageCnt + 1
            (ELSE)
            ENDIF
            Write ItemNo, BeginQty, UnitsSold
            LineCnt = LineCnt + 1
        End
    End Class
End Class
```

When the RetPercentSold method is called, if the number of units sold for the item is greater than 0 (UnitsSold > 0), the percentage of the beginning inventory of that item sold during the month (PercentSold) is computed and returned to the caller. If no units have been sold, or if the number of units returned by buyers of the item exceeded the number of units sold so that the value of UnitsSold is negative, the RetPercentSold method sets PercentSold to 0.0. The variable named Flag acts as a **program switch** to set up the logic needed to handle a special situation: An instance of the inner class Exception needs to be instantiated before a call is made to any methods of that class. When Flag contains a 1, as it does initially, the switch is ON. This causes the THEN clause of the nested IFTHENELSE structure to instantiate an instance of Exception with the statement Exception: ExcepNow and set Flag to 0, or OFF. Because the program switch is OFF from that time forward, the THEN clause is not executed again during this instantiation of Exception. That is the logic we need. (You will learn more about program switches in Chapter 13.)

The RetPercentSold method then invokes the WriteExcep method of ExcepNow to write the item number, beginning quantity, and units sold to an exception file. That method checks first to determine if a heading line needs to be written. Of course, it does need to be written the first time that WriteExcep is called. It is written again (at the top of a second page) after 55 lines have been written. This logic should look familiar to you. We have used it in several solutions in earlier chapters. In any case, the item number, beginning quantity, and units sold are written to the exception file, the line count is incremented by 1, and control returns to the statement following the IFTHENELSE control structure in the RetPercentSold method, which returns the current value of PercentSold to the caller of the RetPercentSold method.

The inner class Exception has three data members that are unique to it: MaxLines, LineCnt, and PageCnt. These data members are not used by the enclosing class MonthlyInventory because it has nothing to do with writing exceptions. The WriteExcep method of the inner class writes out the values of item number, beginning quantity, and units sold each time it is called. It has access to these private variables of MonthlyInventory because it is an inner class of MonthlyInventory.

When the RetEndQty method of the MonthlyInventory class is invoked, it computes and returns the quantity on hand at the end of the month (EndQty) of a particular inventory item.

An inner class has no separate existence outside of the class that encloses it. You can instantiate multiple instances of an inner class if the logic you need to solve a problem requires it. You actually can nest classes as many levels deep as you like, but there is likely to be little reason for deep nesting in most programs.

Benefits of Object-Oriented Design

Many benefits can be gained from object-oriented design and programming. Although some of the benefits were mentioned in previous discussions, it seems advisable at this point to focus specifically on the advantages of an object-oriented approach.

A first major advantage is simpler maintenance. If a class definition is changed, that change will be picked up in any program that incorporates an object of that class. There is no need to scan through many programs searching for all uses of the class and then change each occurrence individually. If classes are carefully designed, with attention to encapsulation and data hiding, many changes can be made internal to a class with no impact at all on users of some methods in the class and with isolated, well-defined impact on others. Functionality added to a class by adding new methods need have no impact at all on existing users. Modular design of procedure-oriented applications also helps to ease maintenance tasks, but there tends to be a "crispness" of separation in an object-oriented approach that is often not obvious in a procedure-oriented approach, even if it is there.

Applications that are built on a well-thought-out object-oriented design are resilient to change. That means that changes can be made without invalidating the current design and without having unforeseen "side effects" on other parts of the applications. (For example, "I did not know you were also looking at or changing the value of that variable," or "I did not know you depended on that value being reinitialized.") It also means that application systems can be evolved over time. There is no need to complete an 18-month project before a part of the application system can be released to users. Suppose a major change in user requirements does surface—it will not cause the system to have to be redesigned completely or abandoned.

Another advantage is higher productivity. Because class libraries containing many prebuilt (and thus working and tested) classes exist, much of the design and programming for an application already may be done. The class libraries often provide many of the technical details needed in a program. Thus, the developer need not work at that level of technical detail; he or she can concentrate on the business requirements of the application. As another example, it is much easier to work with widgets as objects than to define pixel locations on a display screen. Because class libraries provide prebuilt classes, programmers can use these classes as basic building blocks of their applications.

An object-oriented approach encourages and facilitates reuse not only of software but also of entire designs. It leads to the creation of reusable, reliable design patterns and application frameworks. Even if a class definition does not completely fit the developer's needs, the developer can reuse much of the class and tailor it to his or her needs through overriding. Similarly, even if a documented design pattern is not "exactly right" for a particular problem solution, it may be a very good point from which to start.

The availability of well-tested, proven components means less overall new work to be done on a project. The existence of reusable design patterns and application frameworks means fewer unknowns. Less new work and fewer unknowns means less development risk and greater predictability of development schedules.

Finally, an object-oriented approach encourages the developer to work within a scope that he or she can handle easily. As humans, we can think about only so much at one time; we can remember only a limited number of details. We work much more comfortably with a task we can "get our arms around." If a class you are designing seems to you to be getting "too big" or "too complex," chances are it is. Step back. Rethink. This is easy to do within an object-oriented approach to design and programming.

Object-Oriented Languages

A wide variety of programming languages exist. The languages have been defined, supported, and evolved over the past 50 years. Developers often extensively use one language that becomes their "favorite." Java, Smalltalk, C#, C++, ObjectPascal, and Eiffel are all **object-oriented languages**. They can be used to write object-oriented programs, much as C, COBOL, and Pascal have been and are being used to write procedure-oriented programs. Java, Smalltalk, and Eiffel were designed "from the ground up" to be object-oriented. Each of these languages includes a rich set of predefined classes, readily available to all developers. Because languages such as C++ and ObjectPascal have evolved from a procedure-oriented base to also support an object-oriented approach, they are sometimes referred to as **hybrid languages**. From a purist's perspective, Visual Basic and Ada are often described as **object-based languages** rather than object-oriented because neither provides direct support for inheritance. However, since objects and classes are elements of both object-based and object-oriented languages, it is possible and often desirable to use object-oriented design methods for both object-oriented and object-based programming-language implementations.

Key Terms

association	aggregate	composite
binary association	part	iterative design
path	"has-a" relationship	inner class
association name	unfilled diamond	enclosing class
role name	symbol	program switch
self-association	cycle	object-oriented
multiplicity	solid-filled diamond	language
link	symbol	hybrid language
whole-part relationship	composition	object-based language
aggregation		

Exercises

1. (a) What is an association?
   (b) Give some examples of associations within a college or university environment.

2. What do we mean by the multiplicity of a class in a relationship?

3. State the meaning of the following symbols:
   (a) 1..*
   (b) 0..1
   (c) *
   (d) 1
   (e) 6
   (f) 6..8
   (g) 2..*

4. What is the difference between a path and a link, as the terms are used in this chapter?

5. (a) What distinguishes an aggregation relationship from a relationship that is an association but not an aggregation?
   (b) How is an aggregation relationship represented on a UML class diagram?

6. (a) What distinguishes a composition relationship from an aggregation relationship?
   (b) How is a composition relationship represented on a UML class diagram?

7. Define each of the relationships in the situations below as a generalization/specialization, a simple association, an aggregation, or a composition:
   (a) A car has a windshield.
   (b) A family owns a house.
   (c) An instructor teaches a class.
   (d) A husband is married to a wife.
   (e) A boy has a dog.
   (f) Men's basketball is a kind of athletics.
   (g) Dessert is a part of a meal.
   (h) A waitress is an employee of a restaurant.

    (i) A pizza parlor is a business.

    (j) A student's jacket has a hood.

8. When are role names required on a UML class diagram?

9. Recall our definition of a Gradebook class in Chapter 10 (see Figure 10–4). In Chapter 11, we defined a Faculty class that inherited some characteristics from an Employee class (see Figure 11–2). Each Faculty member maintains one or more Gradebooks. Each Gradebook belongs to one and only one Faculty member. Construct a UML class diagram to show both the maintains relationship and the classes that apply.

10. Expand the UML class diagram in Figure 12–1 to show a self-association relationship in which both ends of the relationship are of the Person class. A Person may be a worker or a manager. A manager may supervise any number of employees. A worker may be supervised by one and only one manager.

11. List all of the facts that you can learn from the following UML class diagram:

Person	* PlaysFor▸ 0..1	Organization	* ◂Owns 1..*	Business
	athlete team		team owner	

12. Look at the UML class diagram in Figure 12–9.

    (a) What three facts can you conclude about the relationship between the class Agenda and the class Event?

    (b) How does the relationship between Refreshments and HorsD'Oeuvres differ from the relationship between Refreshments and DinnerMenu?

    (c) Which classes on the diagram are composites?

    (d) Which documents for the event are to be printed off campus?

    (e) Which class is the parent of EntryHallDecorations?

    (f) Which class is the parent of Drinks?

13. Refer to the UML class diagram in Figure 12–9.

    (a) Suggest additional classes that may be needed to handle an initiation ceremony and dinner.

    (b) Add the classes you suggested in Exercise 13(a) to a copy of the UML class diagram in Figure 12–9 and show their relationships to other classes on the diagram.

14. What is meant by iterative design?

15. What advantages does use of an inner class offer?

16. Distinguish between object-oriented languages and object-based languages.

17. List at least five advantages of an object-oriented approach to design and programming.

Part 3

Chapter 13: Array Applications
Chapter 14: Master File Update Processing
Chapter 15: Control-Break Processing

Array Applications

Objectives

Upon completion of this chapter you should be able to

- Design an algorithm to load values into a table.
- Design an algorithm that searches a table using a sequential search.
- Design an algorithm that searches a table using a binary search.
- Design an algorithm that makes use of one or more program switches.
- Design an algorithm that sorts values in a table.

Introduction

In Chapter 9 we saw that it is sometimes useful to read, process, and write collections of similar data items as a data group rather than as separate data items. We saw how subscript notation can be used to refer to specific elements of a group. We developed solution algorithms to find and print the smallest of a set of values, average a set of values, multiply the elements in a table by the elements in corresponding positions of a list, and so on. (In mathematical problem solving, the last of these algorithms is usually referred to as multiplying a matrix by a vector.)

Table Lookups

In programming terminology, reading a table into computer storage is often referred to as **loading a table**. This step usually occurs soon after program execution is initiated, that is, in the initialization portion of the program. After a table has been loaded, it can be referred to repeatedly during subsequent processing steps. Often another input value is read and used in a search of the table. For example, a payroll application may accept a table showing wage classes and corresponding pay rates as one of its inputs. Then, an employee time record that contains the hours worked and wage class of a particular employee may be submitted as a later input. In order to determine what pay rate to use in calculating the employee's pay, the class/rate table must first be searched. This operation is called a **table lookup**.

Sample Problem 13.1 (Table-Lookup Example)

Problem:

To understand how tables and table lookups may be applied in practical situations, consider the following example. A two-dimensional table containing a master list of unique item numbers and corresponding unit prices for registered pharmaceutical products will be input. Once this table is input, additional records will be input, each containing an item number and quantity for one of these products. The item number input will be looked up in the table. If the item number is found, the corresponding price in the table will be used to compute the total price for the particular item. If the item number input is not found in the table, an appropriate message will be output.

Solution:

The structure chart in Figure 13–1 shows a modular design for this problem solution. The overall control module is shown in Figures 13–2 (flowchart) and 13–3 (pseudocode). As the program begins execution, A000 calls another module (B000) to input the table.

The program flowchart in Figure 13–4 and the pseudocode in Figure 13–5 show the processing steps for module B000. The name ITEM_PRICE represents the table shown in Figure 13–6. This table stores the reference data of item numbers and unit prices. In this example, the fifth row, first column of the table [ITEM_PRICE(5,1)] contains an item number of 54387. This item's unit price is $10.95, the amount located in the fifth row, second column of the table [ITEM_PRICE(5,2)].

A DOWHILE control structure in module B000 controls the loading of the table. Nested IFTHENELSE control structures are used because the number of items to be included in the table is not fixed. It is not known beforehand exactly how many times the DOWHILE loop needs to be executed.

Because the pharmaceutical products are registered items, the designer of the solution algorithm has determined that not more than 200 items will have to be identified and priced. If fewer than 200 items are entered, a special input record (trailer record) with an item number of 99999 is to be entered to indicate that the table is complete. (Recall the discussion of trailer records in Chapter 5.) The inner IFTHENELSE checks the value of the subscript ROW, which is used as a count of the number of table entries read into storage. If the maximum number of table entries (200) has been read, a special end-of-file indicator (EOF)

Figure 13–1
Inventory Problem (Structure Chart)

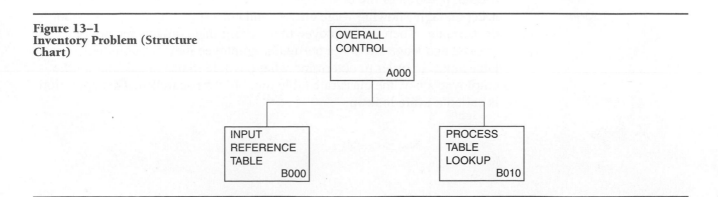

Figure 13–2
Inventory Problem—Overall
Control (Flowchart)

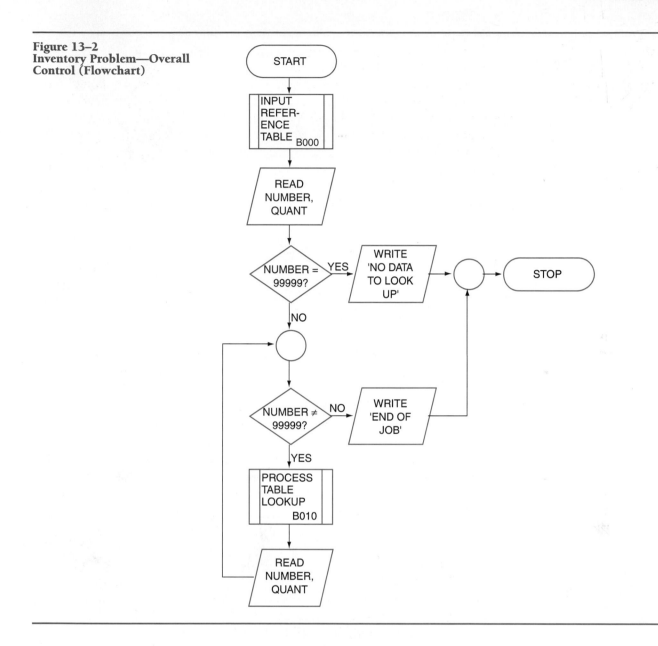

Figure 13–3
Inventory Problem—Overall
Control (Pseudocode)

```
Start
Input reference table (B000)
Read NUMBER, QUANT
IF NUMBER = 99999 THEN
    Write 'No data to look up'
ELSE
    DOWHILE NUMBER ≠ 99999
        Process table lookup (B010)
        Read NUMBER, QUANT
    ENDDO
    Write 'End of job'
ENDIF
Stop
```

Figure 13–4
Inventory Problem—Input
Reference Table (Flowchart)

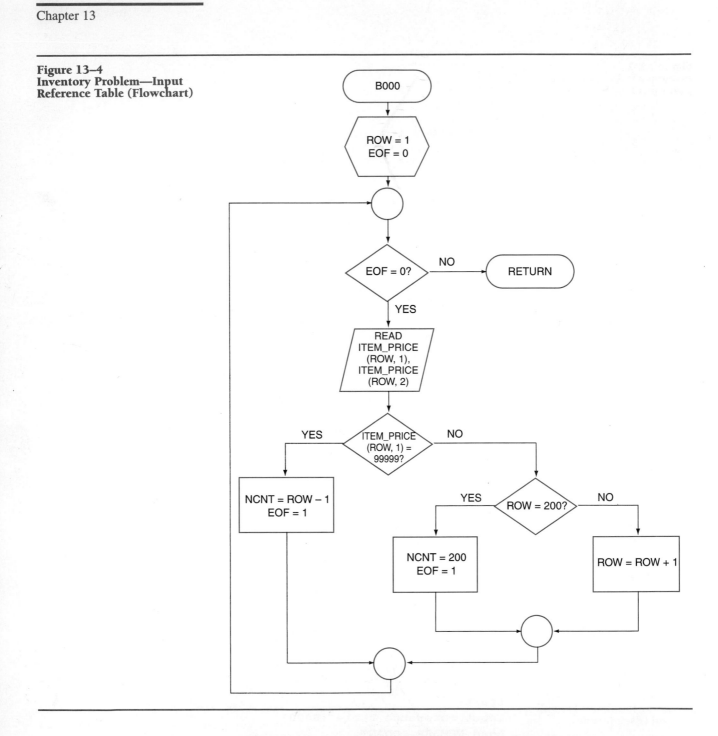

is set to 1 and the number of table entries (NCNT) is set at 200. The outermost IFTHENELSE checks the item numbers of up to 200 entries to watch for the 99999 input record. If the 99999 record is detected, EOF is set to 1. NCNT is set to ROW – 1 to prevent the 99999 record from being included in the table. Whichever of these tested conditions occurs first causes the DOWHILE loop to be exited.

The DOWHILE loop is controlled by the variable named EOF. The use of EOF is an example of how another important programming technique—the use of **program switches**—can be applied. We discussed program switches briefly in Chapter 12. In general, a program switch is used to set up the logic

Figure 13–5
Inventory Problem—Input
Reference Table (Pseudocode)

```
B000
Enter
ROW = 1
EOF = 0
DOWHILE EOF = 0
   Read ITEM_PRICE(ROW,1), ITEM_PRICE(ROW,2)
   IF ITEM_PRICE(ROW,1) = 99999 THEN
      NCNT = ROW - 1
      EOF = 1
   ELSE
      IF ROW = 200 THEN
         NCNT = 200
         EOF = 1
      ELSE
         ROW = ROW + 1
      ENDIF
   ENDIF
ENDDO
Return
```

Figure 13–6
Inventory Problem—
ITEM_PRICE Reference Table

ITEM_PRICE TABLE

		COLUMNS	
		1	2
	1	18337	50.85
	2	35795	5.85
	3	11427	1.50
	4	98547	105.50
ROWS	5	54387	10.95
	6	77378	3.50
	7	48591	14.95
	⋮	⋮	⋮
	199	UNUSED	UNUSED
	200	UNUSED	UNUSED

needed to deal with a special condition that may arise during processing. It may be implemented in any of several ways, depending on the hardware and software characteristics of the computer system in use. In this case, EOF is being used as a switch that contains either a value of 1 or a value of 0. When EOF contains a 1, the switch is ON; when EOF contains a 0, the switch is OFF. The switch EOF is set to 0 (OFF) in the preparation step in Figures 13–4 and 13–5. The switch EOF is set to 1 (ON) within the DOWHILE loop when either of the two conditions indicating no more records are to be read occurs.

Note that in either case of loop termination, the number of table entries read into storage is placed in NCNT before EOF is set. NCNT can then be used as a loop control when the table is referred to in subsequent processing.

Now let's consider the logic in the overall control module (A000) (see Figures 13–2 and 13–3). After the values for the table have been input, module B000 returns control to A000 and another type of input is read: inquiries to be processed against the table. Each inquiry consists of an item number and a quantity. Because the number of inquiries to be processed will vary from one run of this program to another, a special input record containing 99999 as an item number is to be entered as an end-of-input record after all inquiries have been processed. An IFTHENELSE is used to make sure that there are, in fact, some inquiries to process. If not, a special "No data to look up" message is written as output. A DOWHILE structure is used to control the input of the rest of the inquiries. For each record that is input, the reference table is searched to determine if the item number that is currently input (NUMBER) is contained in the table. If so, the unit price corresponding to the item number (the second column in the current row of the table) is multiplied by the quantity entered as input (QUANT) to determine the total cost. The table is searched and the computation is done in a separate module (B010), shown in Figures 13–7 (flowchart) and 13–8 (pseudocode).

A DOWHILE loop controls the main processing in module B010. The loop is executed repeatedly until either one of two conditions exists:

- The item number of the inquiry received as input in A000 (NUMBER) is matched with an item number in the reference table [ITEM_PRICE(ROW,1)].
- All item numbers in the table have been examined without finding an item number that matches the item number in the inquiry.

These conditions are tested for by the nested IFTHENELSE control structures within the DOWHILE. Whichever occurs first causes another switch, called DONE, to be set to 1. Note that DONE is initialized to 0 at the beginning of B010. When the DONE switch is tested and found to be 1, the DOWHILE loop is exited.

The THEN path of the outer IFTHENELSE shows the processing that occurs when an item-number match is found in the table. The DONE switch is set to 1 after this processing occurs. If a match is not found, then the next entry in the item-number table must be checked.

In the ELSE path of the outer IFTHENELSE, the subscript ROW is incremented and also checked to determine if ROW exceeds the number of entries (NCNT) in the table. If this is the case, no item number matching the input item number exists in the reference table. The THEN clause of the innermost IFTHENELSE is then executed to output the message "No data available for" followed by the unmatched item number. The DONE switch is set to 1.

In either case, the DOWHILE loop will be exited and control will return to A000. At this point another record will be read. After this record is tested, either module B010 will be executed again or the DOWHILE loop in A000 will be exited and an "End of job" message will be written.

**Figure 13–7
Inventory Problem—Process
Table Lookup (Flowchart)**

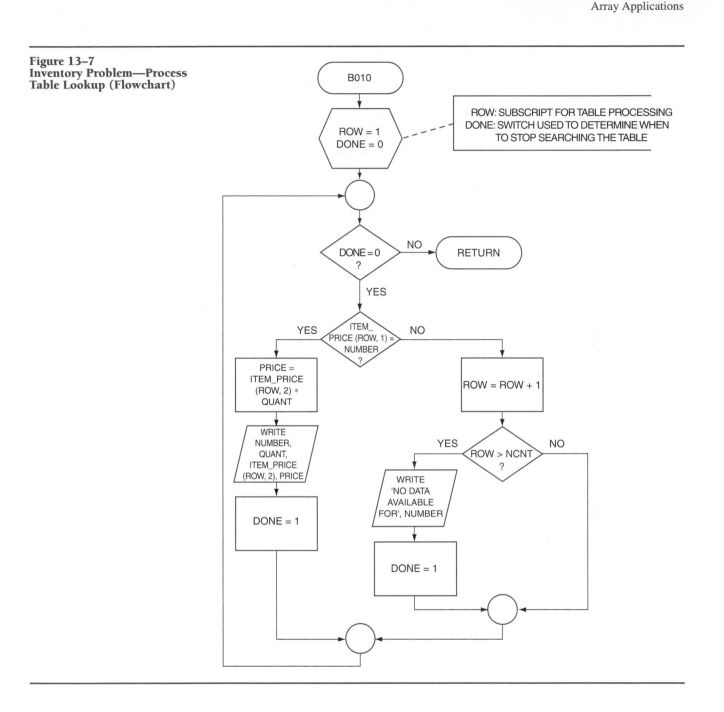

What assumptions have you made about the ordering of entries within the item-number/unit-price reference table? Perhaps you think that the table entries are in ascending item-number sequence. This is not a requirement of the solution algorithm. The item numbers in the table may be in ascending sequence, or descending sequence, or no particular sequence at all (as is the case in Figure 13–6). Because the table-lookup module starts at the beginning of the table for each inquiry, it will work in any of these cases.

However, the sequence of entries in a table can affect the efficiency of table searching in terms of processing time. Because the table-lookup routine in Figures 13–7 and 13–8 always starts with the first entry and proceeds toward the last, the most-used item number should be the first

Figure 13–8
Inventory Problem—Process
Table Lookup (Pseudocode)

```
B010
Enter
ROW = 1
DONE = 0
DOWHILE DONE = 0
  IF ITEM_PRICE(ROW,1) = NUMBER THEN
     PRICE = ITEM_PRICE(ROW,2) * QUANT
     Write NUMBER, QUANT, ITEM_PRICE(ROW,2), PRICE
     DONE = 1
  ELSE
     ROW = ROW + 1
     IF ROW > NCNT THEN
        Write 'No data available for', NUMBER
        DONE = 1
     (ELSE)
     ENDIF
  ENDIF
ENDDO
Return
```

entry in the table, and the least-used item number should be the last. For example, if 60 percent of the inquiries involve item numbers 55040, 30456, and 32045, these item numbers should be the first ones in the table. The rest of the table should also be sequenced according to frequency of use. The total amount of time required for a run will be affected accordingly.

Binary Searches

A more sophisticated table-lookup routine should be considered in cases in which (1) the frequency of use of table entries is evenly distributed over the table, (2) the number of entries in the table is very large, or (3) processing efficiency or high system performance is mandatory. The routine commonly employs a **binary search**. When this table-lookup technique is used, the entries in the table being searched must be in either ascending or descending sequence according to the values of a particular data item or items common to all entries. The portion of an entry that contains that data item or items is called the **key field**. The term *binary* is appropriate for this lookup technique because the portion of the table being searched is halved repeatedly. The search begins with an entry at or near the middle of the table. Based on a comparison, the search continues in either the first half or the second half of the table. The other half of the table is ignored. The next comparison is made against the value at or near the middle of the half just selected. The search continues by successively halving the portion of the table remaining. Eventually, a match occurs between the **search key** of the data being processed against the table and the key field of a table entry. Alternatively, it may be determined that no entry with a matching key field exists in the table. In such a case, an exception routine must be carried out.

Figure 13–9 illustrates how a table of 16 elements is accessed when a binary search is used. As you can see, 5 comparisons, at most, are needed to

Figure 13–9
Binary Search

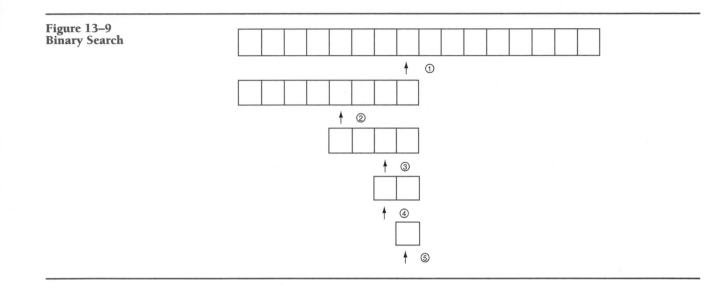

determine whether or not a match is found. As noted previously, the binary search technique is particularly valuable when a table contains a large number of entries. For example, in a table with 1000 entries, the maximum number of comparisons that may be needed is only 10.

Sample Problem 13.2 (Binary Search Example)

Problem:

Assume that 70 company sales offices are identified by unique two-digit codes on a company's internal documents. When reports are printed for external use, however, the sales offices must be identified by location rather than by code number. Each program that provides a report for external use must have access to information associating the office code numbers with the actual office locations. The office codes and corresponding locations will be input in ascending order into two lists. These lists will be used to locate subsequent office codes, also provided as input. Because the codes in the list are assumed to be in order, the office code list will be searched using the binary search technique. It is important to remember that the binary search technique can be used to search an array only if the values in the array are in ascending or descending sequence.

Solution:

The structure chart in Figure 13–10 shows that this algorithm will be implemented as four modules. The overall control module (A000) will call module B000 to input the code numbers and corresponding locations into two tables. Another module (B010) will be used to search the code numbers table, and a third module (B020) will handle the processing that must occur after the table has been searched.

Execution begins at the first step in the overall control module. The flowchart for this module is shown in Figure 13–11, and the corresponding pseudocode is shown in Figure 13–12.

**Figure 13–10
Binary Search (Structure
Chart)**

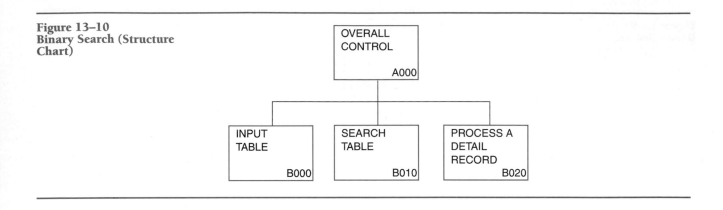

The first step in A000 calls module B000 (see Figures 13–13 and 13–14) to input the code numbers and corresponding locations into two one-dimensional arrays—CODE and LOCATION. The first step in module B000 inputs the first code and location into the first position of each array. A variable SUB will be used as a subscript and is set to 1. Another variable, VALID_TABLE, will be used to denote whether or not the values in the tables are in ascending order. VALID_TABLE is initially set to "YES" because we assume the table is properly ordered. A DOWHILE control structure is used to control the reading of the rest of the two-digit office codes and the corresponding office locations into the arrays CODE and LOCATION. Together, these arrays form a reference table.

Because the office-code/office-location reference table is to be accessed using a binary search technique, the entries in the table must be ordered. A sequence check is performed to make certain they are ordered. An IFTHENELSE control structure is used to perform the sequence-check. To avoid sequence-checking the first code number against a preceding code number (because there is no code number preceding the first one), the first code and location are input in a step preceding the loop. The first code and location that are input within the loop are CODE(2) and LOCATION(2). In this case the value of SUB will be 2 and the value of SUB-1 will be 1. Hence, the value of SUB will never be 0 or less.

Each office code number must be greater than the one that precedes it. The IFTHENELSE control structure will compare the office code number just input (CODE(SUB)) with the previous office code number (CODE(SUB-1)). If the previous code is greater than the new code, then the table values are out of sequence. The YES path of this decision-making step will be taken and the value of the variable VALID_TABLE will be set to "NO". The value of VALID_TABLE can be checked in other modules to determine whether the office-code/office-location table is in sequence. If the table is not in sequence, the reference table that the input-table module loads into storage should not be used in subsequent processing.

The DOWHILE loop in Figures 13–13 and 13–14 is exited when all 70 table entries have been read into storage. Control is then returned to the overall control module (A000), where processing continues.

The next processing step in module A000 is an IFTHENELSE control structure that checks the value of the variable VALID_TABLE. If the value is not "YES", indicating a sequence error, an error message is output and

**Figure 13–11
Binary Search—Overall
Control (Flowchart)**

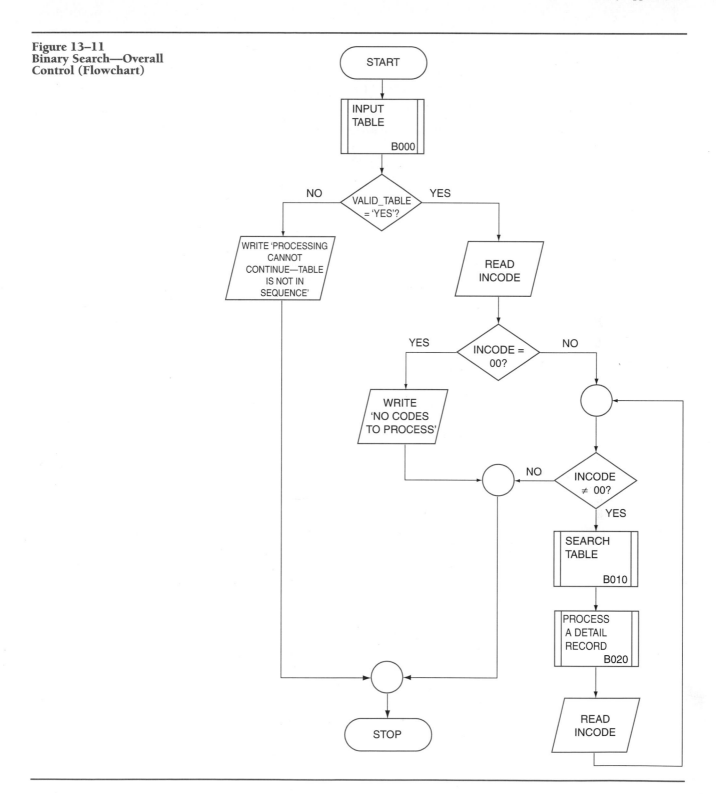

program execution is terminated. If the value is "YES", then normal processing can occur as indicated in the YES path of the IFTHENELSE control structure. Specific office codes will be input and processed until an
office code of 00 is input. An IFTHENELSE control structure is used to

Figure 13–12
Binary Search—
Overall Control
(Pseudocode)

```
A000
Start
Input table (B000)
IF VALID_TABLE = 'YES' THEN
        Read INCODE
        IF INCODE = 00 THEN
                Write 'No codes to process'
        ELSE
                DOWHILE INCODE ≠ 00
                    Search table (B010)
                    Process a detail record (B020)
                    Read INCODE
                ENDDO
        ENDIF
ELSE
        Write 'Processing cannot continue--table is not in sequence'
ENDIF
Stop
```

Figure 13–13
Binary Search—Input Table
(Flowchart)

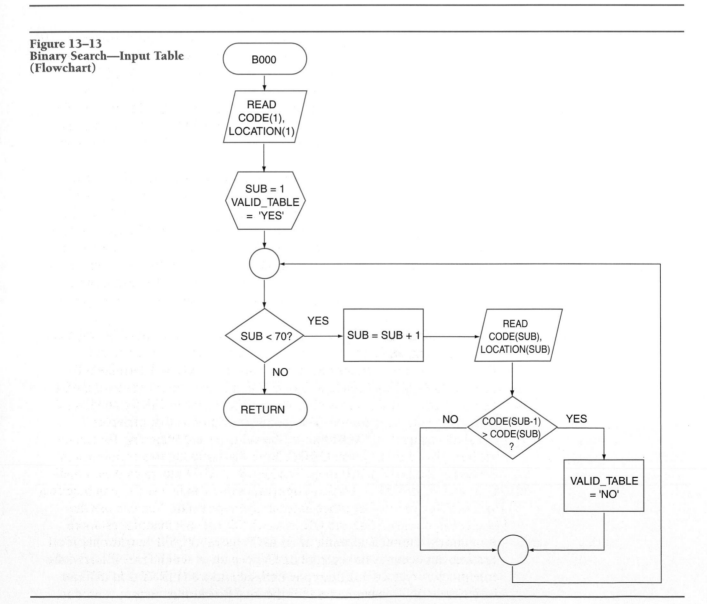

Figure 13–14
Binary Search—Input Table
(Pseudocode)

```
B000
Enter
Read CODE(1), LOCATION(1)
SUB = 1
VALID_TABLE = 'YES'
DOWHILE SUB < 70
    SUB = SUB + 1
    Read CODE(SUB), LOCATION(SUB)
    IF CODE(SUB-1) > CODE(SUB)  THEN
        VALID_TABLE = 'NO'
    (ELSE)
    ENDIF
ENDDO
Return
```

check that at least one office code is input. If not, a "No data" message will be output. A DOWHILE control structure is used to control the processing of each office code input. For each code that is input, module B010 is called to perform a binary search on the reference table. The reference table must be searched to locate the office code in the CODE array so the corresponding location can be retrieved from the LOCATION array. The flowchart and corresponding pseudocode for module B010 are shown in Figures 13–15 and 13–16.

The first step in module B010 initializes three variables. The first variable, FOUND, is used to indicate whether or not the office code input has been found in the array CODE. We initially set this variable to "NO" because the search has not yet begun. A variable called FIRST, used to store the lower bound of the reference table, is set to 1. A variable called LAST, used to store the upper bound of the table, is set to 70 (because there are 70 entries). A DOWHILE loop is used to control whether the search should begin or continue. There are two conditions that must be met to begin or continue the search.

The first condition is that the lower bound of the table must be less than or equal to the upper bound of the table. Therefore the lower bound, FIRST, is compared against the upper bound, LAST, to determine if the current value of FIRST is less than or equal to the current value of LAST. The first time, the answer is obvious, but the values of FIRST and LAST will change, as we shall see. The second condition is that the value FOUND be equal to "NO"; that is, the value we are searching for has not yet been found in the array CODE. Note that there are two conditions in the test of the DOWHILE loop. The keyword AND separates these conditions and is one of two **boolean operators** that can be used to combine two or more conditions. The other boolean operator is OR. You can use the boolean operators AND and OR in a conditional statement to test more than one condition. The result of an AND operation will be true only if all conditions are met. The result of an OR operation will be true if any of the conditions is met. In this example, both conditions (FIRST ≤ LAST and FOUND = "NO") must be true for the loop processing steps to continue.

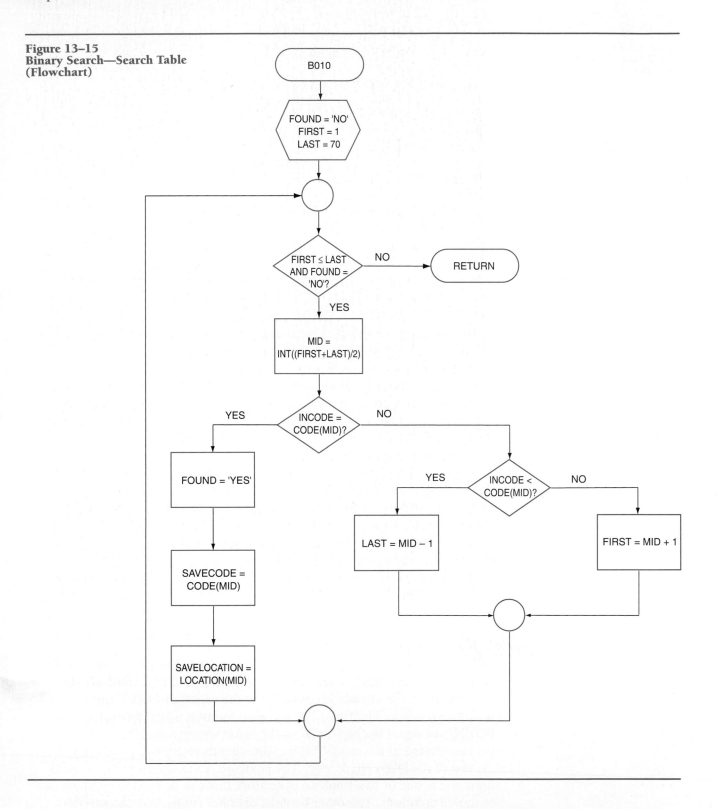

The first time these two conditions are checked they are both true; there-fore the processing steps within the loop are executed. The midpoint of the office-code/office-location reference table is computed by adding the lower-bound FIRST and the upper-bound LAST and then dividing their

Figure 13–16
Binary Search—Search Table
(Pseudocode)

```
B010
Enter
FOUND = 'NO'
FIRST = 1
LAST = 70
DOWHILE FIRST ≤ LAST AND FOUND = 'NO'
    MID = INT((FIRST + LAST) / 2)
    IF INCODE = CODE(MID) THEN
            FOUND = 'YES'
            SAVECODE = CODE(MID)
            SAVELOCATION = LOCATION(MID)
    ELSE
            IF INCODE < CODE(MID) THEN
                    LAST = MID - 1
            ELSE
                    FIRST = MID + 1
            ENDIF
    ENDIF
ENDDO
Return
```

sum by 2. A special built-in function (INT, which is short for INTEGER) is then used to **truncate** (cut off or delete) the decimal portion of the result of the division and assign the truncated result to MID. MID is then used as a subscript during table processing. INT is a keyword that we have used to identify the function that truncates a value to integer form. (The programming language later used when writing the program may use that term or a different one.) The input to the INT function is a single decimal number, and the output is the same number with the decimal portion truncated. For example, INT(5.3) = 5; INT(7.9) = 7; INT(6.0) = 6.

Next, the current value of INCODE (the office-code number read as input) is compared to the midpoint value of the code-number portion of the reference table, CODE(MID). [MID = INT((1 + 70) / 2) the first time through, so we know that MID is equal to 35.] The program execution then proceeds as follows:

- IF the value of INCODE is equal to the value of CODE(MID), the table search is ended because the code numbers match.
- IF the value of INCODE is less than the value of CODE(MID), the lower half of the reference table is searched.
- IF the value of INCODE is greater than the value of CODE(MID), the upper half of the reference table is searched.

These tests are set up in a structured manner as nested IFTHENELSE patterns.

Note that if INCODE is equal to CODE(MID), the variable FOUND is set to "YES" and the values in the two arrays at position MID are saved in the two variables SAVECODE and SAVELOCATION for use later in processing. Because FOUND is now equal to "YES," the condition FOUND = "NO" in the DOWHILE loop will no longer be true; the loop will be exited; and control will be returned to A000.

If INCODE is not equal to CODE(MID), the DOWHILE loop will be executed again. The value of INCODE is tested against the midpoint value of the half of the table that contains the code number, as indicated by previous algorithmic steps. As before, if the compared values are equal, the search is ended. If not, then the appropriate half of the divided table (i.e., the quarter of the table the value may be in) is specified. Halving continues in this manner until the search is successful, or until it is proved that the code number cannot be found in the table.

If adjustments to the lower and upper bounds of the table cause them to overlap (i.e., FIRST > LAST), it is assumed that the code number input has not been found in the table. The DOWHILE loop will be exited and control returned to A000. At this point, module B020 will be executed. This module will handle the processing of the office code and location, as well as alternative action if the office code input has not been found in the table. We will not concern ourselves with the details of this module in this chapter.

Because the binary search algorithm can be somewhat complex, it may be useful to illustrate the details of each step with an example. Figure 13–17 shows a simulation (or trace) of this algorithm with some typical input values. First, a set of arbitrary values is read into the CODE array. Figure 13–17 then illustrates the values of FIRST, LAST, MID, and CODE(MID) for a value of INCODE, say 21, that will eventually be found in the array, and for another value of INCODE, say 30, that will not be found in the array. The search will end when the lower-bound FIRST is found to be greater than the upper-bound LAST (22 is greater than 21).

Sorting Lists

In the previous example, the office codes that were input into the list were checked to ensure they were in ascending order. What if, however, the office codes to be input were not in ascending order? We could still input them into the office-code list and then use another algorithm to sort the list. Before we look at a sorting algorithm, we need to discuss how to exchange two values within the computer's memory. This process is necessary in the sort procedure that follows.

Exchanging Values

Suppose the contents of memory location NUM1 was defined to be 5 and the contents of memory location NUM2 was defined to be 3. Let us look at how we could exchange the values in NUM1 and NUM2. Your first attempt to exchange these two values might be to simply use two assignment statements, as shown in Figure 13–18. This approach will produce an incorrect result because the old value for NUM1 is erased in Step 1 before it can be assigned to NUM2.

The correct procedure for exchanging values is shown in Figure 13–19. One additional storage location, TEMP, is required for use as a temporary

Figure 13–17
Binary Search—Simulation

	CODE			CODE
1	01		25	34
2	02		26	35
3	04		27	36
4	05		28	37
5	06		29	40
6	09		30	41
7	10		31	42
8	11		32	44
9	12		33	45
10	13		34	46
11	15		35	47
12	17		36	48
13	20		37	50
14	21		38	52
15	22		39	53
16	23		40	54
17	25		41	55
18	26		42	58
19	27		43	59
20	28		44	60
21	29		45	61
22	31		·	·
23	32		·	·
24	33		·	·
			70	99

INPUT INCODE = 21

FIRST	LAST	MID	INCODE: CODE(MID)	
1	70	35	21 : 47	(<)
1	34	17	21 : 25	(<)
1	16	8	21 : 11	(>)
9	16	12	21 : 17	(>)
13	16	14	21 : 21	(=)

INCODE = CODE (14)   INCODE FOUND

INPUT INCODE = 30

FIRST	LAST	MID	INCODE: CODE(MID)	
1	70	35	30 : 47	(<)
1	34	17	30 : 25	(>)
18	34	26	30 : 35	(<)
18	25	21	30 : 29	(>)
22	25	23	30 : 32	(<)
22	22	22	30 : 31	(<)
22	21			

FIRST > LAST   INCODE NOT FOUND

storage area. TEMP is used to temporarily hold the value of NUM2 (Step 1) because a new value (NUM1) will replace the old value of NUM2 (Step 2). Once that happens, TEMP (the old value of NUM2) can replace the old value of NUM1 (Step 3).

The previous example illustrated the exchange with simple variables. Figure 13–20 shows the same logical process, used this time with a list of five elements. In this example, the third and fourth elements of the one-dimensional array LIST are exchanged.

Figure 13–18
Exchanging Values
Incorrectly

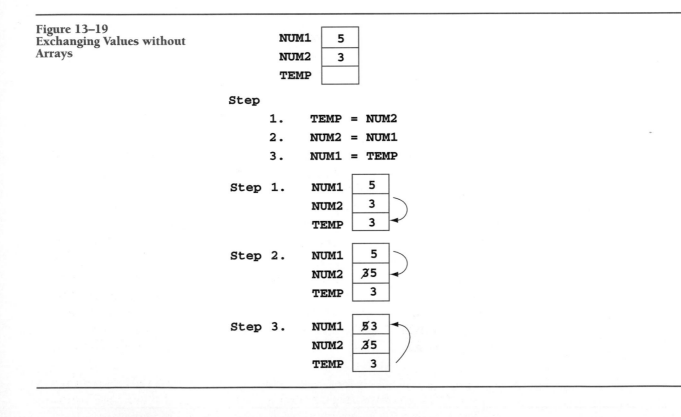

Figure 13–19
Exchanging Values without
Arrays

Sample Problem 13.3 (Sort Example)

Problem:

Sort a list of four values into ascending order and output the sorted list.

Solution:

A common method of sorting data into a required sequence is to read the values to be sorted into storage as a group of data items. The value in the first position in the group is compared to the value in the next position of the group. If the value in the first position is larger than the value in the second, they are interchanged. After the values in positions 2, 3, 4, . . . , n

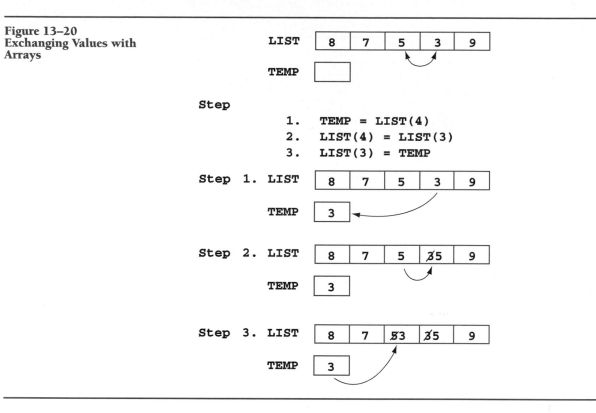

Figure 13–20
Exchanging Values with
Arrays

have been compared with the value currently in the first position, the **first pass** through the values is complete. The first position in the group is definitely known to contain the smallest value.

The first position is then temporarily ignored. The value in the second position of the group is compared with the values in position 3, 4, 5, . . . , n in a **second pass**. At the end of this pass, the second position is known to contain the second-to-smallest value.

The execution of passes continues, each pass requiring one less comparison operation than the preceding pass. After n – 1 passes, the values in the group are in ascending order. This is illustrated using a list of four values, as shown in Figure 13–21. As initially read into the computer, the values are 493, .06, 5, and .015. To arrange these values in ascending order, 4–1, or 3, passes are required.

Figure 13–22 shows a structure chart indicating the modules required to perform the actions diagrammed in Figure 13–21. Figure 13–23 shows the logic within the overall control module. This module calls three lower-level modules to input the array elements (B000), sort the array elements (B010), and output the sorted array (B020). The logic required in B000 and B020 is nearly identical to the algorithms used to input and output array values in Chapter 9. To review those algorithms, see Figures 9–3 and 9–4.

Figure 13–24 shows the flowchart and Figure 13–25 the corresponding pseudocode for module B010, which actually sorts the list. This module utilizes a nested DOWHILE loop. The outer DOWHILE loop controls the number of passes, and the inner (nested) DOWHILE loop controls the number of comparisons done in a single pass. An IFTHENELSE nested

**Figure 13–21
Sorting Four Numbers**

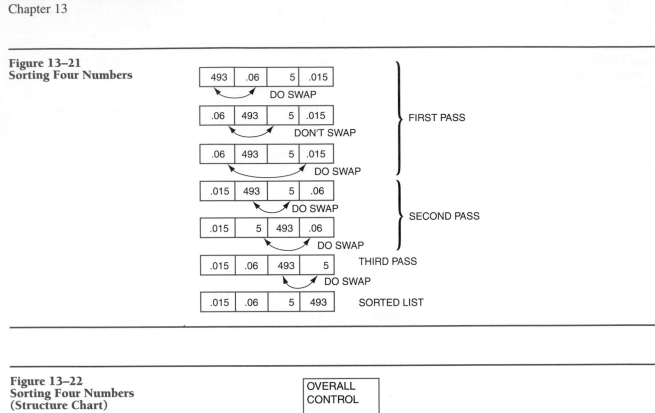

**Figure 13–22
Sorting Four Numbers
(Structure Chart)**

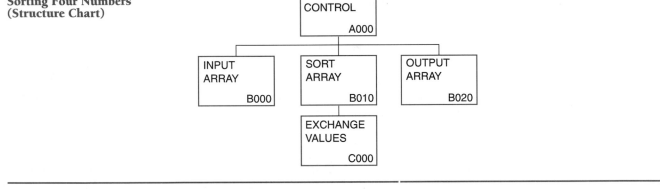

**Figure 13–23
Sorting Four Numbers—
Overall Control**

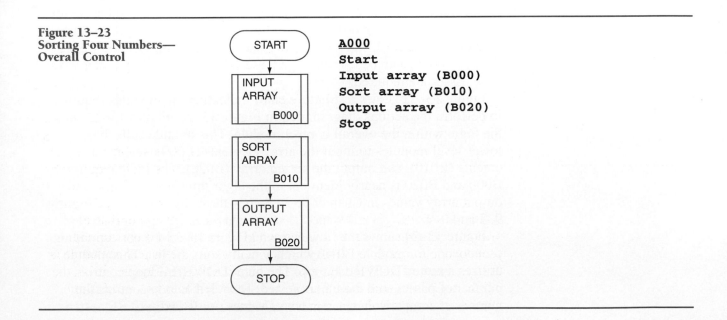

Figure 13–24
Sorting Four Numbers—Sort Array (Flowchart)

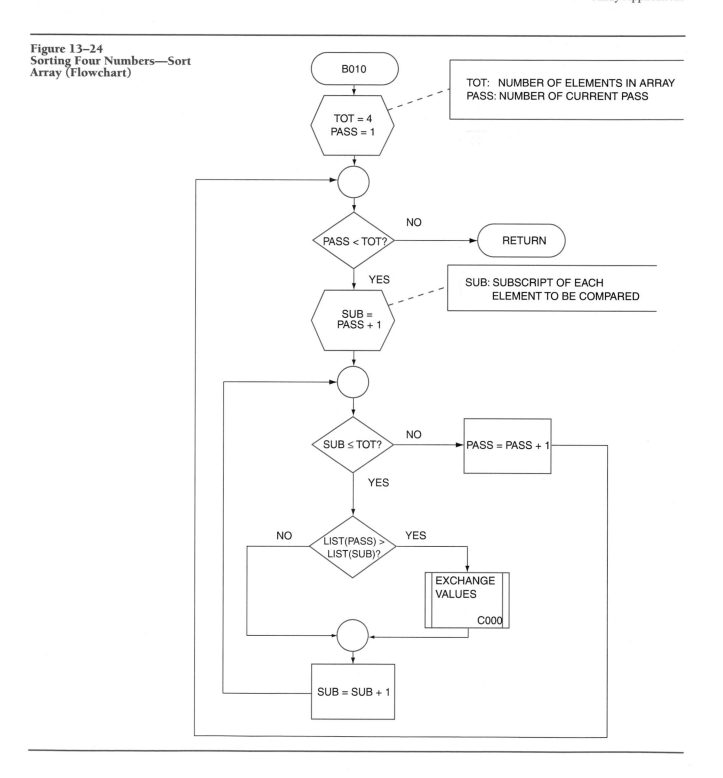

inside the inner DOWHILE loop actually compares the appropriate elements in each pass, causing their values to be exchanged if the current value being compared is larger than the next value. The actual exchange is done in the third-level module C000, which is shown in Figure 13–26.

This algorithm sorts only four values, but it demonstrates a technique that can be used to sort any number of values (see Exercise 13).

Figure 13–25
Sorting Four Numbers—Sort Array (Pseudocode)

```
B010
Enter
TOT = 4
PASS = 1
DOWHILE PASS < TOT
   SUB = PASS + 1
   DOWHILE SUB ≤ TOT
      IF LIST(PASS) > LIST(SUB) THEN
         Exchange values (C000)
      (ELSE)
      ENDIF
      SUB = SUB + 1
   ENDDO
   PASS = PASS + 1
ENDDO
Return
```

Figure 13–26
Sorting Four Numbers—Exchange Values

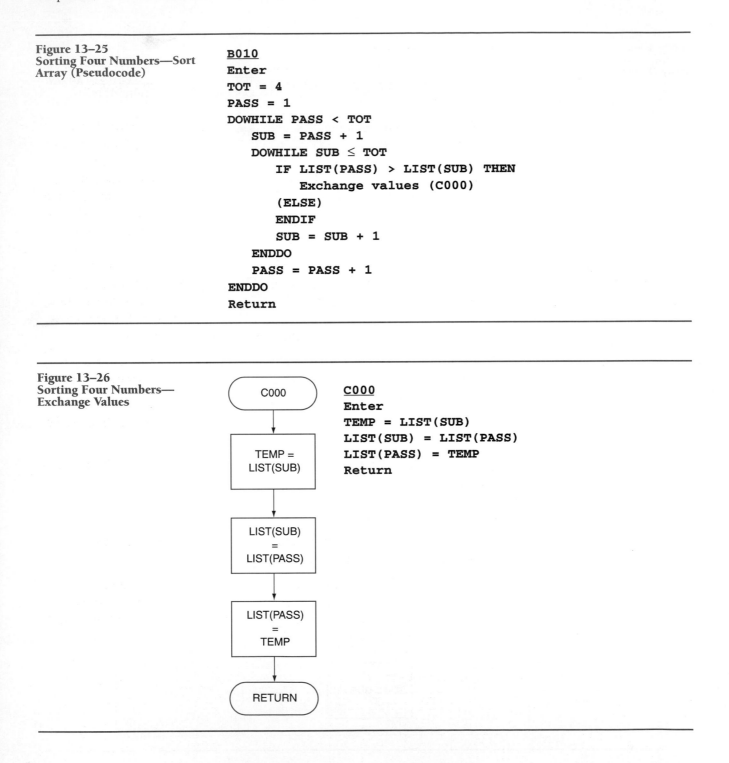

```
C000
Enter
TEMP = LIST(SUB)
LIST(SUB) = LIST(PASS)
LIST(PASS) = TEMP
Return
```

Key Terms

loading a table	key field	first pass (in sorting)
table lookup	search key	second pass
program switch	boolean operator	(in sorting)
binary search	truncate	

Exercises

1. **(a)** What is a table-lookup operation?
   (b) Give some examples (other than those in this book) of common situations where table-lookup operations can be used in problem solving.

2. The table-lookup module in Figures 13–7 and 13–8 is effective, but not particularly efficient. The module always begins searching for a new item number at the beginning of the table. Modify the solution algorithm in Figures 13–2, 13–3, 13–4, 13–5, 13–7, and 13–8 as follows: Assume that the input used to construct the table and the input referring to the table are to be processed in item-number sequence. Include sequence checks in the solution algorithm for verification. Terminate execution if (a) an item number equal to or lower than the preceding item number is provided as table input, or (b) an item number lower than the preceding item number is provided as input referring to the table. Begin each table search at the point in the item-number portion of the table at which the preceding item number was found. Express the solution algorithm in both flowchart and pseudocode forms. Be sure to plan a well-structured program.

3. Explain how a binary search technique can be applied in searching a table of U.S. cities and their populations to find the population of the city of Chicago.

4. For what kinds of problem situations is use of a binary search apt to be especially appropriate?

Refer to Figures 13–11 through 13–17 to complete Exercises 5 through 8.

5. What search key is used in this solution algorithm?

6. What is the key field of a reference table entry?

7. **(a)** Identify two program switches used in this solution algorithm.
   (b) Explain how each of the switches is used.

8. **(a)** Assume the value 11 is submitted as input and assigned to INCODE. Perform a simulation (trace) using the CODE array values specified in Figure 13–17. Use the following table to record the data.

INCODE = 11

FIRST	LAST	MID	INCODE : CODE(MID)

(b) Assume INCODE is equal to 38 and redo Exercise 8(a).

9. Use program flowcharting or pseudocode to plan a well-structured, modular program that reads N and an N-member list called A. Include checks to make certain that the members of A are unique and in ascending order. Program execution should terminate if they are not. If they are, the program should read ARG, an input value to be processed against the list, in subsequent processing steps.

- If ARG is less than A(1), set CODE equal to 0. Print out ARG, A(1), and CODE.

- If ARG is equal to some member of A, set CODE equal to 1. Print out ARG and CODE.

- If ARG is between A(1) and A(N), but there is no member of A equal to ARG, set CODE equal to 2. Print out ARG, CODE, and the member of A that is closest to but not greater than ARG.

- If ARG is greater than A(N), set CODE equal to 3. Print out ARG, A(N), and CODE.

10. Repeat Exercise 9, but allow for a variable number of inputs to be processed against the table. An input value of 999 for ARG should be recognized as an end-of-file indicator. The program should print an end-of-job message as verification that all input has been processed.

11. The program logic in Figures 13–24 and 13–25 causes a list of four values to be arranged in ascending sequence. To sort these same values into descending sequence, only one processing step must be modified. Identify this step by redoing that portion of the program logic shown in Figures 13–24 and 13–25.

12. The algorithm that you created in response to Exercise 11 should show how to perform a descending sort. Assume the following data items are provided as input to the program that you have planned: 52, .091, 708, 10.
    (a) What output should be provided by the program?
    (b) Perform a simulation (trace) to verify that the program will perform as you intend.
    (c) If the results of your simulation in Exercise 12(b) do not match the results you specified in Exercise 12(a), examine both your solution algorithm and your simulation to determine where errors have occurred. Make changes to eliminate the errors.

13. Consider modules B000 and B010 in the sort algorithm in this chapter. Modify the program flowcharts and pseudocode representations for these modules to show how to place any number of values in ascending sequence. An input value of 0 should be recognized as a dummy indicator, signaling that all values to be sorted have been read. When the sort operation is finished, the message "SORT COMPLETED" should be printed for control purposes.

14. Construct a program flowchart and corresponding pseudocode to solve the following problem: Read a series of numbers, one number per record, into a one-dimensional array called X. The number of input values will be specified by a number on the first input record. This first number should not be included in the array. Your algorithm is to compute the mean and the standard deviation of the values in array X. The mean can be computed according to the following formula:

$$\overline{X} = \sum_{i=1}^{N} X_i / N$$

The symbol Σ means the summation of; the mean is represented by the symbol $\overline{X}$. Thus, the formula says that the mean is equal to the summation of the array elements, X_i, where i goes from 1 to N, divided by N.

The standard deviation can be computed according to the following formula:

$$\sigma = \sqrt{\frac{\sum_{i=1}^{N} (X_i - \overline{X})^2}{N - 1}}$$

First, square each of the differences between each element in the array (i.e., X_i) and the computed mean. Each of these differences squared is added to a variable to accumulate their sum. Then divide this summation by the number of items (N) minus 1. Finally, take the square root of the result to find the standard deviation (σ). After the computations are complete, output the array X, the computed mean, and the computed standard deviation.

15. Construct a flowchart and corresponding pseudocode to solve the following problem: Enter the following data into three one-dimensional arrays: PART#, QOH, and UNITCOST. (Note that the data does not need to be input but can be self-generated within the algorithm.)

Part Number	Quantity on Hand	Unit Cost
115	50	90
120	60	91
125	70	92
130	80	93
135	90	94
140	100	95
145	110	96
150	120	97
155	130	98
160	140	99

After the data has been entered, read an arbitrary number of records containing a part number and a quantity ordered. If the entire quantity ordered for the part is available, output a line containing the part number, quantity ordered, cost/unit, and total cost for that part number. Also, update the quantity-on-hand array by subtracting the quantity ordered from the quantity-on-hand for that part. If the quantity ordered was more than the quantity-on-hand, simply output a line containing the part number and a message stating that the part number is understocked and unavailable for shipment. After all records have been processed, output a list of part numbers that must be reordered and the reorder quantity for each. The reorder point for a part number is when the quantity-on-hand falls below 40. For example, if a part number's quantity-on-hand is 25, then 15 additional units must be ordered for that part number. Use automatic EOF to signal the end of the input and assume that all the part numbers input will be contained in the PART# array. Use the following structure chart to guide you in your solution:

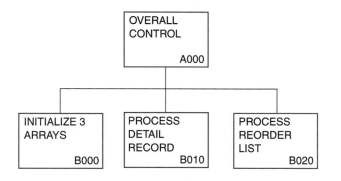

Master File Update Processing

Objectives	***Upon completion of this chapter you should be able to***
	■ Distinguish between online processing and batch processing.
	■ Define the terms *file maintenance* and *file maintenance run*.
	■ Explain the purpose of a turnaround file.
	■ Distinguish between sequential and direct processing.
	■ Design an algorithm that updates a sequential master file.

Introduction	The programmer directs a great deal of attention to the logic within a program, but he or she also must be conscious of the environment in which the program will operate. If the execution of a program is initiated by a user at a terminal (or personal computer or workstation) to meet the user's problem-solving needs, the program is said to operate in an **online-processing environment**. Such a program is often an independent entity. It generates its own input or accepts input directly from the user, without that input being operated on beforehand by another program. It provides output that is routed directly to the user or to other users at terminals (or personal computers or workstations): The output is not used as input to another program.
	Any program designed to solve one specific problem is apt to operate in this fashion. For example, a single program may plot the path of an object fired vertically from the Earth's surface as a function of time, according to a ballistics formula. Another single program may determine when return on investment will begin in a business venture, given principal, rate of interest to be compounded annually, and an accumulated sum to be attained. In effect, such a program is a system in itself.
	In a typical business organization, many computer programs are run on a regularly scheduled basis. Large volumes of input are collected and processed as sets, or **batches**, by the first of a series of interrelated programs, during a single processing run. Each complete series of programs is designed to meet the information-processing requirements of one organizational function—payroll, accounts receivable, billing, or inventory control, for example. The complete series of programs forms a system. Source documents within the system may be used to create inputs for one or more programs, and outputs generated by one program may serve as inputs for

321

**Figure 14–1
Monthly Billing Program
(System Flowchart)**

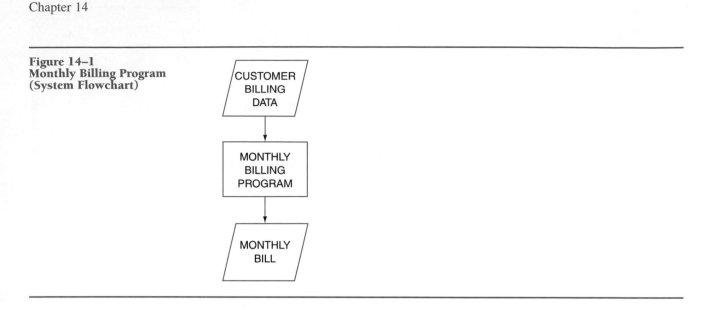

one or more other programs. The programs are said to operate in a **batch-processing environment**.

Whether a system comprises one program or several, it can be described by a system flowchart. The simplest type of system flowchart was introduced in Chapter 2 of this book. It represents one computer program that accepts one form of input and provides one form of output. One of the system flowcharts we discussed in Chapter 2 is shown again in Figure 14–1. Numerous CASE tools support methodologies and tools for describing existing and planned systems in terms of their inputs, outputs, and involved programs or modules. Such information is valuable to people who want to understand the "big picture" of a system.

File Maintenance

Figure 14–2 shows a system flowchart for a system that performs master file updating at a large publishing company. As we learned in Chapter 7, a master file is a collection of related records containing relatively permanent data essential to system processing. However, even permanent data must be changed occasionally; updating of master files for one or more systems is required at almost all information-processing installations. Updating a master file is called **file maintenance**; execution of a program that performs file maintenance is referred to as a **file maintenance run**.

The publishing company using the system described in Figure 14–2 stores the names, addresses, and other pertinent information about all its magazine subscribers in a master name and address file (master N/A file). The file is processed weekly to perform operations such as:

- Add names and addresses of new subscribers.
- Delete names and addresses of subscribers for whom subscriptions have expired.
- Modify records of current subscribers who have renewed their subscriptions or changed their addresses.

Figure 14–2
Master File Update Program
(System Flowchart)

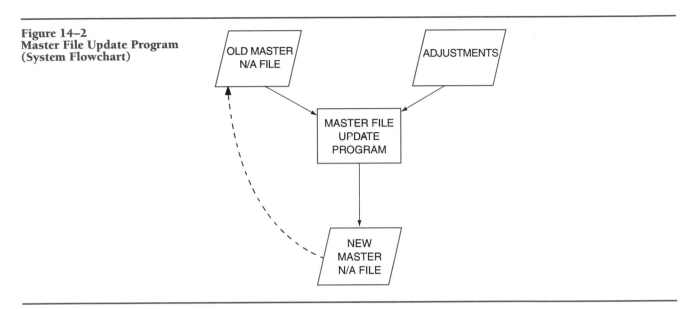

This system flowchart indicates that master-file updating is performed by a master file update program. There are two inputs to the program. One is the current N/A file, called the old master N/A file to distinguish it from the new master N/A file, which will be created by the program. The second input is another file containing adjustments to be made to the master N/A file. This file is the transaction, or detail, file.

The master file update program provides an updated (or new) master N/A file, which will be used as a current (or old) master N/A file the next time this program is executed. Programmers often use broken lines, as shown in Figure 14–2, to indicate a **turnaround file**. The new master N/A file also may be used by other programs (e.g., by a program that prepares address labels for use in mailing).

Sequential Processing

A major consideration when processing records is to determine the sequence in which the records are processed. If a file is stored on magnetic tape, for example, the records must be accessed in the order in which they are stored on the tape. It would take far too long to spin a tape backward or forward looking for a particular record. Instead, a particular record can be accessed only after all preceding records in the file have been accessed. We say that **sequential processing** of the records is required.

So that sequential processing can be performed in an efficient, effective manner, the records in a magnetic-tape master file are arranged in sequence according to a particular data item or items common to all records in the file. The portion of the record that contains this data is called the **key field**. For example, the key field of records in a master payroll file may contain employee number; the key field of records in a master inventory file may contain part number; and so on. All records to be included in the file must have the required key field, and the field must contain valid data.

An additional requirement of sequential processing is that any transaction records to be processed against a sequential master file must be in the same sequence as the master file records. During program execution, the key field value of a transaction record is compared with the key field values of successive master records until an equal comparison or match results. If transaction records and master records were not arranged in the same sequence before the comparisons were made, some master file records would be read long before their matching transaction records were accessed. It is likely that few, if any, matches would occur.

Direct Processing

Master files can be stored on tape, but they also can be stored on disks. Records stored on a magnetic disk can be accessed sequentially or directly. If they are accessed directly, they do not have to be processed in order. This type of processing is called **direct** (or **random**) **processing**. If records are processed directly, they do not need to be sorted first. There is also no requirement to sequence transaction records in the same order as master records. In this case, a **random update procedure** can be used.

Sequential Master File Update Example

Let us direct our attention now to the processing steps within a large program that performs a **sequential update procedure**. Any of several programming techniques can be used to match transaction records with corresponding master records for updating. The remainder of this chapter illustrates one such approach. Before we discuss the problem, it is important to emphasize the assumptions being made and to stress that these assumptions are basic to the successful execution of this program. The assumptions to be made are:

- The master file and the transaction file are assumed to be in the same sequence according to the key fields ID#-M and ID#-T, respectively.
- ID#s read from either file must be in the range from 1 through 10,000.
- Automatic end-of-file processing logic will be used in conjunction with both files.

All such assumptions made during the design stage of program development must be recognized and verified before the design plan is accepted. These assumptions should be stated clearly in the program documentation and pointed out whenever changes to the program or its inputs are discussed.

A structure chart outlining the modular design of a sequential master file update program is shown in Figure 14–3. As you can see, the design is not a simple one. This program consists of many modules, some of which appear in multiple locations on the structure chart. For example, module B020 is called by three different modules (A000, C000, and C010) during program execution. Reuse of a module is far preferable to writing similar code two or even three or more times to implement the same logic. In the past, such duplication of effort occurred often, usually because programmers had no way of knowing what potentially reusable code, routines, and modules already existed, or where to find them. Well-designed CASE tools

Figure 14–3
Sequential Master File
Update Problem (SMFUP)
(Structure Chart)

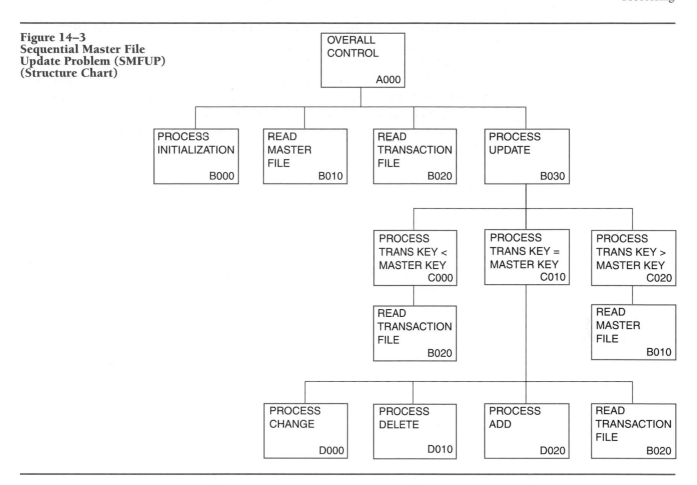

now provide ways to catalog and manage these components in online libraries. The tools promote code reusability and help to increase programmer productivity.

We will now discuss how the program shown in Figure 14–3 will operate in a general manner. Then we will take a look at the specific steps in each module.

After the appropriate initialization is completed, one record is read from the master file and one record is read from the transaction file. The update procedure is begun by comparing the key field from the transaction record to the key field from the master record. Three possibilities exist:

- If the transaction record key is *less than* the master record key, the transaction record is added to the master file and the next transaction is read.

- If the transaction record key is *equal to* the master record key, the master record is either deleted or updated with the information from the transaction record. In this case the next transaction record also needs to be read.

- If the transaction record key is *greater than* the master record key, there are no more transaction records that affect the current master record. Therefore, that master record is output to a new master file and the next master record is read.

This overview of the processing involved in updating a master file is helpful, but many additional details must be taken into account. For example, how do we handle end-of-file processing when we are reading from two separate files and either one may end before the other? What types of errors can occur during processing? These questions and others are answered in the detailed discussion of each module.

The flowchart and pseudocode for the overall control module (A000) are shown in Figures 14–4 and 14–5.

The first module called is B000, which handles all the initialization. The flowchart and pseudocode for B000 are shown in Figure 14–6.

Four program switches, or flags are used in this program, and they are all initially set to 0. Because we are reading from two separate files, each file will reach end-of-file independently of the other. EOF-M will be set to 1 in module B010 when end-of-file has been reached in the master file. Similarly, EOF-T will be set to 1 in module B020 when end-of-file has been reached in the transaction file. This logic is important, because we need to continue to

Figure 14–4
SMFUP—Overall Control
(Flowchart)

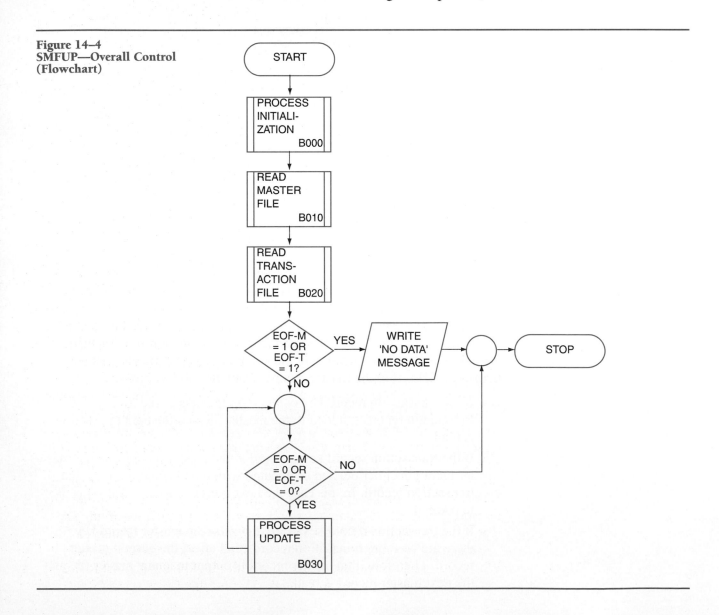

Figure 14–5
SMFUP—Overall Control
(Pseudocode)

```
A000
Start
Process initialization (B000)
Read master file (B010)
Read transaction file (B020)
IF EOF-M = 1 OR EOF-T = 1 THEN
    Write 'No data' message
ELSE
    DOWHILE EOF-M = 0 OR EOF-T = 0
        Process update (B030)
    ENDDO
ENDIF
Stop
```

Figure 14–6
SMFUP—Process
Initialization

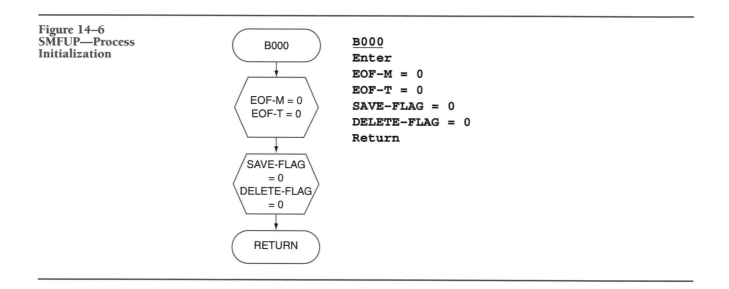

```
B000
Enter
EOF-M = 0
EOF-T = 0
SAVE-FLAG = 0
DELETE-FLAG = 0
Return
```

process the remaining transaction records (valid transactions can only be adds at this point) if we reach end-of-file in the master file first. Similarly, if we reach end-of-file in the transaction file first, we need to read the rest of the master file and copy these records (unchanged) to the new master file. The flowchart and pseudocode for the modules that read the master file and transaction file are shown, respectively, in Figures 14–7 and 14–8.

These two modules are exactly the same, except that one reads the master file (B010) and the other reads the transaction file (B020). Each master record contains an identification number (ID#-M), name, and address of a person subscribing to a magazine. Also included in the record are a magazine code (MAGZ-M) and the expiration date of the magazine (EXPDT-M). Recall from our earlier discussion that the key field being used in this program is ID#-M. Each transaction record contains the same information as the master record, with the addition of a special code field (CODE) specifying the type of transaction. Valid codes are A (add), C (change), and D (delete).

Remember also that the key field for the transaction record is ID#-T. When end-of-file is reached in one of the files (automatically, not trailer record), the EOF flag is set to 1 and the ID# for that file is set to 99999.

Figure 14–7
SMFUP—Read Master File

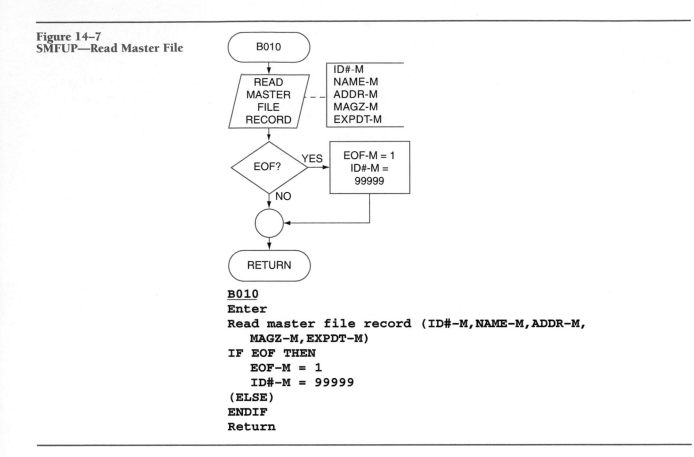

```
B010
Enter
Read master file record (ID#-M,NAME-M,ADDR-M,
   MAGZ-M,EXPDT-M)
IF EOF THEN
   EOF-M = 1
   ID#-M = 99999
(ELSE)
ENDIF
Return
```

Figure 14–8
SMFUP—Read Transaction
File

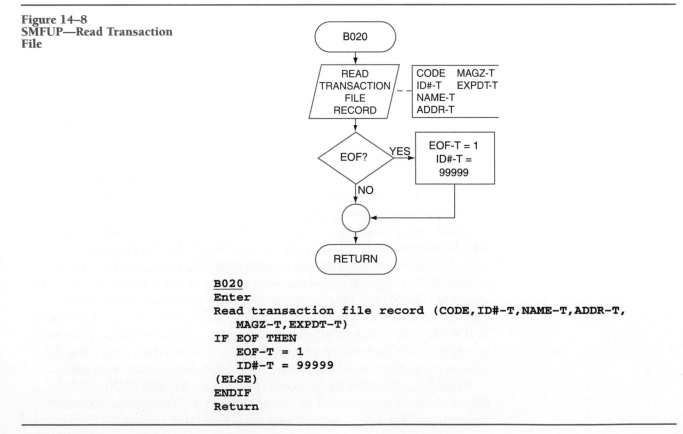

```
B020
Enter
Read transaction file record (CODE,ID#-T,NAME-T,ADDR-T,
   MAGZ-T,EXPDT-T)
IF EOF THEN
   EOF-T = 1
   ID#-T = 99999
(ELSE)
ENDIF
Return
```

Remember the assumption that any ID#s read from either file must be in the range 1 through 10,000.

When end-of-file is finally reached in one of the files, the remaining records in the other file must still be processed. Because this processing involves the comparison of key fields from both the transaction and master files, the ID# from the file that has reached end-of-file no longer exists. If this file is the master file, a key of 99999 in the master file will guarantee that any subsequent transaction key will be less than the master key, and the add operation will be done for each remaining transaction record. If the transaction record reaches end-of-file first, a key of 99999 in the transaction file will guarantee that the transaction key will be greater than any subsequent master key. This will cause the remaining master records to be output unchanged to the new master file.

Two other flags are initialized in B000: SAVE-FLAG and DELETE-FLAG. DELETE-FLAG is used to indicate if the current master record is to be deleted. If DELETE-FLAG is set to 1, the master record will not be output to the new master file. In other words, deleting a record is the act of not writing it to a new master file. SAVE-FLAG is used to indicate if the current master record has been copied to a separate area in memory (save area). This will be done whenever a transaction record needs to be added to the master file. The current master record will be saved in another area and SAVE-FLAG will be set to 1.

The transaction record to be added will then be copied to the area formerly used for the current master record. This is necessary because only a record in the current master record area of memory can be output to the new master file. After this transaction record is actually added to the new master file, the saved master record must be copied back to the current master record area so that it can be processed.

After both files have been read for the first time, a test is made in A000 to determine if the main processing should even be attempted (see Figures 14–4 and 14–5). The first decision step in this module makes sure that neither of the files is empty (end-of-file reached on the first read attempt). If there are no transaction records, the master file does not need to be updated. If there are no master records, the master file does not exist but needs to be created. The creation of a master file should be done in another program. In either case, a "No data" message will be output. (Consider how you might change this module to output either of two messages, depending on which file was empty. See Exercise 5.)

If both files contain at least one record, then the update processing (module B030) will be done inside a DOWHILE loop, as long as either of the files contains records to be processed. Remember, processing must continue with one file even when the other file reaches end-of-file. When end-of-file is finally reached in both files, program execution will terminate.

The flowchart and pseudocode for the process update module (B030) are shown in Figure 14–9. B030 calls three modules to do the actual processing. The first module (C000) gets control if the transaction record key is *less than* the master record key. Module C010 gets control if the transaction record key is *equal to* the master record key. Module C020 handles the writing of a new master record, which will occur when the transaction record key is *greater than* the master record key.

**Figure 14–9
SMFUP—Process Update**

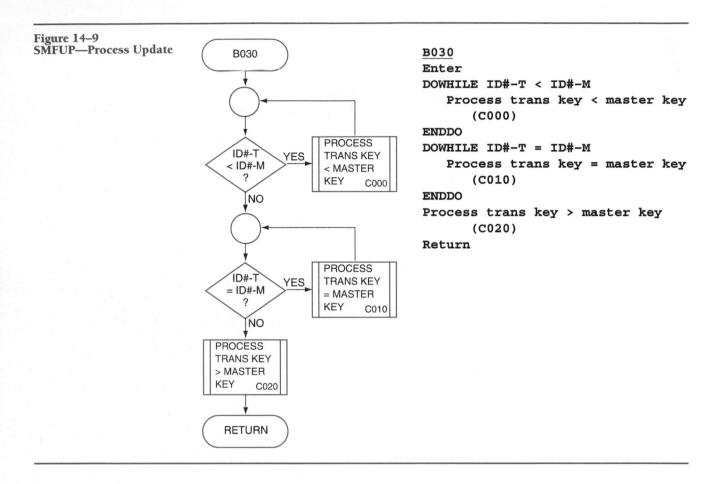

```
B030
Enter
DOWHILE ID#-T < ID#-M
    Process trans key < master key
        (C000)
ENDDO
DOWHILE ID#-T = ID#-M
    Process trans key = master key
        (C010)
ENDDO
Process trans key > master key
        (C020)
Return
```

The flowchart and pseudocode for module C000 are shown in Figures 14–10 and 14–11. A test is made to determine if the transaction code (CODE) is an "A." This is the only valid possibility. If the code is an "A," the current master record will be saved in another area to be processed at a later time, and the save flag will be set to 1. The transaction record to be added then will be copied into the current master record area (see Figure 14–12). If the code is not an "A," either it was invalid or an attempt was made to delete or change a record that does not exist in the current master file. In either case, an error message will be output. (Consider how you might change this module to output one of three specific error messages denoting a bad delete attempt, a bad change attempt, or an invalid code. See Exercise 6.)

After the current transaction record is either copied or determined to be in error, a new transaction record is read and control is returned to B030. The new transaction record's key is then tested against the current master record key. You might wonder why we make this test again. It seems unlikely that a subsequent transaction record key could be less than the master record key, because the current master record key is the previous transaction key. Remember, the keys are assumed to be in sequence. The only time this will occur is if the previous transaction was in error or our assumption is wrong. In either case, the master record key remained unchanged. The new transaction key may still be less than the master record key at this point. Consider the case when a transaction record with an

**Figure 14–10
SMFUP—Process Trans
Key < Master Key
(Flowchart)**

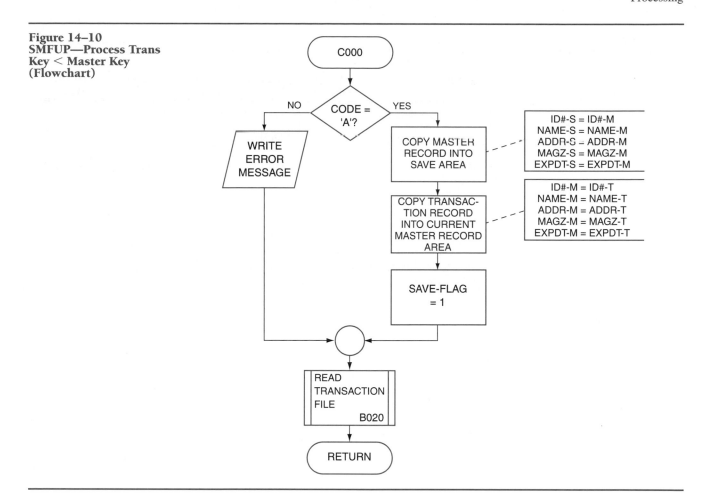

**Figure 14–11
SMFUP—Process Trans
Key < Master Key
(Pseudocode)**

```
C000
Enter
IF CODE = 'A' THEN
    ID#-S = ID#-M
    NAME-S = NAME-M
    ADDR-S = ADDR-M            Copy master record
    MAGZ-S = MAGZ-M            into save area
    EXPDT-S = EXPDT-M
    ID#-M = ID#-T
    NAME-M = NAME-T
    ADDR-M = ADDR-T            Copy transaction record
    MAGZ-M = MAGZ-T            into current master
    EXPDT-M = EXPDT-T          record area
    SAVE-FLAG = 1
ELSE
    Write error message (deleting or changing
        a record that doesn't exist or
        invalid transaction code)
ENDIF
Read transaction file (B020)
Return
```

**Figure 14–12
Adding a Master Record
(Code = "A" and ID#-T
< ID#-M)**

	CURRENT TRANSACTION RECORD			CURRENT MASTER RECORD		SAVE AREA	
	CODE	ID#-T	OTHER FIELDS	ID#-M	OTHER FIELDS	ID#-S	OTHER FIELDS
①	A	000123		000586			
②	A	000123		000586 ———————→		000586	
③	A	000123 ———————→		000123		000586	

invalid code is immediately followed by a valid transaction record with a code of A. Eventually, the transaction record key will no longer be less than the master record key and the second DOWHILE loop test (ID#-T = ID#-M) in B030 is made. If this condition is true, module C010 will be executed. Again, a loop is used, because a single master record may have several transactions associated with it.

The flowchart and pseudocode for module C010 are shown in Figures 14–13 and 14–14. A CASE control structure is used to determine the type of transaction. One of three modules is called if the transaction code is valid, and an error message is output if the transaction code is not valid. In any case, another transaction is read after the previous one is processed, and control is returned to B030. The keys are tested again, and if they are no longer equal, module C020 gets control.

**Figure 14–13
SMFUP—Process Trans Key =
Master Key (Flowchart)**

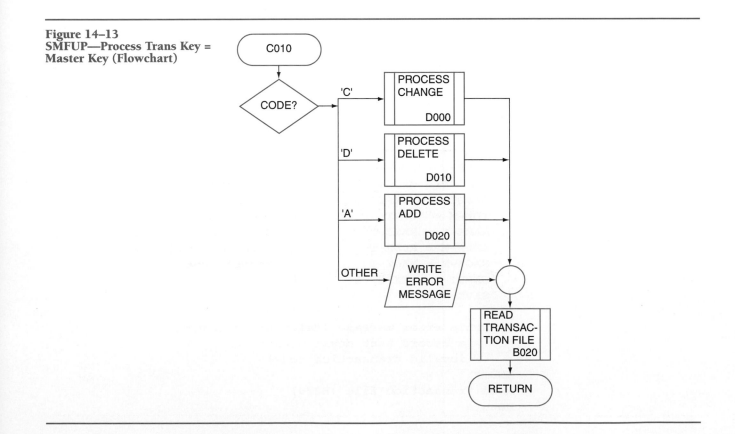

Figure 14–14
SMFUP—Process Trans
Key = Master Key
(Pseudocode)

```
C010
Enter
CASENTRY CODE
    CASE 'C'
        Process change (D000)
    CASE 'D'
        Process delete (D010)
    CASE 'A'
        Process add (D020)
    CASE other
        Write error message (invalid transaction
          code)
ENDCASE
Read transaction file (B020)
Return
```

The flowchart and pseudocode for module C020 are shown in Figures 14–15 and 14–16. This module will be called by B030 only when the transaction record key is greater than the master record key. This indicates that there are no more transactions to be processed against the current master record. At this point the current master record can be written to a new

Figure 14–15
SMFUP—Process Trans
Key > Master Key
(Flowchart)

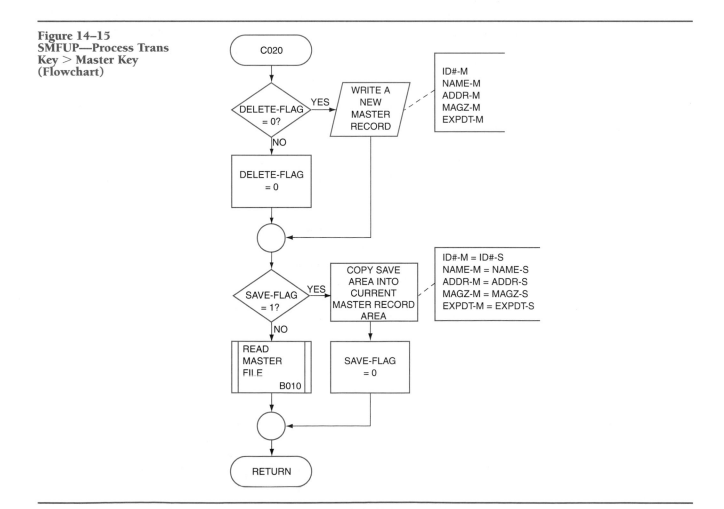

Figure 14–16
SMFUP—Process Trans Key > Master Key (Pseudocode)

```
C020
Enter
IF DELETE-FLAG = 0 THEN
    Write a new master record (ID#-M,NAME-M,ADDR-M,MAGZ-M,
        EXPDT-M)
ELSE
    DELETE-FLAG = 0
ENDIF
IF SAVE-FLAG = 1 THEN
    ID#-M = ID#-S
    NAME-M = NAME-S        Copy save area
    ADDR-M = ADDR-S        into current master
    MAGZ-M = MAGZ-S        record area
    EXPDT-M = EXPDT-S
    SAVE-FLAG = 0
ELSE
    Read master file (B010)
ENDIF
Return
```

master file (another actual file, as shown in the system flowchart in Figure 14–2), unless the delete flag has been set to 1. If a previous transaction specified that this master record was to be deleted, the delete flag was set to 1 (see module D010). In this case, the current master record simply will not be copied to the new master file. The delete flag will be reset to 0 so it does not cause subsequent master records to be deleted.

Because the current master record was either deleted or copied to the new master file, it would seem appropriate to read another record from the master file. Module C020 *may* read a record next, but first it checks the save flag. If the save flag is equal to 0, a master record is read. If the save flag is equal to 1, we know that a master record that has not yet been processed is stored in a save area. (Remember module C000.) In such a case, it is inappropriate to read a new master record, because one exists in the save area and has not yet been processed. This master record in the save area is then copied back into the current master record area, and the save flag is reset to 0 (see Figure 14–17). Control is then returned to B030. B030 returns control to A000 and the end-of-file flags are checked prior to re-entering the main loop.

Recall that module C010 calls one of three modules to process a transaction with a valid code. If the code is equal to "C", module D000 (see Figure

Figure 14–17
Copying Save Area into Current Master Record

	CURRENT TRANSACTION RECORD			CURRENT MASTER RECORD		SAVE AREA	
	CODE	ID#-T	OTHER FIELDS	ID#-M	OTHER FIELDS	ID#-S	OTHER FIELDS
①	C	01234		000123		000586	
②	C	01234		000586 ←		000586	

Figure 14–18
SMFUP—Process Change

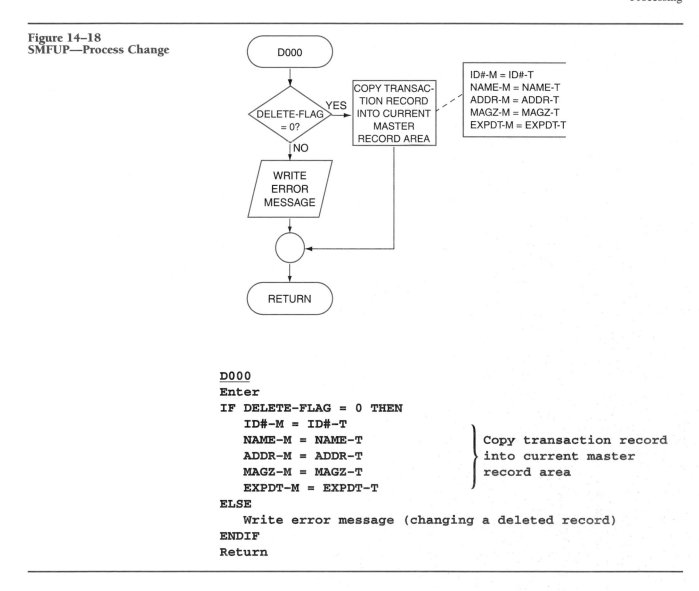

```
D000
Enter
IF DELETE-FLAG = 0 THEN
   ID#-M = ID#-T
   NAME-M = NAME-T        ⎫  Copy transaction record
   ADDR-M = ADDR-T        ⎬  into current master
   MAGZ-M = MAGZ-T        ⎭  record area
   EXPDT-M = EXPDT-T
ELSE
   Write error message (changing a deleted record)
ENDIF
Return
```

14–18) handles the processing. First, the delete flag is checked to make sure that the current master record is not marked for deletion. If it is, an error message is output because it makes no sense to change a deleted record. If the record has not been marked for deletion, the transaction record data is copied into the current master record area, replacing what was there before (see Figure 14–19). Control is then returned to module C010.

If the code is equal to "D," module D010 (see Figure 14–20) handles the processing. Again, the delete flag is checked to determine if the current master record has been marked for deletion. If it has, an error message will be output indicating an attempt has been made to delete an already-deleted record. Although it would not hurt to "delete again" (because we would simply be setting a flag to 1 that was already 1), this action is probably not the intent of the person creating the transaction file. It is more likely a data entry error, so by providing an error message we alert someone to a potential problem. If the current master record has not been marked for deletion, the delete flag is set to 1 and control is returned to module C010.

Figure 14–19
Changing a Master Record
(CODE = "C" and ID#-T =
ID#-M)

	CURRENT TRANSACTION RECORD			CURRENT MASTER RECORD		SAVE AREA	
	CODE	ID#-T	OTHER FIELDS	ID#-M	OTHER FIELDS	ID#-S	OTHER FIELDS
①	C	01234		01234		00586	
②	C	01234 ⟶		01234	one or more of the fields in the current master record will be replaced with data from the current transaction record	00586	

Figure 14–20
SMFUP—Process Delete

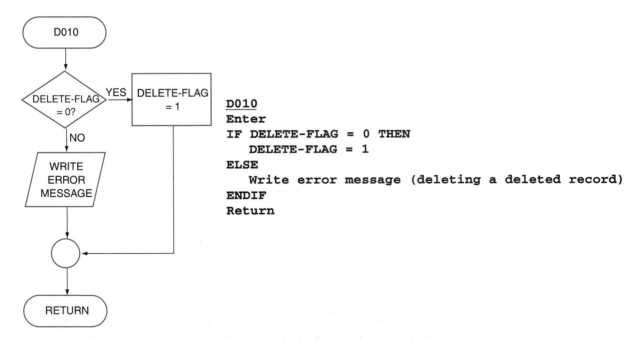

```
D010
Enter
IF DELETE-FLAG = 0 THEN
    DELETE-FLAG = 1
ELSE
    Write error message (deleting a deleted record)
ENDIF
Return
```

If the code is equal to "A", module D020 (see Figure 14–21) handles the processing. In most cases, this module will be called when an attempt to add a record is made and that record already exists. In this case, an error message will be output. It is possible, however, to add a record that has just been deleted. If this happens, ID#-T will be equal to ID#-M and module C010 will be called—not C000, which does the normal add. Therefore, if the delete flag is 1, the current transaction record can be copied right over the current master record (because it was previously deleted) (see Figure 14–22). Do you see why we do not have to save the master record first? The delete flag must be reset to 0, because the current master record should no longer be marked for deletion. An add transaction that immediately follows a delete transaction has the same effect as a single change transaction.

Figure 14–21
SMFUP—Process Add

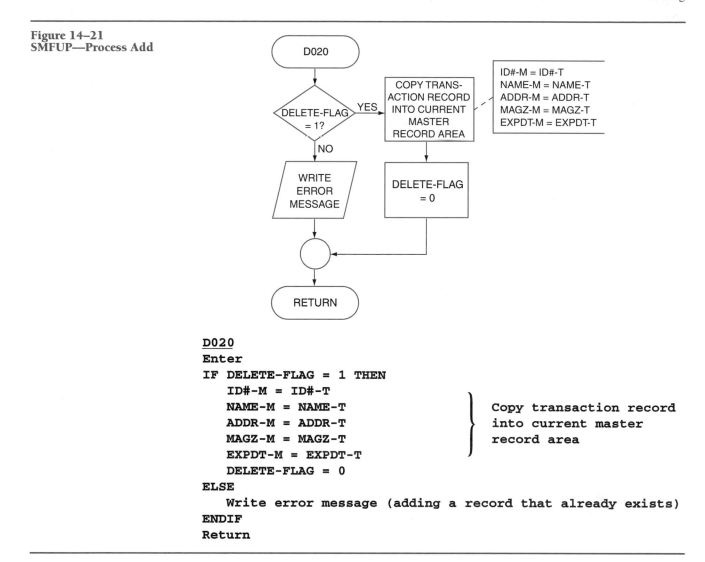

```
D020
Enter
IF DELETE-FLAG = 1 THEN
    ID#-M = ID#-T
    NAME-M = NAME-T        ⎫  Copy transaction record
    ADDR-M = ADDR-T        ⎬  into current master
    MAGZ-M = MAGZ-T        ⎭  record area
    EXPDT-M = EXPDT-T
    DELETE-FLAG = 0
ELSE
    Write error message (adding a record that already exists)
ENDIF
Return
```

Figure 14–22
Adding a Master Record
(CODE = "A" and ID#-T =
ID#-M and DELETE-FLAG = 1)

	CURRENT TRANSACTION RECORD			CURRENT MASTER RECORD		SAVE AREA	
	CODE	ID#-T	OTHER FIELDS	ID#-M	OTHER FIELDS	ID#-S	OTHER FIELDS
①	A	01534		01534		00586	
②	A	01534	⟶	01534	one or more of the fields in the current master record will be replaced with data from the current transaction record	00586	

Key Terms

online-processing
  environment
batch
batch-processing
  environment
file maintenance

file maintenance run
turnaround file
sequential processing
key field
direct (random)
  processing

random update
  procedure
sequential update
  procedure

Exercises

1. Define the following terms:
   (a) online-processing environment
   (b) batch-processing environment
   (c) file maintenance run

2. What is the most significant characteristic of a magnetic-tape file in determining how records in the file must be processed?

3. Explain the purpose of each program switch initialized in Figure 14–6.

4. Modify the solution algorithm presented in this chapter to compute and output a count of the number of master records processed. The count should be (1) initialized to 0, (2) increased by 1 whenever a master record is written as output, and (3) output with an appropriate message before processing is terminated.

5. Redo the flowchart and/or pseudocode in Figures 14–4 and 14–5 to output one of two specific messages, depending on which file contained no data.

6. Redo the flowchart and/or pseudocode in Figures 14–10 and 14–11 to output one of three specific error messages denoting a bad delete attempt, a bad change attempt, or an invalid transaction code.

7. (a) Simulate the execution of the algorithm presented in this chapter using the following input records (make up values for the other fields):

Transaction File		Master File
CODE	ID#-T	ID#-M
C	00002	00002
D	00003	00003
C	00007	00006
C	00007	00007
A	00008	00009
		00010
		00012
		00014

(b) Show the contents of the new master file when processing is completed.
(c) List all the error messages (if any) that will be output.

8. **(a)** Simulate the execution of the algorithm presented in this chapter using the following input records (make up values for the other fields):

Transaction File		Master File
CODE	ID#-T	ID#-M
A	00001	00003
C	00002	00004
A	00002	00005
C	00003	00006
B	00004	00007
C	00005	00008
D	00007	00010
D	00009	
D	00010	
C	00010	

(b) Show the contents of the new master file when processing is completed.

(c) List all the error messages (if any) that will be output.

9. **(a)** Simulate the execution of the algorithm presented in this chapter using the following input records (make up values for the other fields):

Transaction File		Master File
CODE	ID#-T	ID#-M
D	00003	00001
A	00003	00002
C	00005	00003
D	00006	00004
D	00006	00006
C	00007	00007
A	00008	00008
A	00009	
A	00011	
D	00012	
A	00013	

(b) Show the contents of the new master file when processing is completed.

(c) List all the error messages (if any) that will be output.

Control-Break Processing

Upon completion of this chapter you should be able to

- Define the terms *control break* and *control field (key)*.
- Distinguish among a detail-printed report, a group-printed report, and a group-indicated report.
- Design an algorithm that handles single-level control-break processing.
- Design an algorithm that handles multiple-level control-break processing.

Introduction

In most of the problems and solutions discussed in this book, the output contains detail lines, total lines, or both. Recall our discussion of these different types of output lines in Chapter 5. It is sometimes necessary to output intermediate total lines as well as final totals. For example, a sales report may need to show individual employee sales (detail lines), department total sales (intermediate total lines), and accumulated sales from all departments (final total line).

We have already seen many examples of algorithms that output detail lines and/or final total lines. An intermediate total line can also be output; it is normally written when a control break occurs. A **control break** is an interruption of the normal detail processing when the value in a designated key, or control field, of the input records changes. The **key**, or **control field**, is the field that determines when a control break is to occur.

For example, look at Figure 15–1. Assume all records for department 1000 are read first. When the first record for department 2000 is input, an intermediate total line containing accumulated totals for department 1000 can be output. Similarly, when the first record for department 3000 is input an intermediate total line containing totals for department 2000 can be output. Do you see the pattern? When the department number changes, normal processing is interrupted and total information for the previous group of departments is output.

In this example, department number is defined to be the key or control field. In control-break processing, we make one very important assumption: The input must be in sequence according to the values in the control field. When a record with department number 2000 is first input, we assume that all records for department number 1000 have been input.

Figure 15–1
Control Break Processing
Problem (CBPP)—
Sample Input

DEPARTMENT NUMBER	SALESPERSON NAME	SALESPERSON NUMBER	WEEKLY SALES
1000	JOHN WEAVER	3498	5003.00
1000	NANCY SMITH	2281	6154.00
1000	JAMES JOHNSON	3098	4234.50
2000	MARY STEVENS	1154	2213.80
2000	PAUL PRATT	7638	8874.40
3000	THEODORE JONES	5540	9832.30
.	.	.	.
.	.	.	.
.	.	.	.
7000	STEVE BLACK	4554	2394.80
7000	LAURA CUNNINGHAM	1092	5541.70
.	.	.	.
.	.	.	.
.	.	.	.

Only in this way can we be sure that the total information for department 1000 is complete.

In a control-break program, it is possible to output detail lines, intermediate total lines, and final total lines. If the report includes detail lines, it is called a **detail-printed report**. This is a report in which one line is printed for each input record processed. If the report does not include detail lines, it is called a **group-printed report**. This report prints one line of output for each group of input records processed (intermediate total lines), but does not output the detailed information about each input record (see Exercise 10). A group-printed report may be output when only summary information is needed. It is not always necessary or even desirable to output a highly detailed document. Sometimes "less is better," depending on the information in the report and who will be reading it. As discussed previously, the use of CASE tools allows users to review and approve sample outputs early in the program development cycle.

The following problem illustrates the logic behind a control-break algorithm. As you will see, this algorithm produces a detail-printed report.

Sample Problem 15.1 (Single-Level Control Break)

Problem:

ABC Corporation needs a program to produce a weekly sales report. The input to the program consists of department number, salesperson name, number, and weekly sales, as illustrated in Figure 15–1. The input can be assumed to be in ascending order by department number. The report should contain both a report heading and column headings at the top of every

Figure 15–2
CBPP—Detail-Printed Report

```
                        ABC CORPORATION
                      WEEKLY SALES REPORT

   DEPARTMENT        SALESPERSON       SALESPERSON         WEEKLY
     NUMBER             NAME              NUMBER            SALES

      1000          JOHN  WEAVER          3498           5,003.30
      1000          NANCY  SMITH          2281           6,154.00
      1000          JAMES  JOHNSON        3098           4,234.50

               TOTAL  SALES--DEPARTMENT  1000:    15,391.80


      2000          MARY  STEVENS         1154           2,213.80
        .                .                  .                .
        .                .                  .                .
        .                .                  .                .

               TOTAL  SALES--DEPARTMENT  2000:   120,681.30


      3000          THEODORE  JONES       5540           9,832.30
        .                .                  .                .
        .                .                  .                .
        .                .                  .                .

               TOTAL  WEEKLY  SALES: 443,952.90
```

page, and each page should be numbered. No more than 50 detail lines should be output on a single page. Each detail line should contain department number, salesperson name, salesperson number, and weekly sales for one person. There should also be one line showing accumulated sales information for each department. A final total line indicating the total sales volume for all departments should be output at the end of the report. The format of the output is illustrated in Figure 15–2.

Solution:

A structure chart illustrating the modular design of this algorithm is shown in Figure 15–3. All the module names and functions should be familiar to you, with the exception of module B020 (process department change). This module will handle the processing necessary when a control break occurs.

The program flowchart and pseudocode representations of the overall control module (A000) are shown in Figures 15–4 and 15–5.

After the necessary initialization is done, the first record is input. As usual, this record is tested to determine if end-of-file (automatic, not trailer record) has been reached on the first read attempt. If so, a "No data" message is output and processing terminates. If end-of-file has not been

Figure 15–3
CBPP (Structure Chart)

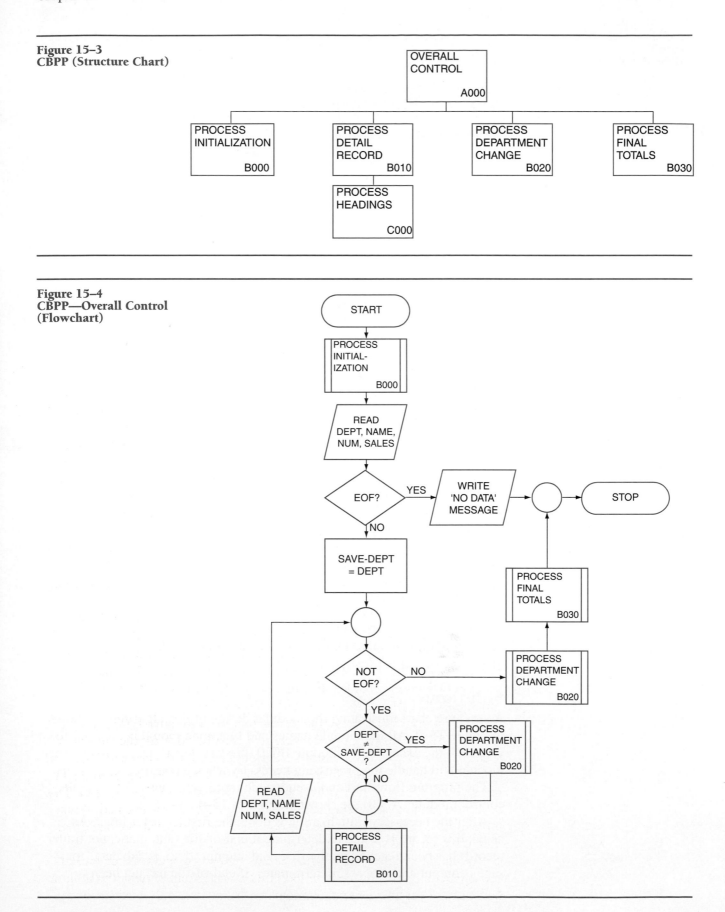

Figure 15–4
CBPP—Overall Control
(Flowchart)

Figure 15–5
CBPP—Overall Control
(Pseudocode)

```
A000
Start
Process initialization (B000)
Read DEPT, NAME, NUM, SALES
IF EOF THEN
   Write 'No data' message
ELSE
   SAVE-DEPT = DEPT
   DOWHILE not EOF
      IF DEPT ≠ SAVE-DEPT THEN
         Process department change (B020)
      (ELSE)
      ENDIF
      Process detail record (B010)
      Read DEPT, NAME, NUM, SALES
   ENDDO
   Process department change (B020)
   Process final totals (B030)
ENDIF
Stop
```

reached, the control field (DEPT) is copied into another area (SAVE-DEPT) before the DOWHILE loop is begun.

You might wonder why. Each time a new record is read, we need to determine if the department number just input is different from the previous department number. It is therefore necessary to save the value of the first department number and check that value against each subsequent department number input. As you can see, this test is the first step that is done in the main processing loop. If the department number just input (DEPT) is not equal to the department number in the save area (SAVE-DEPT), then a control break has occurred and module B020 must be executed. If the two department numbers are, in fact, equal, the normal detail processing (module B010) is executed. Then another record is read.

It is important to see why a null ELSE is used here. If a control break has not occurred, only module B010 is executed—because there are no steps to be done in the NO path of the IFTHENELSE. If a control break has occurred—that is, if the YES path of the IFTHENELSE is taken—module B020 is executed. When module B020 completes its processing, module B010 is then executed. Can you see why?

After the intermediate total line for department number 1000 is output (module B020), the detail processing still must be done for the current record that was input. Remember, a control break for department 1000 was detected because a record for department 2000 was read. That record (for department 2000) must be processed before a new record can be read. This repetitive kind of processing continues until end-of-file is reached. When the end-of-file is finally reached, module B020 is executed one last time; then the final total line is output (module B030).

You might wonder why module B020 is executed again. When end-of-file is reached, the intermediate total for the last group of department numbers has not yet been output. Remember, this is done only when a control break occurs, and a control break can be detected only when the control field in the input changes. An end-of-file is, in effect, a change in the control field. (In this case, it contains a nonexistent department number.)

A program flowchart and corresponding pseudocode for module B000 (process initialization) are shown in Figure 15–6. Two accumulators are set to 0. DEPT-ACCUM is used to keep track of the total weekly sales within one department. FINAL-ACCUM keeps track of the total weekly sales for all departments. MAXLINES is set to 50, the page count (PAGECNT) is set to 1, and the line count (LINECNT) is set to MAXLINES. LINECNT indicates how many detail lines are currently printed on a page of the report. Obviously, no lines have been printed yet, but initializing the line count to MAXLINES makes the computer think it's already at the bottom of a page. Do you remember why this is done?

A program flowchart and corresponding pseudocode for module B010 (process detail report) are shown in Figure 15–7. The first processing step is an IFTHENELSE that determines if headings need to be printed. Remember, when LINECNT reaches MAXLINES or greater, a new page is started and the headings must be printed first. The detail processing in this algorithm is quite simple: The detail line contains information, all of which has been input, so no calculations are necessary. The line is printed and LINECNT is incremented. Finally, the weekly sales amount from the input (SALES) is added to the department accumulator (DEPT-ACCUM) before control is returned to A000.

A program flowchart and corresponding pseudocode for module B020 (process department change) are shown in Figure 15–8. When this module gets control, a control break has occurred and a department total line must be output. Although the actual spacing of output lines is not always a consideration in the design phase, it may help to visualize the final report if we indicate when spacing is to occur. Prior to printing the department total line, a blank line is output; after the total line is printed, two blank lines are output. This will help distinguish the intermediate total line from the detail lines. The department number (SAVE-DEPT) and the accumulated sales for that department (DEPT-ACCUM), as well as identifying text, are output.

Figure 15–6
CBPP—Process Initialization

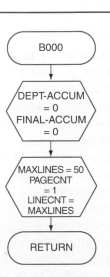

```
B000
Enter
DEPT-ACCUM = 0
FINAL-ACCUM = 0
MAXLINES = 50
PAGECNT = 1
LINECNT = MAXLINES
Return
```

Figure 15–7
CBPP—Process Detail Record

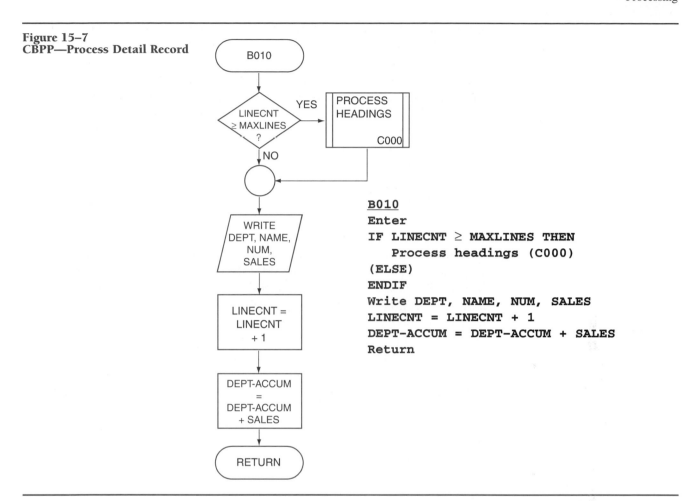

```
B010
Enter
IF LINECNT ≥ MAXLINES THEN
    Process headings (C000)
(ELSE)
ENDIF
Write DEPT, NAME, NUM, SALES
LINECNT = LINECNT + 1
DEPT-ACCUM = DEPT-ACCUM + SALES
Return
```

Why is SAVE-DEPT output and not DEPT? Remember, DEPT contains information from the current input record. Because four physical lines have been printed (three blank lines and one total line), the line count is incremented by 4. The department accumulator is added to the final accumulator and then reset to 0. Do you see why? When the next group of department numbers is processed, we do not want the previous department's total sales to be included in the current department's totals.

Finally, because we are getting ready to process a new group of department numbers, the save area (SAVE-DEPT) needs to be updated. Remember, this area is used to hold the value of the current department being processed. For example, when a control break occurs the first time, the current department number must be changed from 1000 to 2000.

A program flowchart and corresponding pseudocode for module B030 (process final totals) are shown in Figure 15–9. This module simply writes out the value of the final accumulator (FINAL-ACCUM) with an identifying message. Remember, FINAL-ACCUM represents the total weekly sales for all departments.

A program flowchart and corresponding pseudocode for module C000 (process headings) are shown in Figure 15–10. In this module, we are again trying to indicate some degree of spacing. The report heading is followed by one blank line, and the column headings are followed by one blank line.

Figure 15–8
CBPP—Process Department
Change

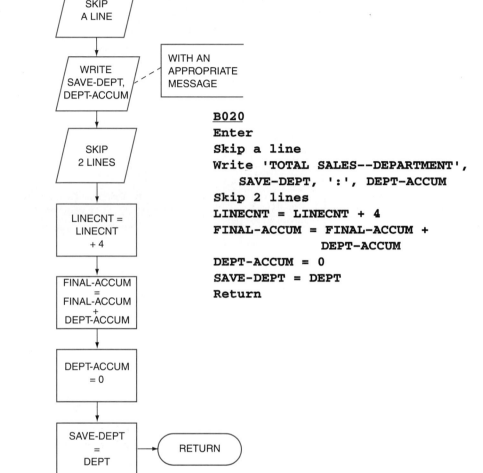

```
B020
Enter
Skip a line
Write 'TOTAL SALES--DEPARTMENT',
    SAVE-DEPT, ':', DEPT-ACCUM
Skip 2 lines
LINECNT = LINECNT + 4
FINAL-ACCUM = FINAL-ACCUM +
                  DEPT-ACCUM
DEPT-ACCUM = 0
SAVE-DEPT = DEPT
Return
```

Figure 15–9
CBPP—Process Final Totals

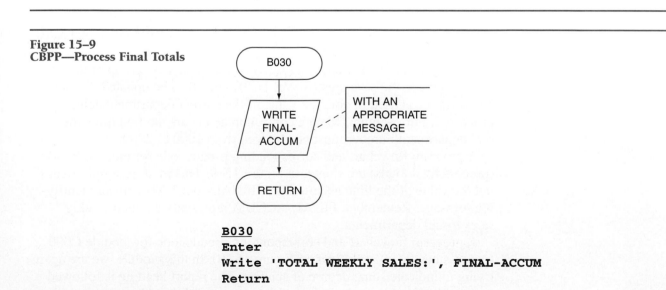

```
B030
Enter
Write 'TOTAL WEEKLY SALES:', FINAL-ACCUM
Return
```

The spacing is primarily for readability; it may or may not be specified by the user. In either case, the programmer must include an indication of spacing requirements somewhere in the documentation.

You might wonder why the line count is not checked prior to writing out either the intermediate total line in B020 or the final total line in B030. It is entirely possible that the line count could be equal to MAXLINES prior to writing out the total lines. Because module B020 does not check the line count, a department total line is always output on the same page on which the detail lines for that same department are printed. This enhances the readability of the report; it would be very confusing to see a department total at the top of a new page without the detail lines preceding it. Similarly, the final total line is always output on the same page on which the last department total line is printed.

With some quick computations, we can figure the maximum number of lines that could be output on a page. Each time B010 is called, the line count is checked. If no intermediate total lines are output, the line count will be incremented by 1 each time B010 executes—until it gets to 50. Then a new page will be started and the line count will be reset to 0. If

Figure 15–10
CBPP—Process Headings

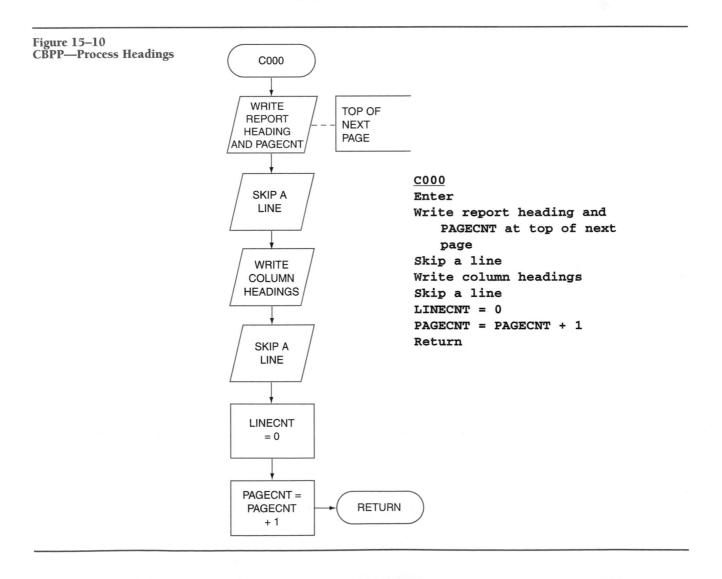

```
C000
Enter
Write report heading and
    PAGECNT at top of next
    page
Skip a line
Write column headings
Skip a line
LINECNT = 0
PAGECNT = PAGECNT + 1
Return
```

B020 is called, let's say midpage (LINECNT about 30), the four intermediate total lines (blank line, actual total line, and two blank lines) are output on the current page and the line count is incremented to 34.

The maximum number of lines would be printed only if the following sequence of steps were executed: LINECNT is 49 on entry to B010 (no new page would be started yet); B010 executes, incrementing LINECNT to 50; B010 returns control to A000 and a new record is read; a control break occurs, so B020 executes next, incrementing LINECNT to 54; B020 returns control to A000, which immediately gives control to B010; B010 checks LINECNT and calls C000, which starts a new page and resets LINECNT to 0. The only time B010 is not executed immediately after B020 is when end-of-file is reached. At this point B030 executes and outputs one last line, the final total line. Because the line count was previously 54, we now know that 55 lines (excluding headings) have been output. If module C000 outputs six lines (two report heading lines, one blank line, two column heading lines, and one blank line), the total number of printed lines on one page would be 61. Because most printers can print 66 physical lines on a page, this should not present a problem. It is important to interpret the specifications properly. The results would have been entirely different if the problem statement had said that a maximum of 50 lines (any type) could be output.

Sample Problem 15.2 (Multiple-Level Control Break)

Problem:

XYZ Computer Company needs a program to produce a weekly sales report. As illustrated in Figure 15–11, the input to the program consists of branch number, salesperson number, customer number, item description, and sales amount of item. The input can be assumed to be in ascending order by customer number, within salesperson number, within branch number.

The report should contain both a report heading and column headings at the top of every page. Each page should be numbered. No more than 40 detail lines should be output on a single page. Each detail line should contain branch number, salesperson number, customer number, description of item, and sales amount of item. There also should be one line showing accumulated sales information for each customer, for each salesperson, and for each branch. A final total line indicating the total sales volume for all branches should be output at the end of the report. The format of the output is illustrated in Figure 15–12. Note that the customer number is printed next to only the first item purchased by each customer. This is an example of a **group-indicated report**. Group-indicated reports generally appear less cluttered and easier to read than detail-printed reports. Note that both the salesperson number and the branch number are also group indicated.

Solution:

A structure chart illustrating the modular design of this algorithm is shown in Figure 15–13. All the module names and functions should be familiar to you. In this example, a control break will occur whenever a customer number, salesperson number, or branch number changes. Note that we have

Figure 15–11
Multiple-Level Control-Break
Processing (MLCBPP)—
Sample Input

BRANCH NUMBER	SALESPERSON NUMBER	CUSTOMER NUMBER	ITEM DESCRIPTION	SALES AMOUNT
10	1122	345	Monitor	400.00
10	1122	345	Printer	200.00
10	1122	345	Modem	100.00
10	1122	412	Copier	500.00
10	1122	412	Printer	600.00
10	1122	567	Keyboard	100.00
10	2334	123	Modem	100.00
10	2334	123	Computer	1000.00
10	2334	246	Hard Drive	350.00
20	1111	137	Hard Drive	350.00
20	1111	137	Monitor	400.00
20	2222	387	Computer	1000.00
20	3333	190	Copier	500.00
20	3333	190	Modem	100.00
20	3333	224	Hard Drive	350.00
20	3333	367	Keyboard	100.00
20	3333	367	Fax Machine	500.00
30	1479	108	Computer	1000.00
.	.	.	.	.
.	.	.	.	.
.	.	.	.	.

included three separate modules (B020, B030, and B040) to handle this processing.

The program flowchart and pseudocode representations of the overall control module (A000) are shown in Figures 15–14 and 15–15.

After the necessary initialization is done, the first record is input. As usual, this record is tested to determine if end-of-file (automatic, not trailer record) has been reached on the first read attempt. If so, a "No data" message is output and processing terminates. If end-of-file has not been reached, then each control field (CUST#, SALES#, and BRANCH#) is copied into another area (SAVE-CUST#, SAVE-SALES#, and SAVE-BRANCH#) before the DOWHILE loop is begun. You probably remember using the same step in the previous problem. Once again, each time a new record is read, we need to determine if the values in any of the control fields just input (CUST#, SALES#, or BRANCH#) are different from the previous values. It is therefore necessary to save the values of the first customer number, salesperson number, and branch number, and to check those values against each subsequent customer number, salesperson number, and branch number input.

The first processing step within the DOWHILE loop blanks out three areas: CUST#-OUT, SALES#-OUT, and BRANCH#-OUT. These variable names represent the customer number, salesperson number, and branch

Figure 15–12
MLCBPP—Group-Indicated
Report

```
                         XYZ COMPUTER COMPANY
                         WEEKLY SALES REPORT

  BRANCH       SALESPERSON      CUSTOMER          ITEM          SALES
  NUMBER         NUMBER         NUMBER         DESCRIPTION      AMOUNT

    10            1122            345          Monitor          400.00
                                              Printer          200.00
                                              Modem            100.00

                 TOTAL CUSTOMER 345 SALES:                     700.00

                                 412          Copier           500.00
                                              Printer          600.00

                 TOTAL CUSTOMER 412 SALES:                    1100.00

                                 567          Keyboard         100.00

                 TOTAL CUSTOMER 567 SALES:                     100.00

                 TOTAL SALESPERSON 1122 SALES:                1900.00

                 2334            123          Modem            100.00
                                              Computer        1000.00

                 TOTAL CUSTOMER 123 SALES:                    1100.00

                                 246          Hard Drive       350.00

                 TOTAL CUSTOMER 246 SALES:                     350.00

                 TOTAL SALESPERSON 2334 SALES:                1450.00

                 TOTAL BRANCH 10 SALES:                       3350.00

    20            1111            137          Hard Drive       350.00
                                              Monitor          400.00
                                                 .
                                                 .
                                                 .
                 TOTAL BRANCH 20 SALES:                       3300.00

    30            1479            108          Computer        1000.00
                                                 .
                                                 .
                                                 .
                 TOTAL BRANCH 30 SALES:                       5000.00

                 TOTAL ACCUMULATED SALES:                    78950.00
```

Figure 15–13
MLCBPP (Structure Chart)

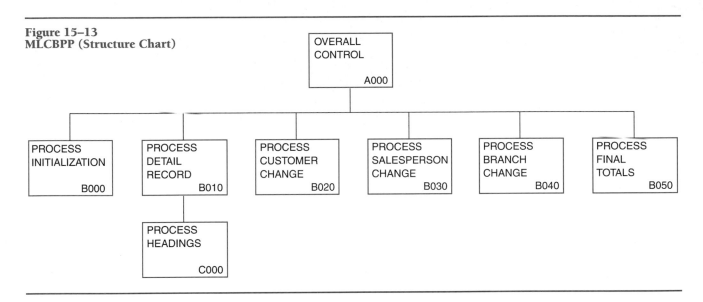

number fields that will be output. This step is necessary because the report is to be group indicated. The only time that the actual customer number, salesperson number, and branch number will be placed in these fields for output will be at the start of a new page (done in the headings module) and when one of the control fields changes (done in the appropriate control-break module).

This step is then followed by three nested IFTHENELSE statements, which determine if one of the control fields has changed. It is very important to notice the order in which these tests are made. A branch number change implies both a salesperson number change and a customer number change. When there is a change in branch number, the module that processes a customer change will be executed first, followed by the module that processes a salesperson change, and then the module that processes the branch change. When there is a change in branch number, the customer total line will be output first, followed by the salesperson total line, and finally the branch total line. Similarly, a salesperson number change implies a customer number change. When there is a change in salesperson number, the module that processes a customer change will be executed first, followed by the module that processes a salesperson change. Again, the customer total line will be output first, followed by the salesperson total line. Lastly, when there is a change in customer number, only the module that processes a customer number change will be executed and a customer total line will be output. If none of the control fields has changed, the normal detail processing (module B010) is executed. Then another record is read. It is important to see that module B010 also is executed when modules B020, B030, and/or B040 complete their processing. Do you remember why?

After one or more total lines are output, the detail processing still must be done for the current record that was input. Remember, a control break was detected as a result of a change in one or more control fields based on information input from a new record. This new record must be processed before another new record can be read. This repetitive kind of processing continues until end-of-file is reached. When the end-of-file is finally

**Figure 15–14
MLCBPP—Overall Control
(Flowchart)**

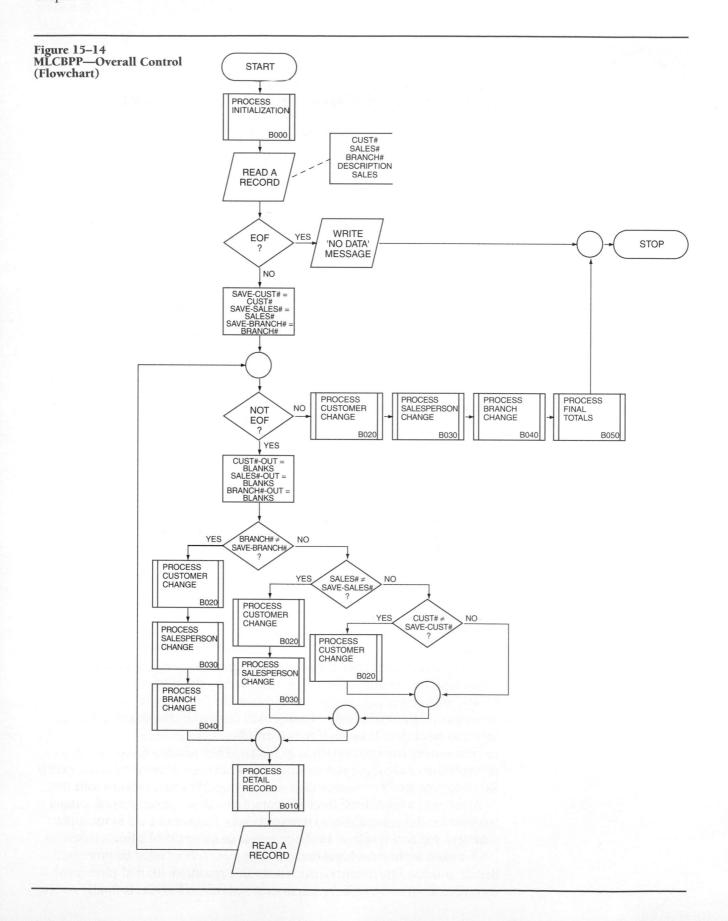

Figure 15–15
MLCBPP—Overall Control
(Pseudocode)

```
A000
Start
Process initialization (B000)
Read CUST#, SALES#, BRANCH#, DESCRIPTION, SALES
IF EOF THEN
   Write 'No Data' message
ELSE
   SAVE-CUST# = CUST#
   SAVE-SALES# = SALES#
   SAVE-BRANCH# = BRANCH#
   DOWHILE not EOF
      CUST#-OUT = Blanks
      SALES#-OUT = Blanks
      BRANCH#-OUT = Blanks
      IF BRANCH# ≠ SAVE-BRANCH# THEN
         Process customer change (B020)
         Process salesperson change (B030)
         Process branch change (B040)
      ELSE
         IF SALES# ≠ SAVE-SALES# THEN
            Process customer change (B020)
            Process salesperson change (B030)
         ELSE
            IF CUST# ≠ SAVE-CUST# THEN
               Process customer change (B020)
            (ELSE)
            ENDIF
         ENDIF
      ENDIF
      Process detail record (B010)
      Read CUST#, SALES#, BRANCH#, DESCRIPTION, SALES
   ENDDO
   Process customer change (B020)
   Process salesperson change (B030)
   Process branch change (B040)
   Process final totals (B050)
ENDIF
Stop
```

reached, modules B020, B030, and B040 are executed one last time, and the final total line is output (module B050).

You might wonder why modules B020, B030, and B040 are executed again. When end-of-file is reached, the intermediate totals for the last group of customer numbers, salesperson numbers, and branch numbers have not yet been output. Remember, this is done only when a control break occurs—and a control break can be detected only when a control field in the input changes. An end-of-file is, in effect, a change in a control field.

A program flowchart and corresponding pseudocode for module B000 (process initialization) are shown in Figures 15–16 and 15–17. MAXLINES is set to 40, the page count (PAGECNT) is set to 1 and the line

Figure 15–16
MLCBPP—Process
Initialization (Flowchart)

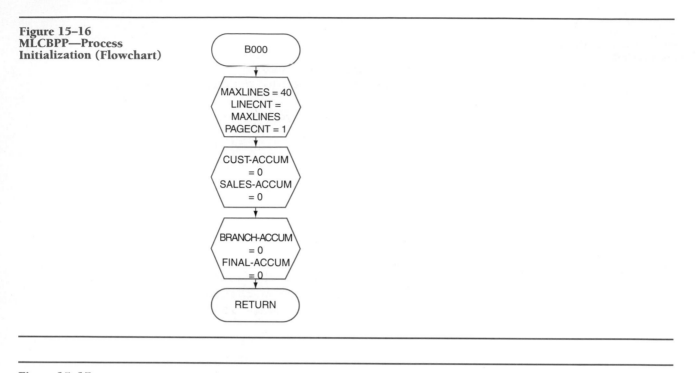

Figure 15–17
MLCBPP—Process
Initialization (Pseudocode)

```
B000
Enter
MAXLINES = 40
LINECNT = MAXLINES
PAGECNT = 1
CUST-ACCUM = 0
SALES-ACCUM = 0
BRANCH-ACCUM = 0
FINAL-ACCUM = 0
Return
```

count (LINECNT) is set to MAXLINES. Four accumulators are set to 0. CUST-ACCUM is used to keep track of the total sales amount for one customer. SALES-ACCUM is used to keep track of the total sales for one salesperson. BRANCH-ACCUM is used to keep track of the total sales for one branch. FINAL-ACCUM keeps track of the total sales for all branches.

A program flowchart and corresponding pseudocode for module B010 (process detail record) are shown in Figures 15–18 and 15–19. The first processing step is an IFTHENELSE that checks to see if headings need to be printed. Remember, when LINECNT reaches MAXLINES or greater, a new page is started and the headings must be printed first. The detail processing in this algorithm is quite simple. The detail line contains information—all of which has been input—so no calculations are necessary. The line is printed and LINECNT is incremented. Note that we are printing the fields BRANCH#-OUT, SALES#-OUT, and CUST#-OUT and not the fields BRANCH#, SALES#, and CUST#. BRANCH#-OUT, SALES#-OUT, and CUST#-OUT will either contain the actual branch number, salesperson number, and customer number or contain blanks— depending on whether the current record contains information on the first

**Figure 15–18
MLCBPP—Process Detail
Record (Flowchart)**

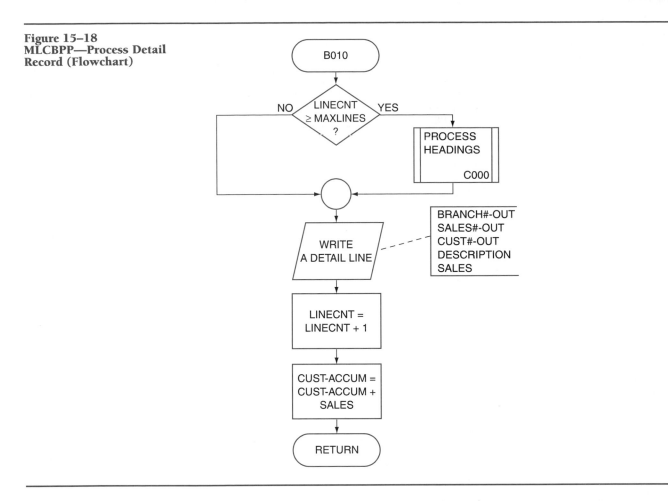

**Figure 15–19
MLCBPP—Process Detail
Record (Pseudocode)**

```
B010
Enter
IF LINECNT ≥ MAXLINES THEN
    Process headings (C000)
(ELSE)
ENDIF
Write BRANCH#-OUT, SALES#-OUT, CUST#-OUT, DESCRIPTION, SALES
LINECNT = LINECNT + 1
CUST-ACCUM = CUST-ACCUM + SALES
Return
```

branch number, salesperson number, or customer number in a group. This
is necessary because the report is to be group indicated. Finally, the sales
amount for one customer from the input (SALES) is added to the customer
number accumulator (CUST-ACCUM) before control is returned to A000.

A program flowchart and corresponding pseudocode for module B020
(process customer change) are shown in Figures 15–20 and 15–21. When
this module gets control, a control break has occurred and a customer total

Figure 15–20
MLCBPP—Process Customer
Change (Flowchart)

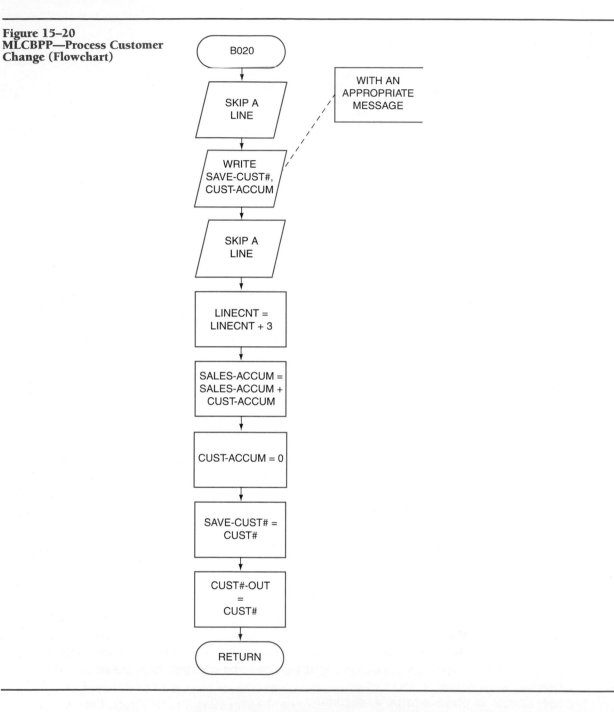

line must be output. Prior to printing the customer total line, a blank line is output; after the total line is printed, another blank line is output. This will help to set the customer total line apart from the other lines in the report. The customer number (SAVE-CUST#) and the accumulated sales amount for that customer (CUST-ACCUM), as well as identifying text, are output.

Remember, CUST# contains information from the current input record; therefore, we need to output SAVE-CUST#. Because three physical lines have been printed (two blank lines and one total line), the line count is incremented by 3. The customer accumulator is added to the salesperson accumulator and then reset to 0. Do you see why? When the next customer is

**Figure 15–21
MLCBPP—Process Customer
Change (Pseudocode)**

```
B020
Enter
Skip a line
Write 'TOTAL CUSTOMER', SAVE-CUST#, 'SALES:',
  CUST-ACCUM
Skip a line
LINECNT = LINECNT + 3
SALES-ACCUM = SALES-ACCUM + CUST-ACCUM
CUST-ACCUM = 0
SAVE-CUST# = CUST#
CUST#-OUT = CUST#
Return
```

processed, we do not want the previous customer's total sales to be in-cluded in the current customer's totals.

Finally, because we are getting ready to process a new customer, the save area (SAVE-CUST#) needs to be updated. Remember, this area is used to hold the value of the current customer number being processed. For example, when a control break occurs the first time, the current customer number must be changed from 345 to 412. CUST#-OUT must also be reset to the new customer number so that the actual number—not a blank field—will be output the next time a detail line is written.

A program flowchart and corresponding pseudocode for module B030 (process salesperson change) are shown in Figures 15–22 and 15–23. When this module gets control, a control break has occurred and a salesperson total line must be output. After the total line is printed, a blank line is output. This will help to set the salesperson total line apart from the other lines in the re-port. The salesperson number (SAVE-SALES#) and the accumulated sales for that salesperson (SALES-ACCUM), as well as identifying text, are output.

Remember that SALES# contains information from the current input record; therefore, we need to output SAVE-SALES#. The line count is then incremented by 2. The salesperson accumulator is added to the branch accu-mulator and then reset to 0. This is the same step we used in the previous module. When the next salesperson is processed, we do not want the previ-ous salesperson's total sales to be included in the current salesperson's totals.

Finally, because we are getting ready to process a new salesperson, the save area (SAVE-SALES#) needs to be updated. Remember, this area is used to hold the value of the current salesperson number being processed. For example, when this control break occurs the first time, the current salesperson number must be changed from 1122 to 2334. SALES#-OUT also must be reset to the new salesperson number, so that the actual number will be output the next time a detail line is written.

A program flowchart and corresponding pseudocode for module B040 (process branch change) are shown in Figures 15–24 and 15–25. When this module gets control, a control break has occurred and a branch total line must be output. After the total line is printed, a blank line is output. This will help set the branch total line apart from the other lines in the report. The branch number (SAVE-BRANCH#) and the accumulated sales for that branch (BRANCH-ACCUM), as well as identifying text, are output.

**Figure 15–22
MLCBPP—Process
Salesperson Change
(Flowchart)**

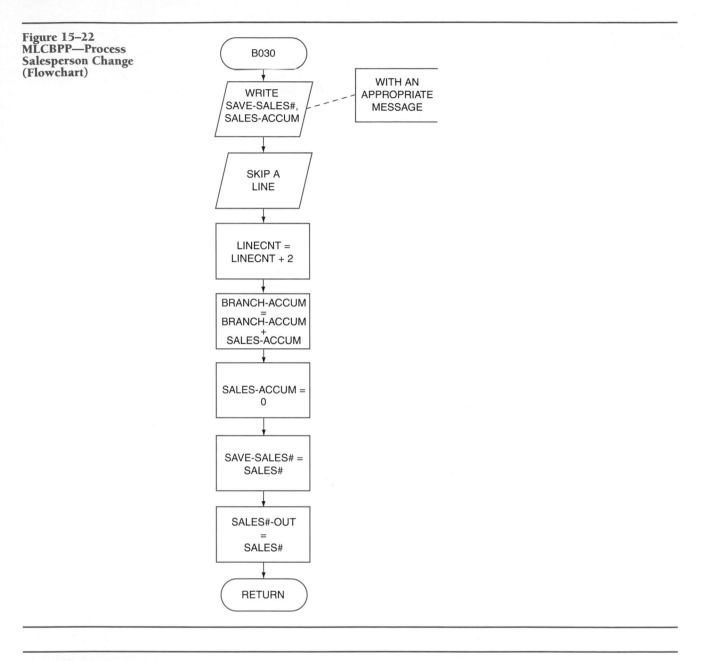

**Figure 15–23
MLCBPP—Process
Salesperson Change
(Pseudocode)**

```
B030
Enter
Write 'TOTAL SALESPERSON', SAVE-SALES#, 'SALES:', SALES-ACCUM
Skip a line
LINECNT = LINECNT + 2
BRANCH-ACCUM = BRANCH-ACCUM + SALES-ACCUM
SALES-ACCUM = 0
SAVE-SALES# = SALES#
SALES#-OUT = SALES#
Return
```

Figure 15–24
MLCBPP—Process Branch
Change (Flowchart)

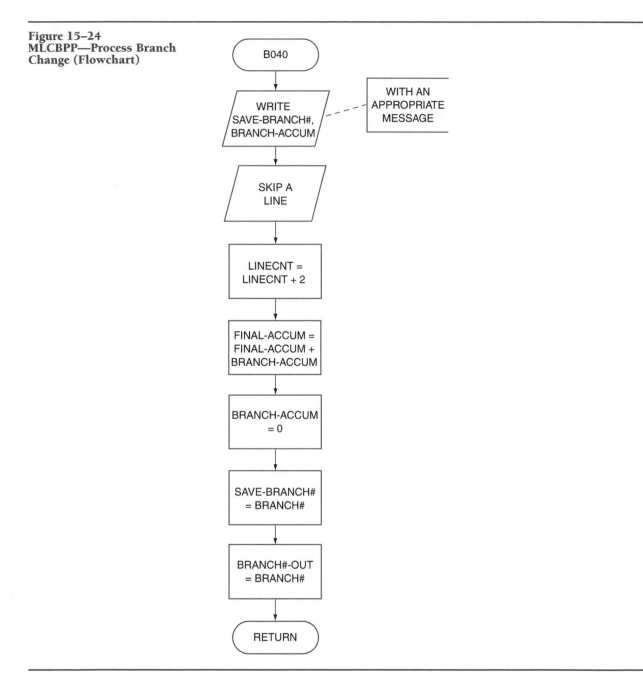

Figure 15–25
MLCBPP—Process Branch
Change (Pseudocode)

```
B040
Enter
Write 'TOTAL BRANCH', SAVE-BRANCH#, 'SALES:', BRANCH-ACCUM
Skip a line
LINECNT = LINECNT + 2
FINAL-ACCUM = FINAL-ACCUM + BRANCH-ACCUM
BRANCH-ACCUM = 0
SAVE-BRANCH# = BRANCH#
BRANCH#-OUT = BRANCH#
Return
```

Remember that BRANCH# contains information from the current input record; therefore, we need to output SAVE-BRANCH#. The line count is then incremented by 2. The branch accumulator is added to the final accumulator and then reset to 0. When the next branch is processed, we do not want the previous branch's total sales to be included in the current branch's totals.

Finally, because we are getting ready to process a new branch, the save area (SAVE-BRANCH#) needs to be updated. Remember, this area is used to hold the value of the current branch number being processed. For example, when this control break occurs the first time, the current branch number must be changed from 10 to 20. BRANCH#-OUT also must be reset to the new branch number, so that the actual number will be output the next time a detail line is written.

A program flowchart and corresponding pseudocode for module B050 (process final totals) are shown in Figures 15–26 and 15–27. This module simply writes out the value of the final accumulator (FINAL-ACCUM) with an identifying message. Remember, FINAL-ACCUM represents the total sales for all branches.

A program flowchart and corresponding pseudocode for module C000 (process headings) are shown in Figure 15–28 and 15–29. In this module, we are again trying to indicate some degree of spacing. The report heading is followed by one blank line, and the column headings are followed by one blank line. The spacing is primarily for readability; it may or may not be specified by the user. In either case, the programmer must include an indication of spacing requirements somewhere in the documentation. Notice also that the three output fields (CUST#-OUT, SALES#-OUT, and BRANCH#-OUT) are set to their corresponding input values (CUST#,

Figure 15–26
MLCBPP—Process Final Totals (Flowchart)

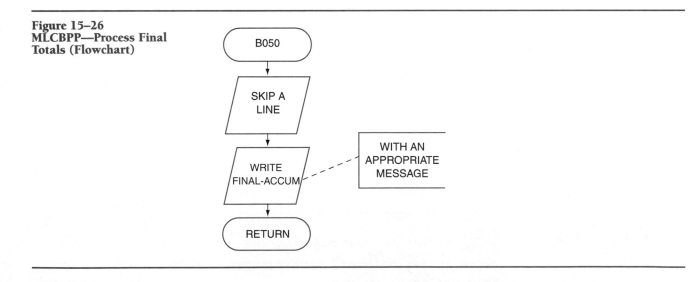

Figure 15–27
MLCBPP—Process Final Totals (Pseudocode)

```
B050
Enter
Skip a line
Write 'TOTAL ACCUMULATED SALES:', FINAL-ACCUM
Return
```

SALES#, and BRANCH#). Can you see why this is necessary? Whenever a new page is started, the actual branch number, salesperson number, and customer number should be output—not blank fields. This step likewise improves the readability of the report.

Figure 15–28
MLCBPP—Process Headings
(Flowchart)

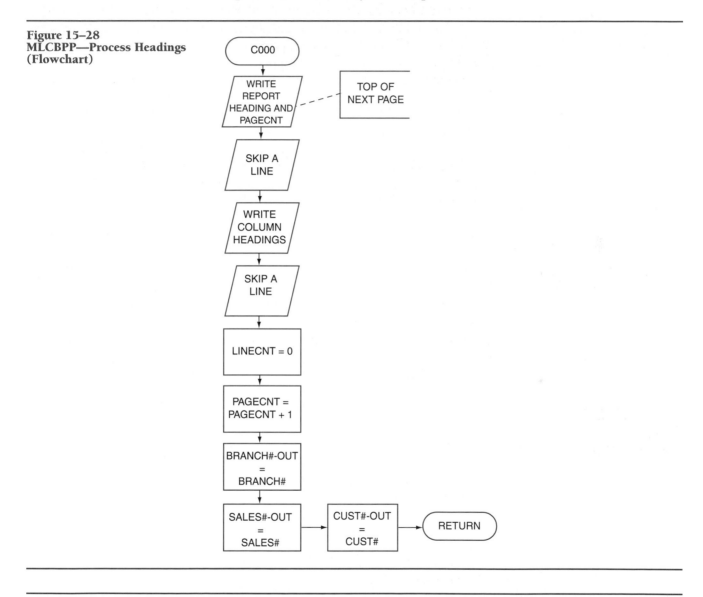

Figure 15–29
MLCBPP—Process Headings
(Pseudocode)

```
C000
Enter
Write report heading and PAGECNT
    at top of next page
Skip a line
Write column headings
Skip a line
LINECNT = 0
PAGECNT = PAGECNT + 1
BRANCH#-OUT = BRANCH#
SALES#-OUT = SALES#
CUST#-OUT = CUST#
Return
```

Key Terms

control break

control field (key)

detail-printed report

group-printed report

group-indicated report

Exercises

1. (a) What is a detail-printed report?
   (b) What is a group-printed report?
   (c) What is a group-indicated report?

2. What is a control field?

3. What does it mean to say that a "control break" has occurred?

4. What would happen if SAVE-DEPT were not reset to DEPT in Figure 15–8 (module B020)?

5. What would happen if DEPT-ACCUM were not reset to 0 in Figure 15–8 (module B020)?

6. What would happen if FINAL-ACCUM were also reset to 0 in Figure 15–8 (module B020)?

7. Explain what would happen if the processing steps inside the DOWHILE loop in Figure 15–4 (module A000) had been written as follows:

8. Harris Wholesale Distributors maintains comprehensive records of the type, volume, and sales price of all merchandise dispatched from its warehouse locations. These records are updated daily with computer help. They provide the basis for numerous management inquiries, as well as for weekly sales reports. As a part of year-end processing, a summary report showing monthly sales volume—the total dollar value of all merchandise dispatched from the warehouses each month—is needed. Even though the report is a summary report, it is a detail-printed report. Do you understand why?

The input to the program that will generate the summary report comprises one record for each warehouse that reported activity during a month. The record tells the name and location of the warehouse, its total sales volume for the month (in dollars), and a numerical indicator of the month: 1 for January, 2 for February, and so on (see the following illustration). Because the input records are to be read from a distribution master file, they can be assumed to be in ascending order by month (all January records first, then all February records, etc.). There will be at least one warehouse record for each month. A special end-of-file record containing 0 in the month-indicator field, 0s in the sales volume field, and no data in the other fields will be provided as the last input record.

WAREHOUSE	LOCATION	SALES	MONTH
WAREHOUSE 1	CANTON, OHIO	67300.00	2
WAREHOUSE 2	CANTON, OHIO	145717.00	2
MAXIM'S	KANSAS CITY, MO.	25800.00	2
J & D	AUSTIN, TEXAS	456098.00	7
KANE	CHICAGO, ILL.	66400.00	7

The summary report is to contain one print line (detail line) for each warehouse record—that is, for each warehouse from which merchandise was dispatched during a particular month. When all the records for a month have been read and printed, a total sales volume for the month is to be printed. A grand total indicating the sales volume for the year is to be printed at the end of the report.

```
                     HARRIS WHOLESALE DISTRIBUTORS
                     YEAR-END TOTAL SALES REPORT

MONTH           WAREHOUSE           LOCATION                        SALES

January         Tompkin's          Aberdeen, Maryland           45,069.00
January         Warehouse A        Des Moines, Iowa             44,036.00
   .               .                  .                             .
   .               .                  .                             .
   .               .                  .                             .
                                                               _____
                JANUARY TOTAL SALES VOLUME                   $3,846,077.00
   .               .                  .                             .
   .               .                  .                             .
   .               .                  .                             .
                                                               _____
                YEAR-END TOTAL SALES VOLUME                 $51,994,223.00
```

Construct a flowchart and corresponding pseudocode to solve this problem, but consider the following requirement: The illustration shows that the name of the month is to be output—not the number of the month as shown in the input. For example, January will be output, not the number 1. The name is clearly more descriptive to anyone reading the report. To make this task easier, assume that you have available a one-dimensional array called MONTHNAME as follows:

MONTHNAME

JANUARY
FEBRUARY
MARCH
APRIL
MAY
JUNE
JULY
AUGUST
SEPTEMBER
OCTOBER
NOVEMBER
DECEMBER

In this case, MONTHNAME(1) = JANUARY, MONTHNAME(2) = FEBRUARY, and so on.

9. Redo Exercise 8 to provide a group-indicated report. The report will contain the name of the month on only the first detail line printed for a particular month, as follows:

```
                    HARRIS WHOLESALE DISTRIBUTORS
                    YEAR-END TOTAL SALES REPORT

  MONTH          WAREHOUSE             LOCATION                      SALES

  January        Tompkin's             Aberdeen, Maryland         45,069.00
                 Warehouse A           Des Moines, Iowa           44,036.00

      .               .                     .                          .
      .               .                     .                          .
      .               .                     .                          .

                 JANUARY TOTAL SALES VOLUME                  $3,846,077.00

  February       Warehouse B           Washington, D. C.          35,072.00
                 Warehouse C           Chicago, Illinois          37,345.00
                 Warehouse D           Boston, Mass.              42,128.00
      .               .                     .                          .
      .               .                     .                          .
      .               .                     .                          .

                 FEBRUARY TOTAL SALES VOLUME                 $8,723,119.00
      .               .                     .                          .
      .               .                     .                          .
      .               .                     .                          .
```

10. Student records are input, each containing student ID, student name, and one test score. There may be several scores for one student but each score will be input on a separate record, as follows:

STUDENT ID	STUDENT NAME	TEST SCORE
111111111	Mary Davidson	80
111111111	Mary Davidson	90
111111111	Mary Davidson	100
222222222	Dean Black	70
222222222	Dean Black	60
333333333	Ginny Smith	85
.	.	.
.	.	.
.	.	.

Assume the records are in ascending order by student ID. Construct a flowchart and corresponding pseudocode to compute and output the average for each student. Each line of output should contain the student ID, student name, and average. Note that a group-printed report is needed. No detail lines or final total lines are required.

11. Redo Exercise 10 to compute two control totals: the number of students for whom averages are computed and the total number of test scores processed for all students. Output these two control totals at the end of the report in a final total line.

12. A rebate report is to be prepared for a major department store. Design an algorithm to produce the required report. The input consists of customer records that contain the customer number, item number purchased, type of purchase code, and amount of purchase. The type of purchase code indicates whether the purchase was made on a sale item (code of 1) or a non-sale item (code of 2). The output consists of a rebate report containing the customer number, item number purchased, type of purchase (ON SALE or REGULAR), and amount of purchase. The messages ON SALE and REGULAR should be extracted from an array SALETYPE (see below) based on the value in the "type of purchase" code field in the input record.

SALETYPE

1	ON SALE
2	REGULAR

In addition, total number of sale items, total number of non-sale items, total amount of sale purchases, total amount of non-sale purchases, total credits issued, and rebate earned are to be printed for each customer.

Total credits for each customer are calculated in the following manner: 0 credits for each regular purchase below $100; 1 credit for each regular purchase between $100 and $500; 2 credits for each regular purchase over $500 but under $1,000; 3 credits for each regular purchase of $1,000 or more. Purchases made on sale items do not count in the credit calculation. Based on the total credits, one of the following messages should be printed:

Rebate Earned
Message

	0	NO REBATE
Credits Issued	1–5	$20 REBATE
	6–20	$50 REBATE
	Over 20	$100 REBATE

After all records are processed, final totals should be printed on a separate page. Final totals should include total number of sale items, total number of non-sale items, total amount of sale purchases, total amount of non-sale purchases, total number of $20 rebates earned, total number of $50 rebates earned, and total number of $100 rebates earned.

A sample of the output follows:

```
                    DEPARTMENT STORE REBATE REPORT          Page 1

CUSTOMER            ITEM            TYPE OF          AMOUNT OF
NUMBER              NUMBER          PURCHASE         PURCHASE

1111                123             ON SALE          $ 150.00
                    234             REGULAR          $ 300.00
                    364             REGULAR          $ 200.00
                    334             ON SALE          $ 950.00

SALE ITEMS          - TOTAL 2       SALE PURCHASES   - TOTAL $1100.00
NON-SALE ITEMS      - TOTAL 2       NON-SALE PURCHASES - TOTAL $ 500.00

TOTAL CREDITS ISSUED    - 2         REBATE EARNED - $20 REBATE

2222                456             REGULAR          $1750.00
                    249             REGULAR          $80.00
                    987             REGULAR          $1650.00

SALE ITEMS          - TOTAL 0       SALE PURCHASES   - TOTAL   $ 0.00
NON-SALE ITEMS      - TOTAL 3       NON-SALE PURCHASES - TOTAL $3480.00

TOTAL CREDITS ISSUED    - 6         REBATE EARNED - $50 REBATE

                         (Separate Page)

                    DEPARTMENT STORE                     Page X
                      FINAL TOTALS

TOTAL NUMBER OF SALE ITEMS          -          125
TOTAL NUMBER OF NON-SALE ITEMS      -          300
TOTAL SALE PURCHASES                -       $ 50,925
TOTAL NON-SALE PURCHASES            -       $142,855

TOTAL REBATES EARNED
  $ 20 REBATES EARNED               -           20
  $ 50 REBATES EARNED               -           10
  $100 REBATES EARNED               -            6
```

This report is to be group indicated; that is, the customer number is to be printed only for the first record in the group. Include the "No data" message, report headings, and column headings on every page. Use automatic EOF to indicate the end of the input. Records are assumed to be in ascending order by customer number.

13. A sales report is to be prepared for the appliance department in a major department store. Design an algorithm to produce the required report. The input consists of salesperson records that contain the salesperson number, salesperson name, type of appliance code, color code, and sales price. The output consists of a kitchen appliance sales report containing the salesperson number, salesperson name, appliance sold, color, and sales price. Assume that you have available two one-dimensional arrays called TYPE and COLOR as follows:

TYPE

1	REFRIGERATOR
2	DISHWASHER
3	COMPACTOR
4	STOVE

COLOR

1	WHITE
2	GOLD
3	AVOCADO

In addition, subtotals and final totals for three control-break levels—salesperson, appliance, and color—are to be output. Design the output as a group-indicated report. A sample of the output follows on the next page.

```
                    KITCHEN APPLIANCE SALES INC.              Page 1

    SALESPERSON        SALESPERSON      APPLIANCE        COLOR        SALES
    NUMBER             NAME             SOLD                          PRICE

       10              John Black       Refrigerator    Avocado      700.00
                                                                     600.00
                                                                     500.00

                                        COLOR Avocado SALES          1800.00
                                                     UNITS SOLD             3

                                                        Gold          500.00
                                                                      600.00

                                        COLOR Gold      SALES         1100.00
                                                     UNITS SOLD             2

                                        TOTAL Refrigerator SALES      2900.00
                                                     UNITS SOLD             5

                                         Dishwasher     White          500.00
                                                                       400.00

                                        COLOR White      SALES          900.00
                                                     UNITS SOLD              2

                                        TOTAL Dishwasher   SALES        900.00
                                                     UNITS SOLD              2

                              TOTAL FOR SALESPERSON NUMBER  10        3800.00
                                                     UNITS SOLD             7
                                             .
                                             .
                                             .
                                        GRAND TOTAL SALES   150560.00
                                        TOTAL UNITS SOLD          300
```

Assume the input data is properly sequenced, and include the "No data" message. Include report headings and column headings on every page, as well as page numbers, with 30 detail lines per page. Use automatic EOF to indicate the end of the input.

14. Redo Exercise 13 to include a group-printed report (on a new page) as follows:

```
            SUMMARY OF SALES BY APPLIANCE SOLD              Page X


APPLIANCE                        NO. OF UNITS               SALES PRICE
                                    SOLD

Refrigerator                        XXX                     XXXXXXXXX.XX
Dishwasher                          XXX                     XXXXXXXXX.XX
Compactor                           XXX                     XXXXXXXXX.XX
Stove                               XXX                     XXXXXXXXX.XX


                        ***GRAND TOTALS ***                 XXXXXXXXXXX.XX
```

15. Redo Exercise 14 to include an alternative group-printed report (on a new page) as follows:

```
            SUMMARY OF SALES BY COLOR                       Page X


COLOR OF                         NO. OF UNITS               SALES PRICE
APPLIANCE                           SOLD

White                               XXX                     XXXXXXXXX.XX
Gold                                XXX                     XXXXXXXXX.XX
Avocado                             XXX                     XXXXXXXXX.XX


                        *** GRAND TOTALS ***                XXXXXXXXXXX.XX
```

Program Flowcharting Symbols

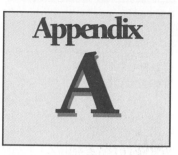

Appendix A

American National Standards Institute (ANSI) recommendations for use of symbols on program flowcharts are presented in this appendix. The shape of each recommended symbol, its meaning, and one or more examples are given. The symbols that you are most apt to find useful in your design work are explained in greater detail in one or more chapters in this book. The chapter in which each symbol is introduced is given in parentheses following the explanation of the symbol below.

Input/Output Symbol

Generalized input/output function; reading data from an input medium or writing data to an output medium (Chapter 2)

OPEN IFILE

WRITE OUTREC

Process Symbol

Any processing step; an operation or group of operations causing change in value, form, or location of data (Chapter 2)

TOTAL = ASUM * .05

Flowline Symbol

Sequence of operations and direction of data flow, arrowheads are required if linkage is not top to bottom or left to right (Chapter 2)

Annotation Symbol	Additional explanation; comments (Chapter 6)

TEST FOR
END-OF-FILE
INDICATOR

ITEM =
99999?   YES

NO

Connector Symbol	Exit to, or entry from, another part of the flowchart; if the *to* or *from* step is on another page, a page reference should be stated (Chapter 3)

B3

FROM PAGE 3

C2

Terminal Interrupt Symbol	Terminal point in a flowchart—start, stop, or break in the line of flow (Chapter 2)

START   STOP

Decision Symbol	Decision-making operation, usually based on a comparison, that determines which of two or more alternative paths should be followed (Chapter 3)

NUMB
= 0?   NO

YES

ACTION
CODE?

1   2   3   OTHER

Preparation Symbol

An operation performed on the program itself for control, initialization, overhead, or cleanup; examples are to set a program switch, to place a limit value in a loop-control variable, and to initialize an accumulator (Chapter 4)

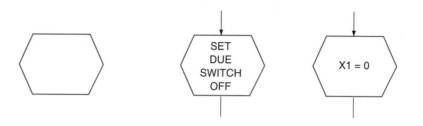

Predefined-Process Symbol

One or more operations specified in detail elsewhere, such as in a reference manual or on a different flowchart, but not on another part of the flowchart where this symbol appears (Chapter 6)

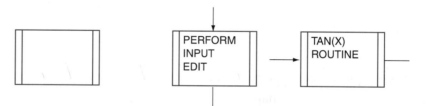

Structured-Programming Control Structures

Appendix B

The three basic patterns of structured programming—SIMPLE SEQUENCE, IFTHENELSE, and DOWHILE—are summarized in this appendix. Two additional control structures, CASE and DOUNTIL, which represent frequently used combinations of these basic patterns, also are summarized.

First, the general form of the control structure is given. Then an example is expressed in both flowchart and pseudocode forms. The chapter in which each structure is introduced is given in parentheses following the explanation.

SIMPLE SEQUENCE

The execution of one processing step after another, in normal execution sequence (Chapter 2)

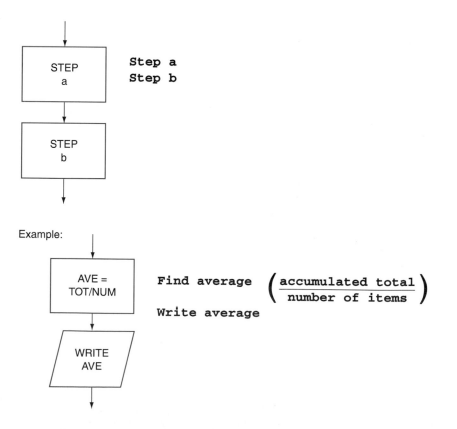

IFTHENELSE

The selection of one of two alternatives (Chapter 3)

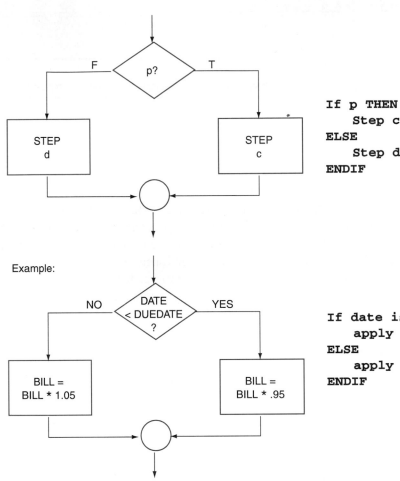

```
If p THEN
    Step c
ELSE
    Step d
ENDIF
```

Example:

```
If date is less than due date THEN
    apply 5% discount to bill
ELSE
    apply 5% interest to bill
ENDIF
```

DOWHILE

The execution of processing steps within a program loop as long as a specified condition is true; a leading-decision loop (Chapter 4)

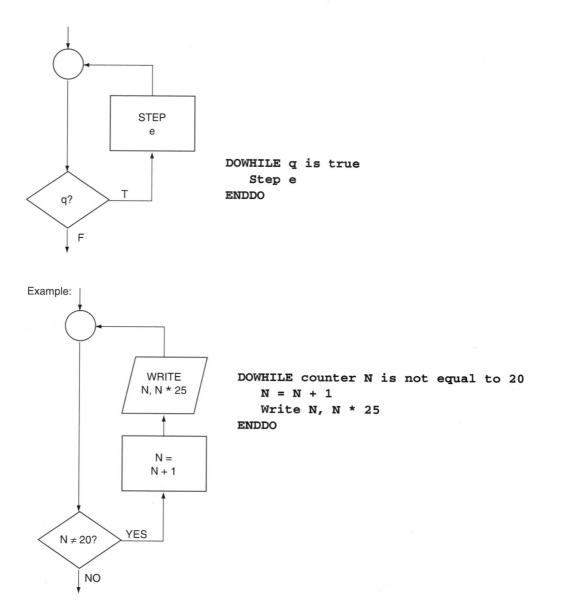

```
DOWHILE q is true
    Step e
ENDDO
```

Example:

```
DOWHILE counter N is not equal to 20
    N = N + 1
    Write N, N * 25
ENDDO
```

CASE

The selection of one of more than two alternatives (Chapter 7)

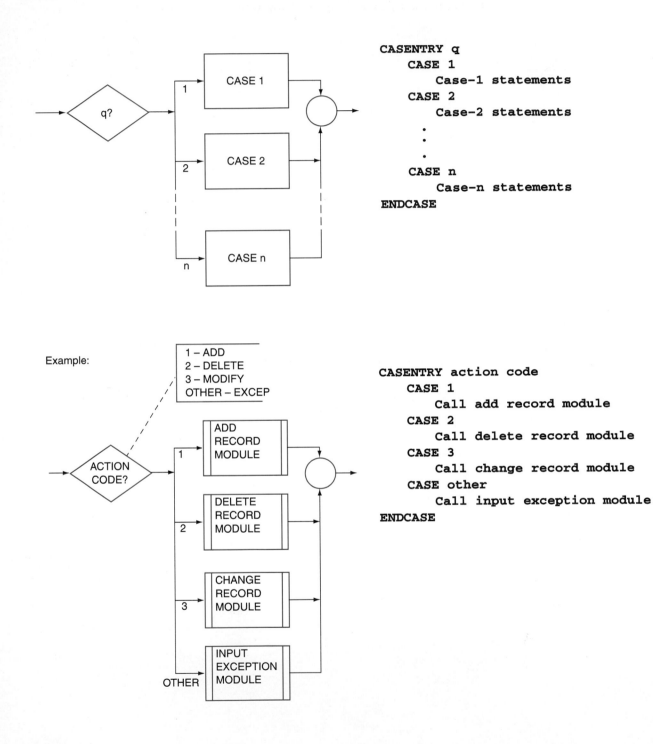

```
CASENTRY q
    CASE 1
        Case-1 statements
    CASE 2
        Case-2 statements
        .
        .
        .
    CASE n
        Case-n statements
ENDCASE
```

Example:

```
CASENTRY action code
    CASE 1
        Call add record module
    CASE 2
        Call delete record module
    CASE 3
        Call change record module
    CASE other
        Call input exception module
ENDCASE
```

DOUNTIL

The execution of processing steps within a program loop until a specified condition is true; a trailing-decision loop (Chapter 8)

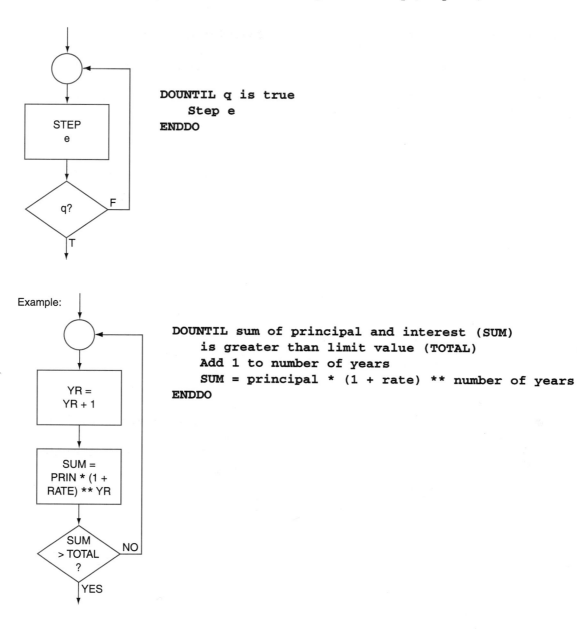

```
DOUNTIL q is true
    Step e
ENDDO
```

```
DOUNTIL sum of principal and interest (SUM)
    is greater than limit value (TOTAL)
    Add 1 to number of years
    SUM = principal * (1 + rate) ** number of years
ENDDO
```

Answers to Selected Exercises

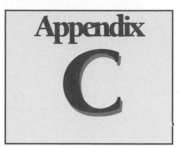

Appendix C

Responses to selected exercises from each of the chapters in this book are given in this appendix. For some of these exercises, there is no one correct answer. In such cases, the responses are representative answers to the problems.

Chapter 1	**1.** The steps in the system development life cycle are:
	▪ Analyze the current system.
	▪ Define the new system requirements.
	▪ Design a new system.
	▪ Develop a new system.
	▪ Implement the new system.
	▪ Evaluate the new system.
	2. The steps in the program development cycle are:
	▪ Review the program requirements.
	▪ Develop the program logic.
	▪ Write the program.
	▪ Test and debug the program.
	▪ Complete the program documentation.

Chapter 2	**1.** An algorithm is a step-by-step procedure to solve a problem.
	4. (a) file, record, field, character
	6. A system flowchart shows general information about an application. The information shown includes the major inputs, processes, and outputs for each program in the application. A program flowchart shows the specific details of how one program works. Both types of flowcharts give us information graphically: The system flowchart gives general information and can involve several programs; the program flowchart gives specific information about only one program.

7. **(a)** The normal direction of flow on both system and program flow-charts is top-to-bottom and left-to-right.
9. **(a)** –49
10. 310.40

Chapter 3

1. A decision-making step provides for a choice among alternative paths, which is a variation in processing sequence dependent on the data entering the system or situations that arise during processing.
3. **(a)** Pseudocode is an English-like description of an algorithm that uses indentation to more clearly identify the three basic control structures.
5. **(a)** A null ELSE indicates that no processing steps are to occur within the false path of an IFTHENELSE statement.
7. In a sequential IFTHENELSE pattern, all the tests are executed regardless of the outcome of previous tests. In a nested IFTHENELSE pattern, a test is either executed or not executed depending on the results of the previous test.

Chapter 4

1. A program loop permits a sequence of processing steps to be done repeatedly (i.e., re-executed or reused) during processing.
5. Because each basic pattern has only one entry point and only one exit point, it can be treated as a SIMPLE SEQUENCE pattern. Further, a series of these basic patterns can be treated as a SIMPLE SEQUENCE. (We say that the contained patterns are nested.) The combining of patterns and building up of logic can continue until a complete program is constructed. A program containing only basic patterns and combinations thereof can have only one entry point and only one exit point. It can itself be thought of as a SIMPLE SEQUENCE, or basic building block.
7. The three basic patterns are SIMPLE SEQUENCE, IFTHENELSE, and DOWHILE.

9. **(a)** Each time the loop is executed, COUNT will be reset to 0. Because the last step in the loop adds 1 to COUNT, the value of COUNT will continue to be equal to 1, never reaching the maximum value of 6. When the test (COUNT < 6) is made to determine if the loop should be executed, the answer will always be yes, thus creating an infinite loop.

15.

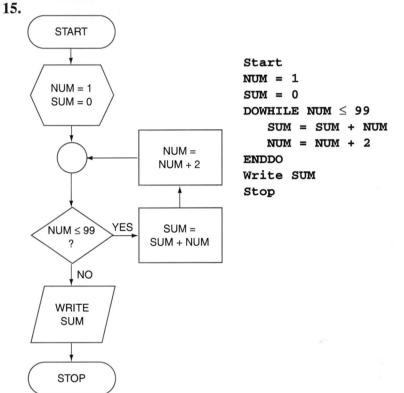

```
Start
NUM = 1
SUM = 0
DOWHILE NUM ≤ 99
    SUM = SUM + NUM
    NUM = NUM + 2
ENDDO
Write SUM
Stop
```

Chapter 5

1. In header record logic, the first record in the input specifies how many records will follow. The loop must be controlled by a counter that gets its initial value from the first record (header record). In trailer record logic, the last record in the input specifies that no more records will follow. The loop is not controlled by a counter. A test is made, following each READ statement, to determine whether or not the last record (trailer record) has been read.

5.

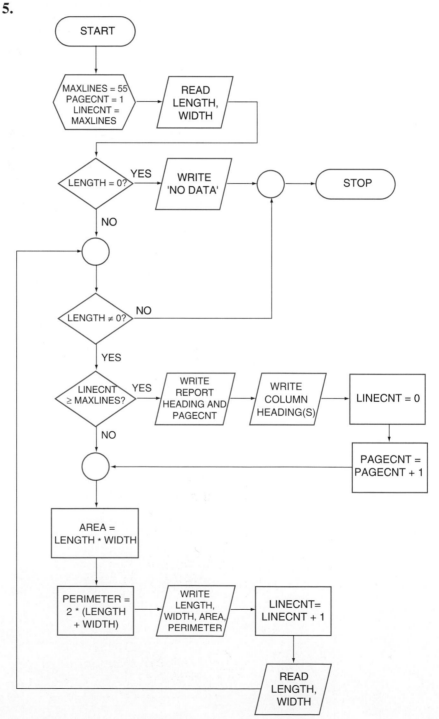

Chapter 6   2.

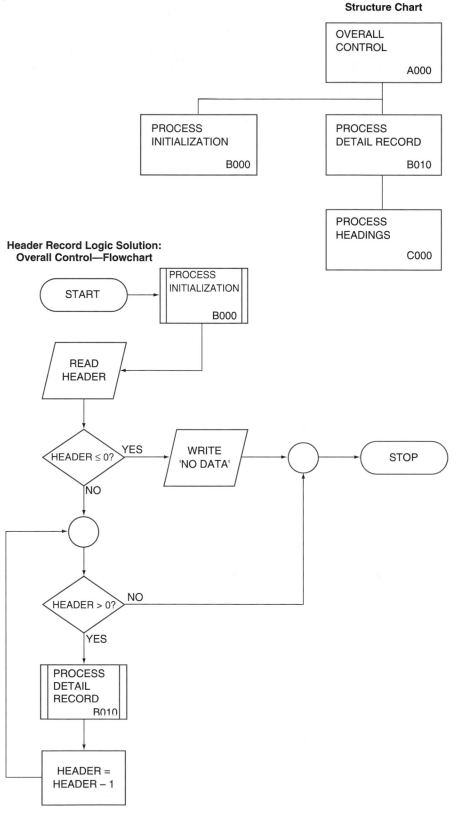

Structure Chart

**Header Record Logic Solution:
Overall Control—Flowchart**

2. (continued)

Process Detail Record—Flowchart

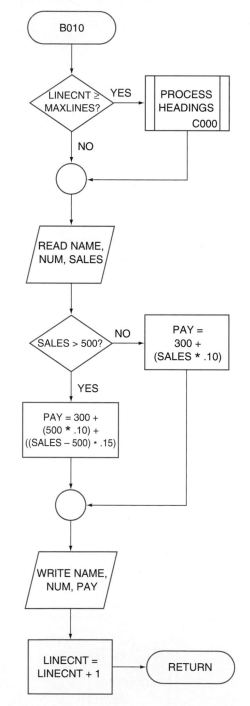

2. (continued)

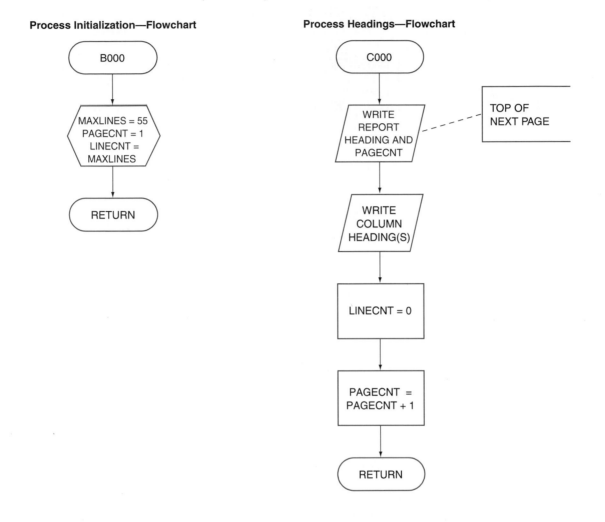

Process Initialization—Flowchart

B000

MAXLINES = 55
PAGECNT = 1
LINECNT = MAXLINES

RETURN

Process Headings—Flowchart

C000

WRITE REPORT HEADING AND PAGECNT ---- TOP OF NEXT PAGE

WRITE COLUMN HEADING(S)

LINECNT = 0

PAGECNT = PAGECNT + 1

RETURN

Chapter 7

1. (a) A master file contains a large volume of relatively permanent data kept for reference purposes. A transaction file contains current activities, or transactions, that will be used to update a master file.

5. (a)

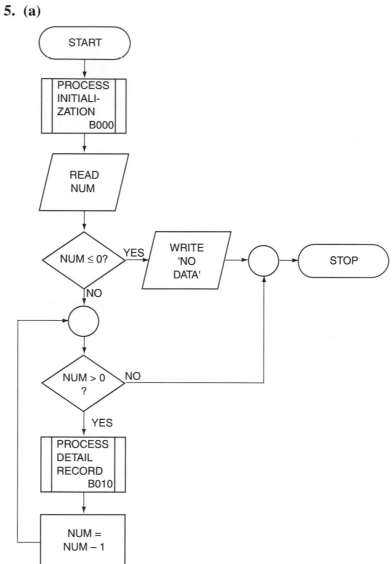

5. (a) (continued)

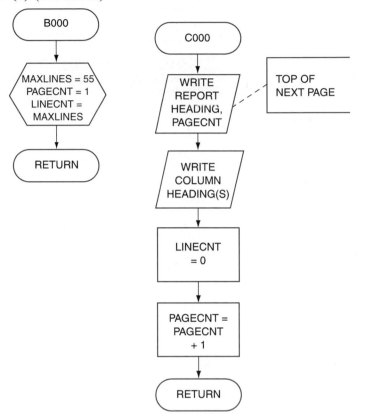

5. **(a)** (continued)

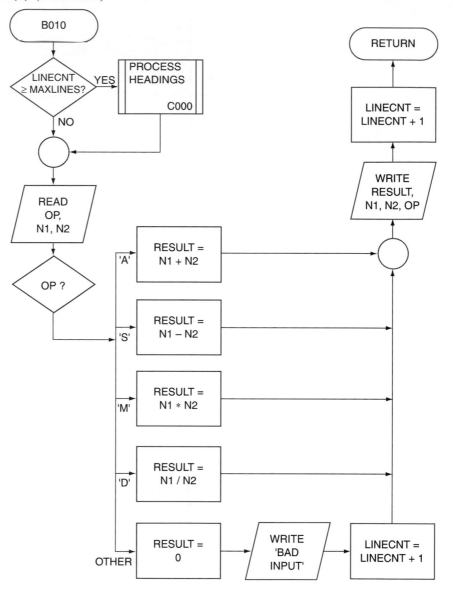

Chapter 8

1. (a) A leading-decision loop is formed in a DOWHILE pattern.

6.

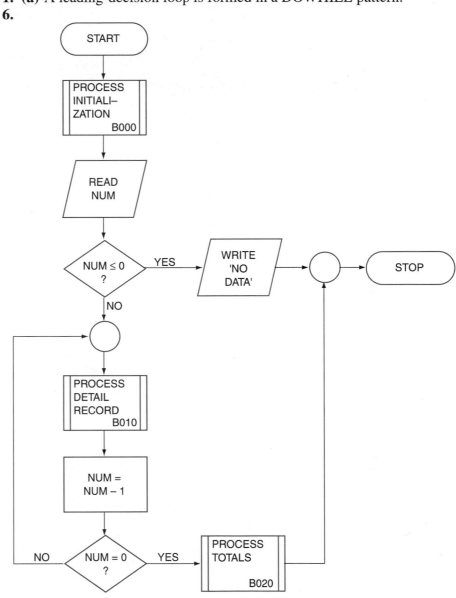

6. (continued)

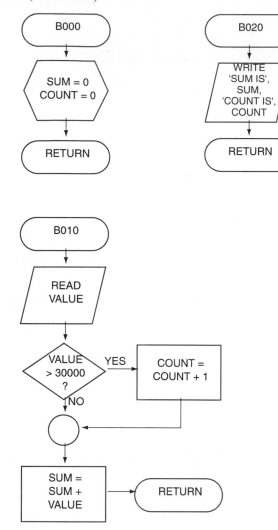

Chapter 9

1. (a) A data group is a collection of similar data items for which a single storage area large enough for all the items is reserved. The individual items are not assigned names; instead, one name is assigned to the entire group of items.

3. (a) E(4) contains 48
   (b) E(1)
   (c) E(R)
   (d) 1 and 8; 0 to 8 (answer can vary depending on the subscript range allowed by the programming language)

Chapter 10

1. In procedure-oriented design of programs, the emphasis is on doing things, that is, on performing actions and on the sequence of those actions; the data is secondary. In object-oriented design of programs, the emphasis is on the data to be manipulated and the operations to be performed on the data.

2. A class represents a template from which individual objects can be created. An instance of a class is called an object. An object is a tangible representation of the data members and methods of the class to which it belongs.

3. A class definition contains both data and operations. The data describes what an object looks like and the operations describe what an object does. The data that defines a class also can be referred to as properties, data members (of the class), or instance variables. The operations also can be referred to as methods or member functions.

6. Encapsulation means to package data members and methods into a single well-defined programming unit (in our case, a class or an object created of that class). Encapsulation is most often achieved through data hiding. The adjective "well-defined" implies that other developers can use the programming unit without having to know its implementation.

9. One purpose of a driver program is to create and manipulate the objects of a particular class.

10. Instantiation is the process of creating an object or, more precisely, of creating an instance of a class.

11. Overloading is the ability to use the same method name to invoke different methods that perform different actions based on the number or type of arguments in the method invocation.

12. Jane Rynn   0
    Any Student   0

13. Jane Rynn   83.33 (rounded to two decimal places)
    Any Student   0

Chapter 11

1. Inheritance is a mechanism by which one class acquires the data and methods of an existing, more general class.

3. (a) A base class also may be called a parent class or a super class.
   (b) A derived class also may be called a child class or a subclass.

4.

5.

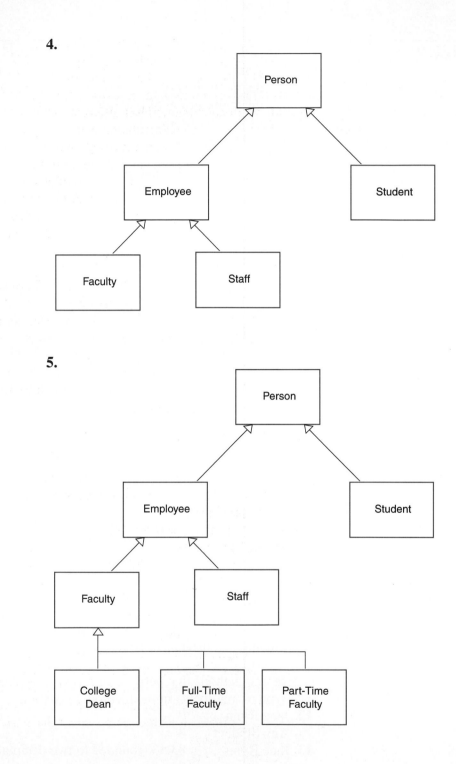

8. Overloading is the ability to use the same method name to invoke different methods that perform different actions based on the number or type of arguments in the method invocation. Overriding is the process of reimplementing in a subclass a method inherited from a base class.

9. An overloaded method has different implementations in the same class; which method implementation is invoked depends on the number or types of arguments in the method invocation. A polymorphic method has the same name as a method in one or more other classes of the same family but has different implementations for the various classes.

10. The calls to the NewFaculty object provide the following output:
Harry Bell   5567   Instructor   25000
The calls to the NewStaff object provide the following output:
Joan Mason   7833   825

11. Invalid rank
Any Employee   9999   No rank assigned   0
Any Employee no longer works here

13. (a) The first term *float* in the method heading indicates that the value of NewBalance returned from the method is a floating-point data item.
(b) The second term *float* in the method heading indicates that the value of an argument passed in a call to the Deposit method must be a floating-point data item.

Chapter 12

1. (a) An association is a relationship among two or more specific classes that describes connections among objects (instances) of those classes.

3. (a) 1..* means 1 or more, without limit.
(b) 0..1 means 0 or 1.
(c) * means 0 or more, without limit.
(d) 1 means 1.
(e) 6 means 6.
(f) 6..8 means 6 or 7 or 8.
(g) 2..* means 2 or more, without limit.

5. (a) In an aggregation relationship, an object may not directly or indirectly be part of itself. That is, there can be no cycles.
(b) An aggregation is shown on a UML class diagram by a line that has an unfilled diamond at the end of the line connected to the class that is the aggregate in the relationship.

7. (a) aggregation
(b) association
(c) association
(d) association
(e) association
(f) generalization/specialization
(g) composition
(h) generalization/specialization
(i) generalization/specialization
(j) aggregation

9.

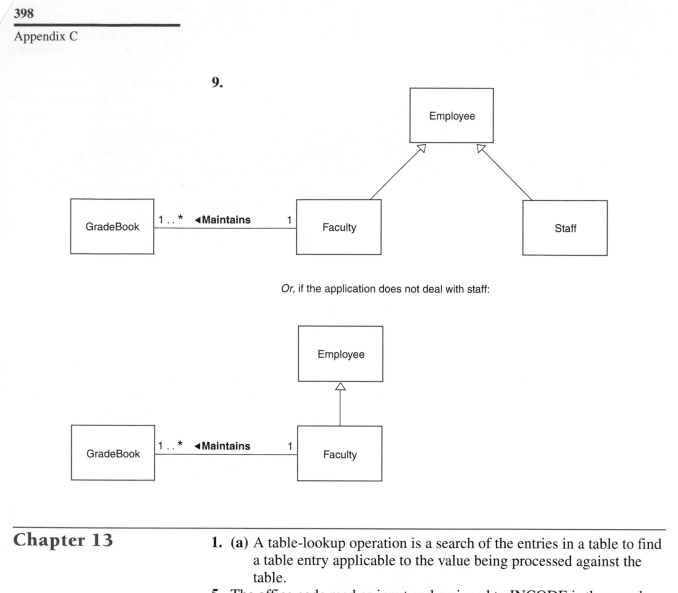

Or, if the application does not deal with staff:

Chapter 13

1. (a) A table-lookup operation is a search of the entries in a table to find a table entry applicable to the value being processed against the table.

5. The office code read as input and assigned to INCODE is the search key.

7. (a) Two program switches used in this solution algorithm are VALID_TABLE and FOUND.

12. (a) 708   52   10   .091

Chapter 14

1. (a) In an online-processing environment, the execution of a program is initiated by a user at a terminal, personal computer, or workstation. The user is in direct communication with the computer.

5.

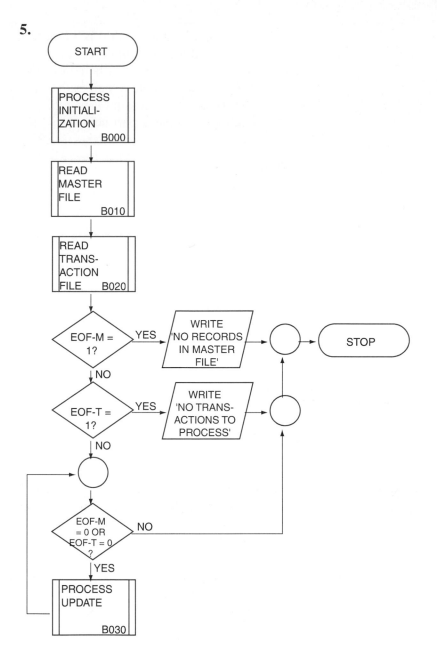

7. (b) The new master file contains the following records:

ID#-M	Status
00002	changed
00006	unchanged
00007	changed (2nd change)
00008	added
00009	unchanged
00010	unchanged
00012	unchanged
00014	unchanged

(c) No error messages will be output.

Chapter 15

1. **(a)** A detail-printed report is a report in which one line of output is printed for each input record processed.

3. A control break occurs when the value in a designated field in the input (the control field) changes. Normal processing is interrupted and control-break processing is begun.

5. Each subsequent department sales total would reflect the sales totals from all previous departments. DEPT-ACCUM would in effect be an accumulation of the sales from all the input records.

Index